FOCUS ON CELL APOPTOSIS RESEARCH

FOCUS ON CELL APOPTOSIS RESEARCH

DANIEL W. CHO
EDITOR

Nova Biomedical Books
New York

For permission to use material from this book please contact us:
Telephone 631-231-7269; Fax 631-231-8175
Web Site: http://www.novapublishers.com

LIBRARY OF CONGRESS CATALOGING-IN-PUBLICATION DATA

Focus on apoptosis research / Daniel W. Cho (editor).
p. cm.
Includes bibliographical references and index.
ISBN 13: 978-1-60021-433-2
ISBN 10: 1-60021-433-9
1. Apoptosis. I. Cho, Daniel W.
[DNLM: 1. Apoptosis. 2. Apoptosis Regulatory Proteins--therapeutic use. 3. Cytoprotection--physiology. QU 375 F6517 2006]
QH671.F63 2006
571.9'36--dc22 2006028758

Published by Nova Science Publishers, Inc. ✢ New York

Contents

PREFACE

Apoptosis is the regulated form of cell death. It is a complex process defined by a set of characteristic morphological and biochemical features that involves the active participation of affected cells in a self-destruction cascade. This programmed cell death plays a critical role in physiological functions such as cell deletion during embryonic development, balancing cell number in continuously renewing tissues and immune system development. Additionally, a dysregulation of apoptosis is underlying in numerous pathological situations such as Parkinson, Alzheimer´s disease and cancer. A number of studies have pointed out an association between consumption of fruits and vegetables, and certain beverages such as tea and wine, which are rich in polyphenols, with reduced risk of chronic diseases, including cancer. Apoptosis is also the regulatory mechanism involved in the removal of unnecessary cells during development and in tissue homeostasis in a wide range of organisms from insects to mammals. The aim of this book is to provide untouched studies in this new and exciting field.

Chapter I - Increased bone resorption is associated with diseases such as osteoporosis, inflammation and arthritis, and lytic bone tumours. Bone resorption is performed by specialised large multinuclear cells called osteoclasts. These cells enter the osteoclast lineage and express the osteoclast marker tartrate-resistant acid phosphatase (TRAP), then differentiate and fuse under the influence of the cytokine RANK ligand (RANKL). On reaching a critical size, they assume a spread morphology, strongly adhered to the underlying surface, which represents the mature state of an osteoclast prepared to resorb bone. In cell cultures these spread cells rapidly lose their TRAP staining and show cytoplasmic condensation and fragmentation, and pyknotic nuclei, suggestive of apoptosis. These early "phantoms" have activated intracellular caspases, further indicating that they die by apoptosis. Both the live spread cells and the dead cells remain strongly adhered and are not influenced by experimental procedures. Phantoms remain identifiable for at least 2 days, and given their large size, they can be counted and their morphology analysed under light microscopy.

The cell density in the culture wells is an important factor influencing osteoclastogenesis, and increases in cell density also accelerate the apoptosis of osteoclasts. RANKL is also a survival factor for osteoclasts and when it was removed, or blocked by receptor antagonists on the last 2 days of 7 day cultures, not only did no more osteoclasts develop, but existing cells entered into apoptosis with a marked increase in phantoms with respect to live cells.

Calcitonin effectively inhibited osteoclastogenesis when present throughout cultures, but when included only at the end of cultures, the rate of spread cell apoptosis increased, suggesting that it also has an inhibitory effect on mature osteoclasts.

Estradiol is a major hormonal control of bone resorption and has been shown to have direct effects on osteoclasts, including increased apoptosis. However, in these cultures estradiol had no consistent effects. TNFα and IL-1β, cytokines associated with inflammation, both strongly promoted the development of spread cells. If induced by a higher concentration of RANKL this would have been associated with a marked increase phantoms, limiting the numbers of live cells. However, both cytokines decreased the number of phantoms, despite the marked increase in spread cells, suggesting a direct effect in decreasing apoptosis. Both cytokines induced a similar response when administered only at the end of cultures, and they protected against the marked increase in apoptosis induced by RANKL withdrawal. PGE2 or TGFβ also induced more modest increases in osteoclastogenesis, including spread cells. With PGE2 this was associated with decreased phantoms, suggesting a direct decrease in apoptosis. If included only at the end of cultures, TGFβ increased osteoclastogenesis and decreased phantoms, suggesting that it too has a weak anti-apoptotic effect. In the absence of RANKL, TGFβ was able to maintain the levels of live spread cells with a transient decrease in phantoms (on day 6 but not day 7). PGE2 had no consistent effect on spread cell apoptosis when added at the end of cultures.

These responses are discussed with respect to the proposed molecular action of each mediator and the convergence of intracellular signalling pathways in the control of osteoclast apoptosis.

Chapter II - Previous studies have shown that apoptosis (programmed cell death) plays a crucial role in the pathomechanism of neurodegenerative disorders. It is a general consensus that apoptosis can be triggered by dysregulation of nerve growth factor physiology and/or oxidative stress processes. Since both of these pathological changes are characteristic to AD, it is conceivable that influencing apoptosis can be a valuable tool in the treatment of neurodegenerative disorders.

Glucosaminoglycans (GAGs) exhibit neuroprotective properties in several animal models of AD. Low molecular weight GAG, C3, attenuates Aβ(25-35)-induced tau-2 immunoreactivity, and AF64A-induced cholinergic lesion in rat brain. Recent studies have also revealed that C3 influences p75 nerve growth factor receptor expression. However, the exact mechanism of the neuroprotective attributes of C3 is not elucidated yet. Since apoptosis is believed to play a pivotal role in numerous neurodegenerative disorders, C3 may induce neuroprotection/neurorepair via affecting apoptotic processes.

The neuroprotective effect of glucosaminoglycans including C3 thus may involve multiple mechanisms, including an influence on processes that lead to programmed cell death. Since apoptosis is a crucial event not only in neurodegenerative diseases but also in the general process of aging, C3 may be a valuable therapeutic agent in Alzheimer's disease-related senile dementia and in other age-related malfunctions.

Chapter III - Alzheimer's disease (AD) and Parkinson's disease (PD) are the most common progressive neurodegenerative disorders that not only affect a large population in the world, but also affect a genetically isolated group of individuals known as the "paisa community" living in Antioquia region, Colombia. Both disorders are neuropathologically characterized by specific neuronal and synaptic loss of the hippocampal pyramidal neurons and dopaminergic neurons of the substantia nigra, respectively. Unfortunately, no effective and/or definitive therapeutic treatment aimed to reduce or to retard the clinic and pathologic symptoms induced by both devastating diseases are presently available. Thus, the biochemical mechanism(s) involved in neural loss has mainly been studied in biological models. In this article the authors have proposed lymphocytes as a cellular model to study the molecular signalling mechanism(s) of cell death-apoptosis- induced by oxidative stress and cytoprotection in AD and PD. Remarkably, lymphocytes and neurons are cellularlly and biochemically alike. Indeed, lymphocytes express at least six different neurochemical systems considered to exclusively pertain to neural cells such as catecholaminergic, serotonergic, acetylcholinergic, glutamatergic, noradrenergic and GABAergic systems. Given that lymphocytes are post-mitotic cells (i.e. G_o cell cycle), possess the synthesis, transport and processing systems of amyloid-beta precursor protein (AβPP), presenilins (PS) and parkin proteins, but most importantly, expresses universal cell death machinery, these cells constitute not only an ideal biomodel to better understanding of the pathological signalling in those neurological entities, but also represent a meaningful tool to design innovative and rational therapeutic strategies focused to arrest the neuronal death in Alzheimer's and Parkinson's patients. Specifically, the authors present a unified molecular mechanism of cell death induced by dopamine and Aβ, and survival signalling mechanism triggered by insulin-like growth factor and cannabinoids against oxidative stress stimuli.

Chapter IV - Non-enzymatic modification of proteins by reducing sugars, a process that is also known as Maillard reaction, leads to the formation of advanced glycation end products (AGEs) *in vivo*. There is a growing body of evidence to show that formation and accumulation of AGEs progress during normal aging, and at an extremely accelerated rate under diabetes, thus being involved in the pathogenesis of various diseases such as diabetic vascular complications, neurodegenerative diseases and cancers. Recently, engagement of their receptor, RAGE, is shown to activate its down-stream signaling and subsequently evoke proliferative or proapoptotic responses in various types of cells. Therefore, in this study, the authors summarize the role for AGEs-RAGE system in the pathogenesis of various disorders, especially focusing on cell growth and death. They further review here the potential therapeutic strategy that may specially target the AGEs-RAGE system in these devastating disorders.

Chapter VI - L-ascorbic acid (LAA) at concentrations of 0.25–1.0mM induced a dose- and time-dependent inhibition of proliferation in acute myeloid leukemia (AML) cell lines and also in leukemic cells from peripheral blood specimens obtained from three patients with AML. LAA induced oxidation of glutathione to oxidized form (GSSG) and subsequent H_2O_2 accumulation in a concentration-dependent manner, in parallel to induction of apoptosis. Induction of apoptosis in LAA-treated AML cells involved a dose-dependent increase of Bax protein, release of cytochrome C from mitochondria to cytosol, activation of caspase 9 and caspase 3, and cleavage of poly [ADP-ribose] polymerase. LAA treatment of cells induces a dose-dependent phosphorylation of extracellular signal regulated kinases (ERK) and results in activation of its catalytic domain. The small G protein Raf1 and MAPKactivated protein kinase 2 are activated by LAA as an upstream and a downstream regulator of ERK, respectively. Although the ERK pathway has been known to activate cell proliferation, pharmacologic inhibition of ERK reduces LAA-dependent apoptosis and growth inhibitory response of acute myeloid leukemia cell lines, suggesting that this signaling cascade positively regulates induction of apoptotic response by LAA.

Arsenic trioxide (As_2O_3) could induce *in vitro* growth inhibition and apoptosis of AML cells. As_2O_3 or LAA suppressed DNA-binding activity of NF-κB composed of p65/p50 heterodimer through preventing the degradation of IκB-α and the nuclear translocation of p65 subsequently as well as interrupting the binding of NF-κB with their consensus sequences. Inhibitory effect of As_2O_3 or LAA on NF-κB DNA activity was dependent upon intracellular glutathione (GSH) and H_2O_2 level, but not superoxide anion.

The addition of LAA to cells resulted in an immediate increase in intracellular total GSH content and GST activity accompanied by the uptake of cysteine. These results suggest a new role of LAA in intracellular sulfur condition and glutathione regulation. Early protein targets of LAA in leukemia cells were identified using a differential proteomics approach in NB4 human leukemia cells treated with LAA. A subunit of protein-disulfide isomerase (a thiol/disulfide exchange catalyst) and immunoglobulin-heavy-chain binding protein (BiP, identical to Hsp70 chaperone) showed quantitative expression profile differences. A myeloid leukemia associated antigen protein (a tropomyosin isoform) showed changes in pI as a result of phosphorylation. The addition of LAA to cells results in an immediate change in the intracellular thiol/disulfide condition and modulates disulfide bond formation in proteins.

Chapter VII - Chronic lymphocytic leukemia (CLL) of B cells is the most common type of leukemia in the Western world and is characterized by the abnormal expansion of CD5+ mature B lymphocytes. The study of the pathways leading to apoptosis resistance in these cells is critical both in understanding the disease process and in devising novel therapeutical strategies. In CLL the accumulation of monoclonal mature B-cells may result from an extended lifespan of B cells, their increased capacity to proliferate, and/or defective apoptosis. Growth of CLL B cells might be a direct consequence of defects of genes regulating apoptosis, but it could be also indirectly affected by abnormal interactions between leukemic cells and the surrounding bystander non-tumoral cells. On CLL B cells levels of expression of various molecules representing major components of apoptotic pathways, such as Bcl-2/Bax, Bid, or the DISC complex (FAS, FADD, Caspase-8) have a key role in regulating the ability of neoplastic cells to undergo apoptosis. In addition, it has been demonstrated that particular cytokines, and specific ligand-receptor interactions, can sustain overexpression of molecules regulating apoptosis, as in the case of IL-4 for Bcl-2 and of CD40-CD40L for FAS and Bid expression. These observations may thus suggest that in CLL, imbalanced production of IL-4 could prolong survival through increased Bcl-2 expression, while CD40 triggering, of leukemic B cells, might sensitize neoplastic B cells to apoptosis through upregulation of FAS and of Bid. However the apoptotic pathway that is triggered by FAS/ FAS ligand (FASL) interaction may be not functional in CLL, and this despite apparently normal levels of expression of the molecules composing the DISC complex both in resting and in CD40-activated cells.

The authors further described in CLL, expansion of a CD3+/CD8+ T cell subset with unique phenotypic features of Tc2, capable of releasing IL-4, that may support CLL survival.

Moreover also IL-2 and IL-15, sharing the common γ chain and the IL-2Rβ chain capable of transducing signals, appear to induce proliferation of CLL B cells. They recently demonstrated that IL-21, utilizing the common γ chain as IL-2 and IL-15, induces, instead, apoptosis of CLL B cells, especially after CD40 triggering, which upregulates IL-21 receptor expression in all the cases studied. In this review they will discuss how CD40 triggering of CLL B cells may modulate expression of cytokine receptors and of molecules implicated in apoptosis of CLL B cells and the mechanisms that regulate these phenomena.

Chapter VIII - Acute pancreatitis has numerous causes, culminating in a common pathway of changes within the pancreatic acinar cell. The pancreatic acinar cell being the functional unit of the exocrine pancreas, it synthesizes, stores, and secretes digestive enzymes. Evidence has accumulated of the significance of apoptotic cell death in the systemic manifestation associated with acute pancreatitis and studies suggest that apoptotic cell death may play a considerable role in affecting mortality and morbidity in severe acute pancreatitis. Apoptosis in pancreatitis is mediated by caspase activation and mitochondrial dysfunction in acinar cells. It is suggested that caspases play an important role in apoptosis. The expression of caspases is a frequent event in chronic pancreatitis and its distribution pattern may reflect the functions of this protease in its role in the apoptotic pathway in atrophic acinar. Also acute pancreatitis modifies mRNA level of both pro and anti-apoptotic genes simultaneously. Numerous studies suggest that pancreatic bcl2, bax and p53 mRNA levels increased extensively in the mild acute pancreatitis and the levels were unchanged during severe acute pancreatitis. This article presents a brief overview of the apoptotic machinery in pancreatitis, including specific and indirect death signals and the underlying molecular events in cell.

Chapter IX - Apoptosis is an active physiological process of genetically programmed cell death. It is a unique process maintaining the homeostasis of variety of physiological systems in the body including the skin. In skin, cell death by apoptosis is an essential way of regulating keratinocyte proliferation and epidermal growth. Terminal differentiation of keratinocyte represents simply a special form of apoptotic death. The process by which keratinocytes undergo apoptosis is a multistep program that includes several mechanisms. One is induced by binding of specific death ligands to death receptors on target cells such as binding of FasL expressed on activated cytotoxic T cells to Fas expressed on keratinocytes. Another is mediated by the release of effectors cell granules such as cytolytic molecule perforin and serine proteases named granzymes. The apoptotic process is not only involved in maintaining skin homeostasis but also has a role in the pathogenesis of a variety of skin diseases. Moreover, failure of cells to undergo apoptosis might be the main cause of some dermatosis. In skin, different apoptotic stimuli may simultaneously trigger multiple apoptotic pathways. Apoptotic keratinocytes along with dysfunctional apoptotic process occur in some skin diseases such as psoriasis and lichen planus. The evidence is accumulating about psoriasis as a T cell-mediated disease. In psoriasis, conflicting results are obtained regarding the role of apoptosis in development of psoriatic lesions. There is a decreased apoptosis of lesional keratinocytes with decreased level of p53 activation. Moreover, it seems that psoriatic keratinocytes has increased resistance to apoptosis due to as yet not well defined mechanisms. Multiple studies revealed increased level of antiapotoptic Bcl-x_L protein in psoriasis. However, several authors demonstrated enhanced up regulation of perforin apoptotic pathway, both, in peripheral blood of psoriatic patients and in psoriatic lesions. A hallmark of lichen planus is the formation of colloid bodies representing apoptotic keratinocytes. Infiltrating T cells seems to be responsible for damage of basal keratinocytes in lichen planus. Experimental data provided evidence for a role of perforin in the apoptosis of lesional keratinocytes. In addition, enhanced Fas-mediated death of epidermal keratinocytes in lichen lesions is also actively involved in the apoptosis of adjacent basal keratinocytes. Here, the authors review the complex role of several apoptotic mechanisms in the etiopathogenesis of papulosquamous diseases, psoriasis and lichen planus.

In: Focus on Cell Apoptosis Research
Editor: Daniel W. Cho, pp. 1-67
ISBN: 1-60021-433-9

Chapter I

Morphological Studies of Osteoclast Apoptosis and the Effects of Endogenous Mediators

Colin R. Gardner[1,*], Jeannette Soria[2] and Miriam Gardner[3]
[1]Associate Professor, Groupe de Recherche MERCI, EA2122 Laboratoire DIFEMA, Faculty of Medicine and Pharmacy, 22, Boulevard Gambetta, 7600 Rouen, France;
[2]Sainte Marie Biochemistry Laboratory, Hotel Dieux, Paris, France;
[3]Centre René Huguenin, 35 rue Dailly, St.-Cloud, Paris, France.

Abstract

Increased bone resorption is associated with diseases such as osteoporosis, inflammation and arthritis, and lytic bone tumours. Bone resorption is performed by specialised large multinuclear cells called osteoclasts. These cells enter the osteoclast lineage and express the osteoclast marker tartrate-resistant acid phosphatase (TRAP), then differentiate and fuse under the influence of the cytokine RANK ligand (RANKL). On reaching a critical size, they assume a spread morphology, strongly adhered to the underlying surface, which represents the mature state of an osteoclast prepared to resorb bone. In cell cultures these spread cells rapidly lose their TRAP staining and show cytoplasmic condensation and fragmentation, and pyknotic nuclei, suggestive of apoptosis. These early "phantoms" have activated intracellular caspases, further indicating that they die by apoptosis. Both the live spread cells and the dead cells remain strongly adhered and are not influenced by experimental procedures. Phantoms remain identifiable for at least 2 days, and given their large size, they can be counted and their morphology analysed under light microscopy.

The cell density in the culture wells is an important factor influencing osteoclastogenesis, and increases in cell density also accelerate the apoptosis of

* Correspondence concerning this article should be addressed to Colin Gardner, Groupe de Recherche MERCI, EA2122 Laboratoire DIFEMA, Faculty of Medicine and Pharmacy, 22, Boulevard Gambetta, 7600 Rouen, France. E-mail : colgard2003@yahoo.fr

osteoclasts. RANKL is also a survival factor for osteoclasts and when it was removed, or blocked by receptor antagonists on the last 2 days of 7 day cultures, not only did no more osteoclasts develop, but existing cells entered into apoptosis with a marked increase in phantoms with respect to live cells.

Calcitonin effectively inhibited osteoclastogenesis when present throughout cultures, but when included only at the end of cultures, the rate of spread cell apoptosis increased, suggesting that it also has an inhibitory effect on mature osteoclasts.

Estradiol is a major hormonal control of bone resorption and has been shown to have direct effects on osteoclasts, including increased apoptosis. However, in these cultures estradiol had no consistent effects. TNFα and IL-1β, cytokines associated with inflammation, both strongly promoted the development of spread cells. If induced by a higher concentration of RANKL this would have been associated with a marked increase phantoms, limiting the numbers of live cells. However, both cytokines decreased the number of phantoms, despite the marked increase in spread cells, suggesting a direct effect in decreasing apoptosis. Both cytokines induced a similar response when administered only at the end of cultures, and they protected against the marked increase in apoptosis induced by RANKL withdrawal. PGE2 or TGFβ also induced more modest increases in osteoclastogenesis, including spread cells. With PGE2 this was associated with decreased phantoms, suggesting a direct decrease in apoptosis. If included only at the end of cultures, TGFβ increased osteoclastogenesis and decreased phantoms, suggesting that it too has a weak anti-apoptotic effect. In the absence of RANKL, TGFβ was able to maintain the levels of live spread cells with a transient decrease in phantoms (on day 6 but not day 7). PGE2 had no consistent effect on spread cell apoptosis when added at the end of cultures.

These responses are discussed with respect to the proposed molecular action of each mediator and the convergence of intracellular signalling pathways in the control of osteoclast apoptosis.

Keywords: apoptosis, osteoclast morphology, cell density, RANKL, OPG, calcitonin, estradiol, cytokines, TGFβ, PGE2, intracellular signaling.

INTRODUCTION

Increased bone resorption occurs in several diseases including osteoporosis, inflammation and arthritis, and bone lytic tumours such as multiple myeloma or bone metastases of other tumour types. The cells which resorb bone are osteoclasts, which are large multinuclear cells specialised for this role. Mononuclear monocytic precursor cells become committed to this lineage under the influence of macrophage colony-stimulating factor (M-CSF) and the TNF superfamily cytokine TNFSF-11/receptor activator of nuclear factor-kappaB (NF-κB) ligand (RANKL), which was originally discovered on T cells and named TNF-related activation-induced cytokine (TRANCE) [1,2,3]. These stimulated precursors express the enzyme tartrate-resistant acid phosphatase (TRAP) which acts as a marker for osteoclasts during all stages of development and maturation. RANKL also stimulates the development and fusion of these cells to become large multinuclear cells. When these differentiated cells reach a critical size they become activated mature osteoclasts and adhere to bone surfaces, forming a lacuna underneath in which to resorb the bone.

Much research has used models of osteoclastogenesis in culture *in vitro*. As it was initially found that the presence and contact of osteoblasts with the osteoclast precursors was necessary for the development of osteoclasts, co-cultures of osteoblastic cell populations and sources of osteoclast precursors (mostly bone marrow cells) were used. The most used endogenous stimulators of bone resorption were vitamin D3 and parathyroid hormone, both of which stimulated osteoclastogenesis via actions on osteoblasts. However, other endogenous mediators such as interleukin-1 (IL-1) tumor necrosis factor α (TNFα), IL-6 in the presence of its soluble receptor, IL-11, IL-17 and E series prostaglandins could also induce osteoclastogenesis. With the discovery of RANKL expressed on the surface of osteoblasts as the major factor in the stimulation of osteoclastogenesis, this molecule was used in its soluble recombinant form to stimulate the osteoclast precursors in the absence of osteoblasts/stromal cells. Such cultures have used either extracted osteoclasts or several sources of precursor cells, including bone marrow cells and spleen cells in animal models (mouse, rat, rabbit and chicken cells have been particularly used). Peripheral blood mononuclear cells, both from animals and from humans, have also been used as a source of osteoclast precursors. In addition, several cell lines have been induced to form osteoclasts, particularly including RAW-264.7, a monocytic/macrophagic cell line. This had several advantages, including the study of the effects of endogenous mediators directly on the osteoclast precursors and the developing osteoclasts, without the complication of indirect effects via osteoblasts, as well as the possibility of studying the morphology of the developing osteoclasts, without the obscuring effect of a layer of osteoblasts.

Microscopic Observation of Osteoclast Apoptosis

Initial morphological indices of death of mature osteoclasts in culture had come from observations in co-cultures. In general, such cultures were optimised to obtain large numbers of osteoclasts in the shortest period possible. This was particularly the case if new compounds with potential actions on osteoclasts were tested routinely. In co-cultures, osteoclast precursors develop on the surface of, or within the the osteoblast cell layer, but the mature cells penetrate the layer and adhere to the bare or coated plastic surface of the bottom of the culture wells. The routine duration of these cultures was 4-5 days but in some experiments the cultures were continued for 6-7 days. In these cultures the osteoblast cell layer became less adhered to the bottom of the well and could easily be removed using cold medium. In this way, the cell layer and the developing osteoclasts were removed, leaving only the mature osteoclasts which were spread and adhered to the bottom of the well. It appeared that these cells were strongly adhered to the plastic as this procedure could be routinely performed without loss or damage to the adhered cells. In some experiments, after staining for TRAP and fixing these cells, it appeared that some cells were dead or dying, but were still strongly adhered to the plastic. This was particularly true in longer duration cultures. These cells did not stain for TRAP, and showed evidence of cytoplasmic condensation and fragmentation. Groups of pyknotic nuclei could be observed, mostly associated with a thickened peripheral ring. The concentrated cytoplasm around these nuclear groups also lost TRAP staining, but more slowly. Eventually the whole cell form becomes black, but still recognisable, particularly the peripheral ring, which was still strongly adhered to the plastic (Fig. 1). Throughout this chapter these cells are called phantoms.

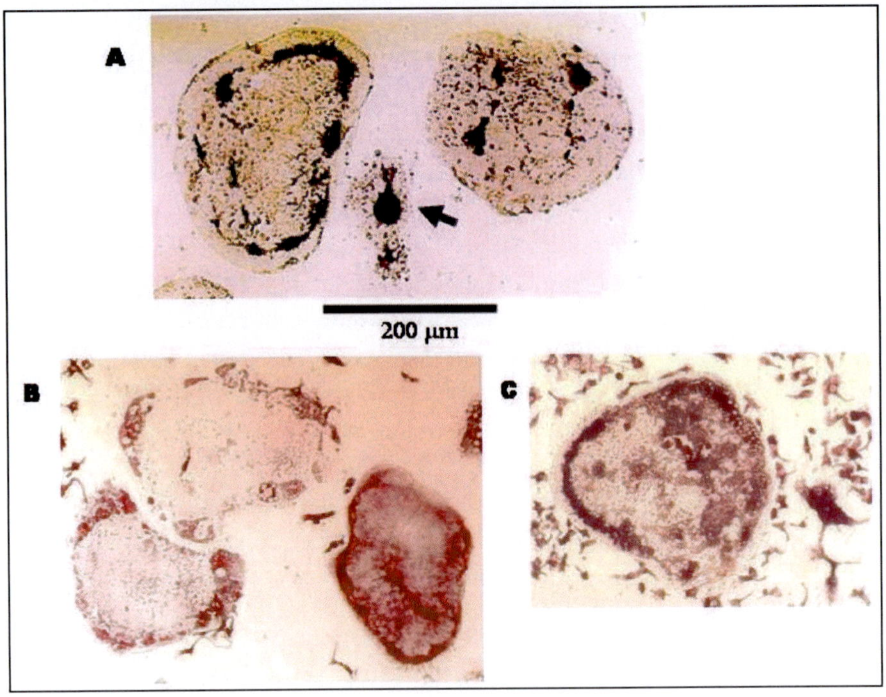

Figure 1. Initial observations of live and dead (phantom) spread cells adhered to the plastic surface of the bottom of culture wells. After 7 days co-culture of mouse calvarial osteoblasts and mouse bone marrow cells the overlaying cell layer was detached by gentle agitation in cold medium and removed, revealing spread osteoclasts adhered to the bottom of the wells. Light microscopic images were recorded at 10 times magnification using Histolab software. A. The majority of cells were dead (phantoms) but were still strongly adhered to the plastic. Note an intermediate (non-spread) osteoclast phantom indicated by the arrow. B. A live spread cell showing full structural integrity and TRAP staining, with two phantoms to the left. C. An image obtained on day 7 of a spleen cell culture showing the condensation of the cytoplasm of a spread cell in the process of apoptosis.

This observation led to studies of the death of mature spread osteoclasts in cultures without osteoblasts. Mouse bone marrow cells (BMC) or spleen cells, stimulated by RANKL in the presence of M-CSF, have been used throughout the studies reported in this chapter.

Culture Methods

The methodology used is based on generally-used standard culture procedures, and is described in brief below.

Adult female CD-1 mice of 8-12 weeks of age (Charles River) were used as the source of cells for all experiments. Immediately after sacrifice, the four long bones of the tibias and femurs were surgically removed and cleaned of other tissues. The spleen was also removed.

For several culture plates performed in parallel, the bone marrow or spleen cells from several mice were pooled before seeding the cells into the plate wells.

Bone Marrow Cell Cultures

Bone marrow cells (BMC) were isolated from the cut shafts of mouse femurs and tibias, by fluid pressure applied with a syringe. BMC were washed and the volume made up to 40 ml with culture medium. A sample of cells was counted, with live cells being identified by Trypan blue exclusion. BMC were cultured with standard seeding cell densities previously found to be optimum for osteoclast development (1 or 2 million cells per well in 24-well plates (NUNC) and 60 thousand cells per well in 96-well plates). However, a series of experiments was performed with a range of seeding cell densities to investigate the effect of cell density on osteoclast apoptosis. Cultures were maintained in αMEM medium (Gibco) (without phenol red, but with added 10% foetal bovine serum (Hyclone), 1% penicillin/streptomycin (Sigma) and 0.1% fungizone), 500μL per well in 24-well plates and 200μL in 96-well plates, for periods up to 7 days, with a change of medium every 2-3 days.

Spleen Cell Cultures

The excised spleen was homogenised by hand in 1 ml culture medium in a glass homogeniser. Remaining cells were washed from the homogeniser and the cell solution made up to 40 ml. A sample of cells was counted, with live cells being identified by Trypan blue exclusion. Spleen cells were cultured with 2 million cells per well in 24-well plates, 100 thousand cells per well in 96-well plates (NUNC). Cultures were maintained in 500μL αMEM medium (Gibco, as for BMC) in 24-well plates or 200μL per well in 96-well plates, for periods of 5 or 7 days, with a change of medium every 2-3 days.

Induction of Osteoclastogenesis

All cultures were performed in the presence of 25 ng/ml M-CSF and osteoclast development was initiated by the addition of recombinant soluble RANKL (Peprotech). In the majority of experiments, a single concentration of RANKL was used (in the range 15-100 ng/ml), depending on the culture cell type and the culture duration. However, in some experiments RANKL concentration-effect curves were performed. In the spleen cell preparations used here, it has been shown that TNFα induced osteoclastogenesis independent of RANKL [4,5]. In some spleen cell cultures TNFα (20-40 ng/ml) has also been used to induce osteoclastogenesis.

The great majority of recombinant proteins used (mediators and their soluble receptors) were purchased from R&D Systems.

Osteoclast Counting

At the end of cultures, the culture medium was removed and cells were fixed by adding 50µL of a solution containing 37% formaldehyde (Sigma) for 30 sec. Plates were then washed twice with physiological saline. Tartrate resistant acid phosphatase (TRAP) was stained with 200µL of a solution containing Fast Garnett (leukocyte acid phosphatase kit, Sigma Diagnostics) in the presence of tartrate, for 30 min at 37°C. The plates were then washed with physiological saline, centrifuged for 2 min at 2000 rpm, and left to dry by evaporation.

The numbers of osteoclasts in different morphological categories representing major stages in their development, maturation and death, were counted under the light microscope (Nikon) at 10-fold magnification. Counting was aided by the use of Histolab and Pegasus image analysis systems and the cell size distribution was analysed using the Sesame system (Microvision Instruments, Paris). The different morphological categories were separated and counted automatically after they had been identified manually. Images were also registered and stored using the Histolab software.

Where the number of cells in different morphological categories were counted in several replicate wells, data are represented as the mean cell count and the variability is represented by the standard error of the mean (SEM).

Morphology of Osteoclastogenesis and Osteoclast Death

The progression of osteoclast differentiation, maturation and death has been represented in these studies by counting osteoclasts in several morphological categories which represent major stages in this progression; 1.Small TRAP-positive cells with 3-6 nuclei. 2. Intermediate osteoclasts with more than 6 nuclei. These cells could have up to more than 30 nuclei as the consequence of several fusions but retained the more indefinite morphology suggestive of motility. 3. Spread osteoclasts which tended to have a definite peripheral ring structure where most of the nuclei were located in groups. This structure may be associated with the reorganisation of actin fibres within the cell into a peripheral ring structure, although it has been suggested that this actin ring does not form so readily in cells cultured on plastic [1]. These cells therefore represent a more mature form of osteoclasts. 4. Phantom osteoclasts which have lost their TRAP staining and which show morphological signs of intracellular condensation and fragmentation, with pyknotic nuclei suggestive of apoptosis.

During osteoclastogenesis, mononuclear TRAP+ cells were the first to develop, and then with fusion, small osteoclasts (3-6 nuclei) were seen. Following further fusion, intermediate cells (more than 6 nuclei) developed and when they had attained a critical size they formed the spread structure. These spread cells are short-lived in these cultures and produce dead or dying phantom cells within a few days (Fig. 1).

There are two main experimental designs which demonstrate the successive arrival of these morphological types. Several culture plates can be incubated at the same time but stopped at regular time intervals. However, changing cell density within the culture wells with time also has an influence on osteoclast development, and this must be taken into account (see *Cell density* section). An alternative approach is to stop the cultures at an appropriate time point which allows the development of all morphological classes (5-7 days), but to vary the concentration of RANKL. In this way the osteoclastogenic stimulus is varied in intensity. Even if all stages of osteoclastogenesis occur at higher RANKL concentrations, only small and some intermediate cells develop at lower, near-threshold concentrations (Fig. 2).

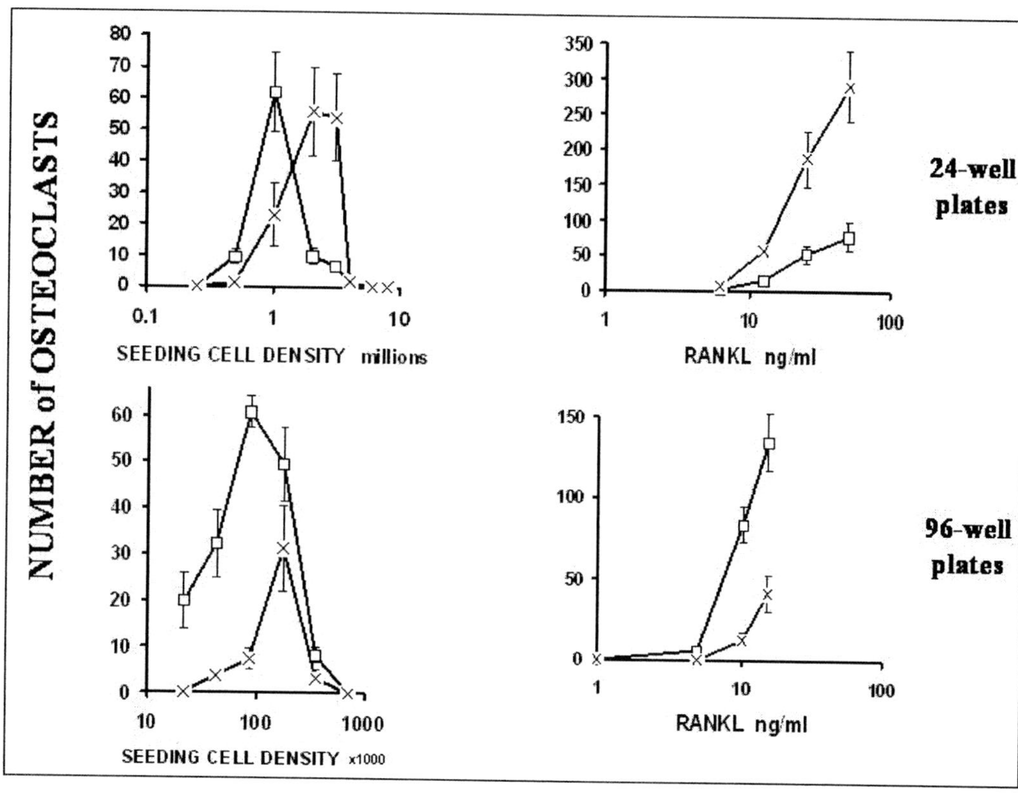

Figure 2. The effects of seeding cell density and well size on the apoptosis of spread osteoclasts on day 7 of mouse spleen cell cultures. Live spread cells and phantoms were separately identified and counted using Histolab software from microscopic images at 10 times magnification, displayed via a colour camera on a computer screen. 2-4 replicate wells were counted for each group. In 24-well plates n=4 for 0.5, 1, 2 and 4 million cells/well, n=2 for all other groups. In 96-well plates n=4 for all groups except the 3 lowest densities tested, where n=3. The variability is represented by the standard error of the mean (SEM). Osteoclastogenesis was induced by RANKL (100 and 30 ng/ml in 24 or 96-well plates, respectively) in the presence of M-CSF (25 ng/ml) in experiments in which the seeding cell density was varied (graphs on the left). The phantoms (crosses) which appeared were in wells with higher cell densities than the peak of live spread cells (squares), in experiments using either 24 or 96-well plates. In experiments with 7-day spleen cell cultures, where the seeding cell density was constant (1 million or 100 thousand cells/well in 24 or 96-well plates, respectively), concentration-response curves for RANKL (graphs on the right) showed concentration-related increases in both spread cells (squares) and phantoms (crosses). However, in 24-well plates there were more phantoms than spread cells, even at low RANKL concentrations, whereas the reverse was true in 96-well plates.

This analysis also indicates that the number of phantoms is dependent on the rate of formation of spread cells. When only low numbers of smaller spread cells have developed there may be no phantoms present, depending on the culture conditions. In spleen cell cultures in 96-well plates it is possible to have relatively large numbers of spread cells but few phantoms. In contrast, in cultures in 24-well plates it appeared that the majority of spread cells died, even though they were only present in small numbers and tended to be smaller in size (Fig.2).

Initially, the standard methods optimised for routine pharmacological studies of osteoclastogenesis were used, but were prolonged, even when there were large numbers of mature spread cells already present. It soon became clear that, as soon as significant numbers of spread cells were present, a proportion of them began to die, despite the continued presence of RANKL and M-CSF, which are considered to be important survival factors for osteoclasts [6]. For example, in initial spleen cell cultures in which osteoclasts tend to develop more slowly, spread cells became numerous on days 5-6 but a large proportion of spread cells were dead on day 7 (Fig. 2). Again, these dead cells were strongly adhered to the bottom of the well, as were their live counterparts. Therefore, detailed studies of the nature of this osteoclast cell death and the experimental factors influencing it, were undertaken. A model was developed in which cell death of osteoclasts could be easily observed and analysed with light microscopy, and the effects of endogenous mediators could be studied.

Caspase Activation During Osteoclast Death

Spread cell death was also observed in BMC cultures (in 96 well plates), there being a mixture of live and phantom cells in the wells on day 6. The fluorescent caspase substrate from the Enzolyte AMC Caspase-3 assay kit (TEBU) was used to detect activated caspases in *in situ* adhered spread cells. The peptide caspase substrate Ac-Asp-Glu-Val-Asp-AMC is weakly fluorescent at 354 nm, but when cleaved by caspases, becomes strongly fluorescent at 442 nm. It is particularly well cleaved by caspase-3 and also by caspase-7. This substrate was added for the first 3 hours (8-11am) in the continued presence of RANKL and M-CSF on day 6. Cells were observed under a fluorescent microscope with the filters centered around 440 nm. The coordinates of the cells studied were recorded and the plates were then removed and stained for TRAP and fixed for later microscopical analysis. The morphology of neither the fluorescent phantoms nor the live spread cells was significantly influenced by these procedures (Fig. 3). Live spread cells or non-spread multinucleated cells showed no fluorescence at all (Fig. 3A). However, the phantom cells consistently showed fluorescence, mostly around the groups of nuclei and in the peripheral ring (Fig. 3B).

However, it is likely that osteoclasts can enter into apoptosis at any stage of their development. RANKL is also a survival factor for osteoclasts and its withdrawal from BMC cultures for the last 16 h of day 5, during the incubation with the fluorescent caspase substrate, stopped further osteoclastogenesis so that there were virtually no spread cells present when the plates were observed in the morning of day 6. RANKL is also a survival factor for osteoclasts and its withdrawal is likely to induce the apoptosis of the osteoclasts present. There were many multinuclear cells with intermediate osteoclast morphology, some of which showed high levels of fluorescence, indicating the activation of caspases (Fig. 4C, D). It is difficult to countermark these cells with TRAP staining as they are motile and often change their form and migrate during the observation period and the processing of the plate prior to fixation. It is also possible that some of these cells detach during processing of the plate [7], and were therefore not accounted for in the analysis of the fixed cell preparation. In situations where many spread cells die in a short period of time, phantoms with the morphology of intermediate cells were also be seen in fixed and stained cultures, and they also appeared to be strongly adhered to the plastic surface (Fig. 4A, B). They represent a small minority of the total number of phantoms, the highest level observed being about 20% in the experiment analysed in figure 5. Where these intermediate cell phantoms occurred, they were included in the total phantom count.

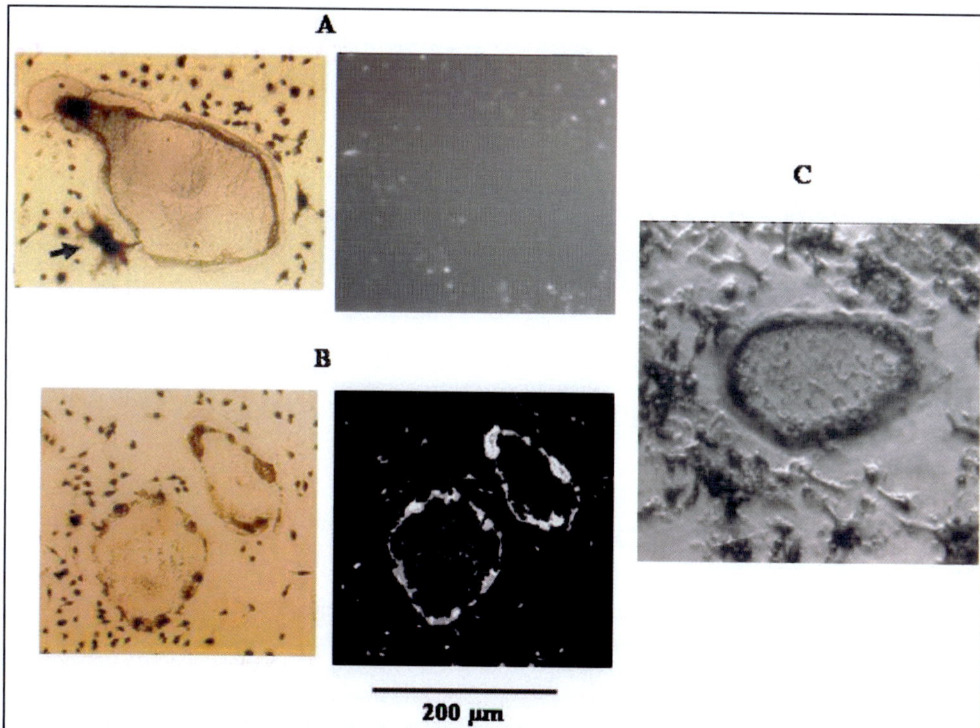

Figure 3. Caspase activation during the death of spread osteoclasts. Osteoclastogenesis was induced in mouse BMC cultures in 96-well plates by RANKL (40 ng/ml) in the presence of M-CSF (25 ng/ml). The fluorescence-inducing caspase substrate from the Enzolyte AMC caspase-3 assay kit (TEBU) (8-fold dilution) was added to the culture medium in the continued presence of RANKL and M-CSF, for the first 3 h (8 am-11am) of day 6. Following the fluorescent recordings the co-ordinates of the recorded cells were noted and the plates were removed for fixation of cultures and staining for TRAP. After drying of the plates, the same spread cells/phantoms were located and images of their TRAP staining and integrity recorded. A. A large live spread cell showing TRAP staining (left), with a live intermediate osteoclast next to it (arrowed). Neither of these cells showed any fluorescence, even after a long exposure (right). B. Two cells identified as phantoms (left) which showed strong fluorescence, particularly around groups of nuclei around the periphery. C. A spread cell identified in an untreated control well in these cultures (shown to be live by subsequent TRAP staining and imagery), was observed with a microscope objective giving the releif of the cells. Note that the peripheral ring of the spread cell is raised but the central cytoplasmic region is flat.

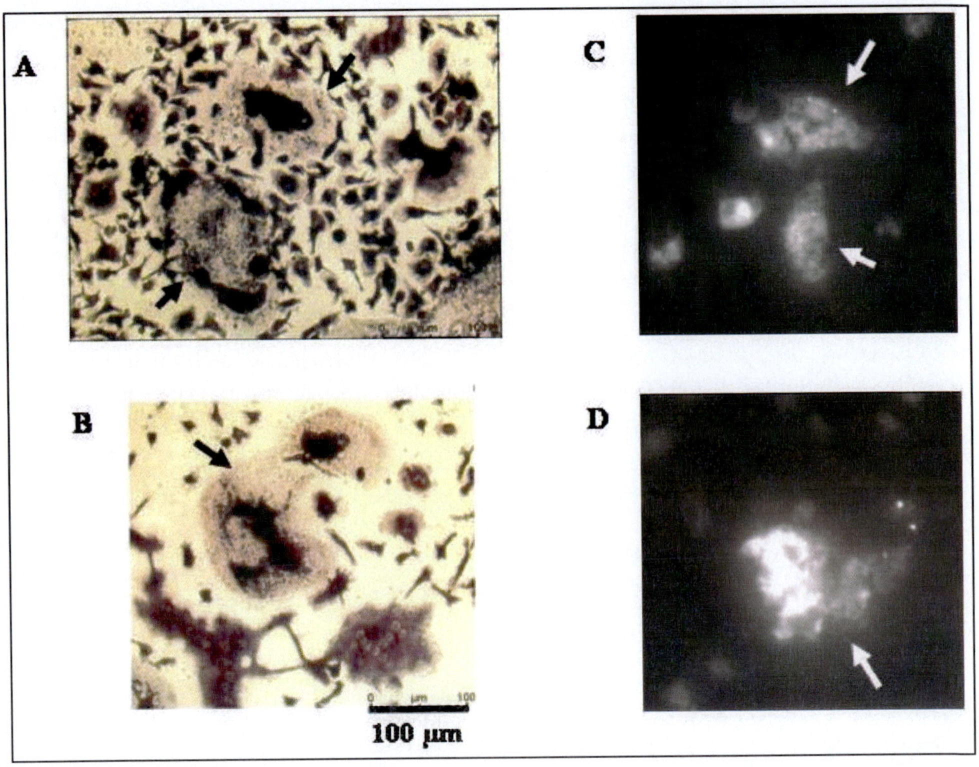

Figure 4. Intermediate (non-spread) osteoclast phantoms. Osteoclastogenesis was induced in mouse BMC cultures in 96-well plates by RANKL (40 ng/ml) in the presence of M-CSF (25 ng/ml). A and B. Phantoms with morphology suggesting that they were intermediate osteoclasts (arrowed) observed on day 7 in wells where there were large numbers of spread cell phantoms. There are several small TRAP-positive osteoclasts present in A, and a live intermediate osteoclast is clearly visible at the bottom of B. By day 5 of such cultures there were numerous osteoclasts with more than 6 nuclei (intermediate) but few with spread morphology. RANKL is a survival factor for osteoclasts, and its removal from the cultures (at 4 pm on day 5) would potentially induce the apoptosis of osteoclasts. Caspase activation in cells with the morphology of intermediate osteoclasts was demonstrated by the addition of an 8-fold dilution of the fluorescence-inducing caspase substrate from the Enzolyte AMC caspase-3 assay kit (TEBU) to the culture medium for 16 h from 4 pm on day 5. On the morning of day 6 the plates were observed under a fluorescent microscope at 20 times magnification with the emission filters centred around 440 nm. Cells resembling intermediate osteoclasts were identified by their size and multinuclear morphology. The fluorescence emission of these cells was then observed and images recorded. C. and D. Cells with the morphology of intermediate osteoclasts showing strong fluorescence (arrowed) suggesting that there are effector caspases activated within the cells and the the cells are undergoing apoptosis. No similar fluorescence was identified in cells with similar morphology in control wells where RANKL was still present.

Some small osteoclasts with 3-6 nuclei also showed strong fluorescence in the cultures from which RANKL had been removed (Fig. 4C). Although there were problems of background fluorescence in mononuclear cells, it appeared that there were more strongly fluorescent mononuclear cells in the absence of RANKL. Therefore, it is possible that the numbers of osteoclasts can be regulated by apoptosis at all stages of their development.

In general, for all the morphological types of osteoclast, the observed activation of caspase enzymes, probably involving those known to be effector caspases in osteoclast apoptosis (e.g. caspase-3, see below), supports the morphological evidence that this cell death is a form of apoptosis.

Most studies on osteoclast apoptosis have been performed using relatively small multinuclear (non-spread) osteoclasts. It cannot be assumed that the processes underlying apoptosis in these cells are the same as those in spread cells. There have been many studies of the intracellular mechanisms involved in osteoclast apoptosis. Effector caspases are always involved and collected evidence suggests that caspase-3 is the effector caspase which plays the major role [7,8]. However, different upstream mechanisms can be involved. Mitochondrial mechanisms are also involved under many circumstances [9,10,11,12], although it has also been suggested that apoptosis induced by detaching osteoclasts from supporting cells involves the activation of an initiator caspase, probably caspase-8, and the direct activation of effector caspases, with little involvement of mitochondrial mechanisms [13]. It seems likely that, as in other cell types, different stimuli activate the core apoptosis mechanisms via different intracellular signalling pathways [14], and that there are several ways to arrive at the final common path of activation of effector caspases and the initiation of the irreversible process of dismantling intracellular structures and functions, that is called apoptosis.

Time Course of Osteoclast Apoptosis

In cultures where many cells were dying (such as day 6 of BMC cultures in 96 well plates), occasionally cells were observed with one region TRAP-negative with disintegrating cytoplasm, yet the rest of the cell apparently alive and normal (Fig. 6A). Such cells were not many in numbers in a given culture well but were regularly seen. This suggests that the cells were fixed during the progression of the process of cell death within the cell, and that this process may be relatively slow. The use of a microscope objective which gave the relief of the cell surface, indicated that the central cytoplasmic region of live spread cells is flat (Fig. 3C) and therefore the layer of cytoplasm within this region is very thin. Some molecular mechanisms of apoptosis rely on the intracellular diffusion of molecules to activate the following steps in the process (e.g. the release of cytochrome c from mitochondria to activate the apoptosome complex and caspase-9). The thin layer of cytoplasm, apparently containing few organelles, may represent a limitation to the progression of diffusion-dependent mechanisms of apoptosis within these large spread cells. Some spread cells were fixed with the cytoplasm in the process of condensing, before complete fragmentation (Fig. 1C).

It is difficult to precisely measure the time course of apoptosis when each plate is fixed at one time point. But judging from the time of generation of spread cells in plates in the same cultures stopped at daily intervals, it appears that the spread cells die rapidly after their generation. This occurs despite the continued presence of survival factors RANKL and M-CSF. The survival effect of RANKL has been demonstrated in these studies (see *RANKL as a survival factor* section and Figs.10,11). The rapid death of spread cells was particularly obvious in spleen cell cultures in 24-well plates, where there were practically always more phantoms than live spread cells in all experiments (see Fig. 2 and the *Well size* section). The spread cells which died tended to be the larger ones. This was particularly clear in BMC cultures where the maximal dimension of spread cells and phantoms were analysed (Fig. 5).

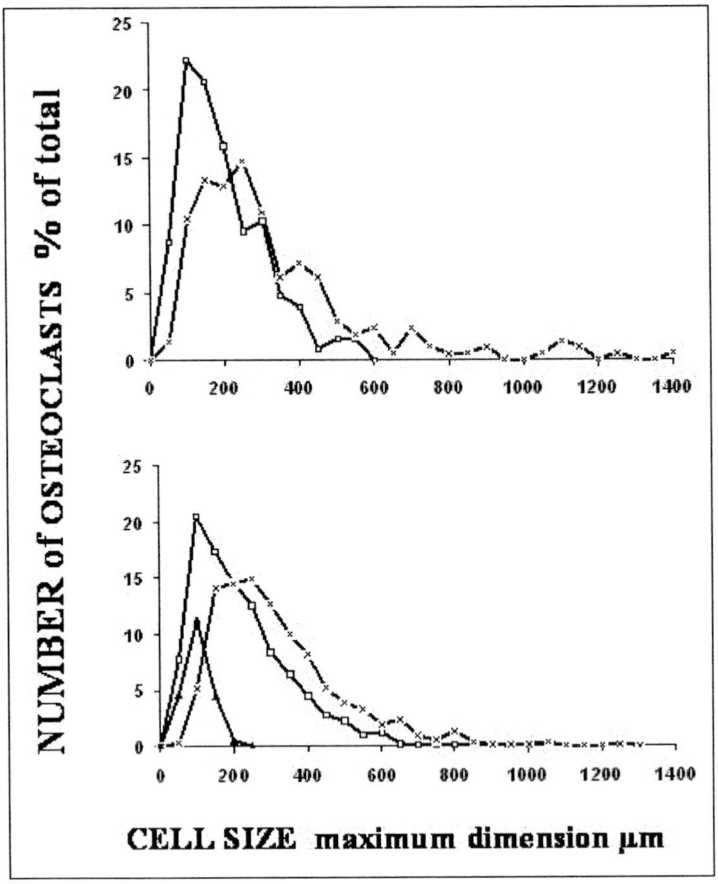

Figure 5. Size analysis of spread cells and phantoms. Osteoclastogenesis was induced in mouse BMC cultures in 96-well plates (seeding density 60 thousand cells /well) by RANKL (40 ng/ml) in the presence of M-CSF (25 ng/ml). On day 6 (upper graph) a significant number of phantoms had developed and the maximal dimension of both live spread cells (squares) and phantoms (crosses) was marked for all the cells in a well and automatically measured using Sesame software. Data from duplicate wells were pooled for analysis. A population curve was constructed for the size of both spread cells and phantoms and it is clear that the early population of phantoms resulted particularly from the apoptosis of larger spread cells. There are several phantoms greater than 600µm but no live spread cells this large. The mode of the spread cell population was 100-200µm whereas that of the phantom population was slightly greater than 200µm, indicating that the whole population of phantoms is larger than the remaining live spread cells. On day 7 (lower graph) when there were more phantoms than live cells, this size difference was still clear. In addition, in the well analysed, there was a significant number of smaller phantoms with non-spread morphology (triangles), which represented 20.9% of the total population of phantoms in this experiment.

The phantoms remain adhered in the wells, uncomplicated by invasion of other cells, for at least 2 days. They remained clearly identifiable and could be easily counted and their dimensions measured. Although monocytic and macrophagic cells were present in the cultures, macrophagic phagocytosis of disintegrating phantoms was never observed. The phantoms disintegrated over a period of 2-3 days, often leaving an empty hole if the spread cell had been surrounded by a continuous cell layer (Fig. 6). Only when the phantom debris had mostly dispersed was there repopulation by mononuclear cells and small developing osteoclasts (Fig. 6). This period of 2-3 days during which phantoms can be identified, counted and analysed, makes the counting of phantom numbers a realistic representation of the rate of apoptosis, given the duration of the cultures.

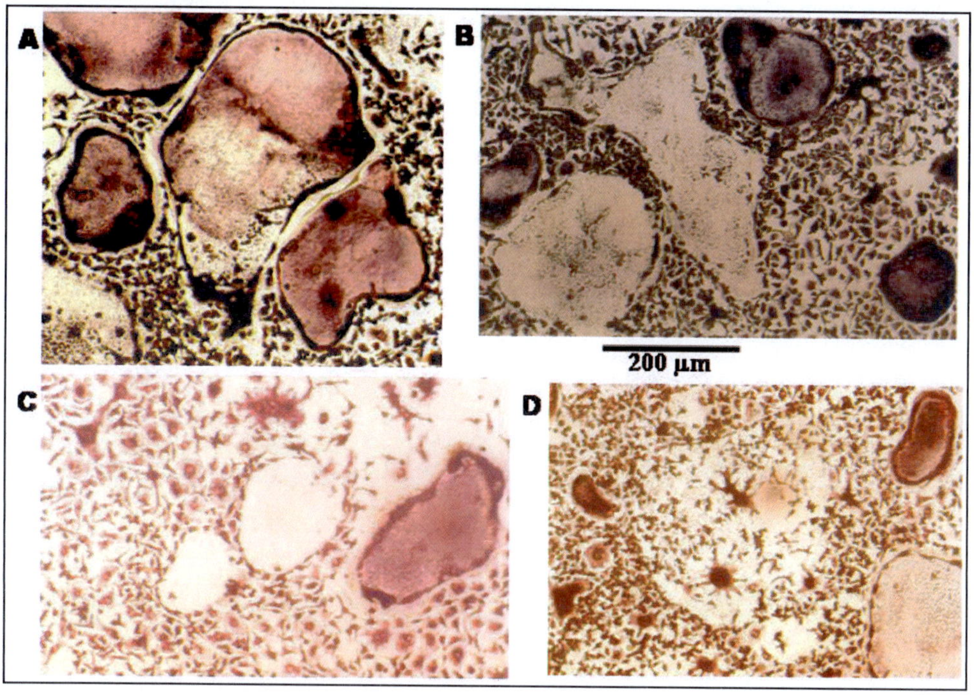

Figure 6. Time course of spread osteoclast apoptosis. All images were taken on day 7 of mouse BMC cultures in 96-well plates (seeding density 60 thousand cells /well) stimulated by RANKL (40 ng/ml) in the presence of M-CSF (25 ng/ml). Light microscopic images were recorded at 10 times magnification using Histolab software. A. A spread cell which appears to be alive and normal in the upper part, but shows no TRAP staining and has fragmented cytoplasm in the lower part. This cell has 3 normal live spread cells around it. B. Large phantoms with nearby small, live spread cells. The image was taken under low light to improve the contrast of the fragmented cytoplasm. C. Holes in a continuous cell layer where phantoms had been. A small amount of debris remained on the plastic in these areas. There is a large live spread cell to the right and several small osteoclasts above these empty areas. D. The repopulation of an area where a phantom had disintegrated, by small cells, including 2 small osteoclasts. In the top right corner of the image is a small live spread cell and in the bottom right corner is part of a more-recent phantom.

Osteoclasts are reported to be much more long-lived on bone surfaces or in vivo [1,15], and the observed rapid apoptosis, together with the flat morphology (Fig. 3C) may be related to their generation on non-bone surfaces. For example, the reorganisation of actin fibrils into a peripheral actin ring structure in spreading osteoclasts occurs much less in osteoclasts generated on plastic surfaces [1]. Cultures on collagen-coated plates gave very similar results to uncoated plastic (see Fig. 24). However, an important role for integrins on the surface of osteoclasts has been proposed. If integrins (primarily $\alpha v\beta 3$ in osteoclasts) are not occupied by a ligand (such as vitronectin) then strong pro-apoptotic signals are sent inside the cells [12,16,17]. It had been established that adhesion of non-spread osteoclasts was necessary for their survival, and disrupting their integrin binding with the disintegrin echistatin induced apoptosis [18]. Apoptosis due to unoccupied integrins was shown to primarily involve caspase-8, in contrast to the spontaneous background apoptosis which primarily involved caspase-9 [13]. As there are few ligand molecules for the osteoclast integrins in these culture plates or medium (collagen is not a ligand for the major integrins on osteoclasts, see [12]), this process of apoptosis might be released rapidly.

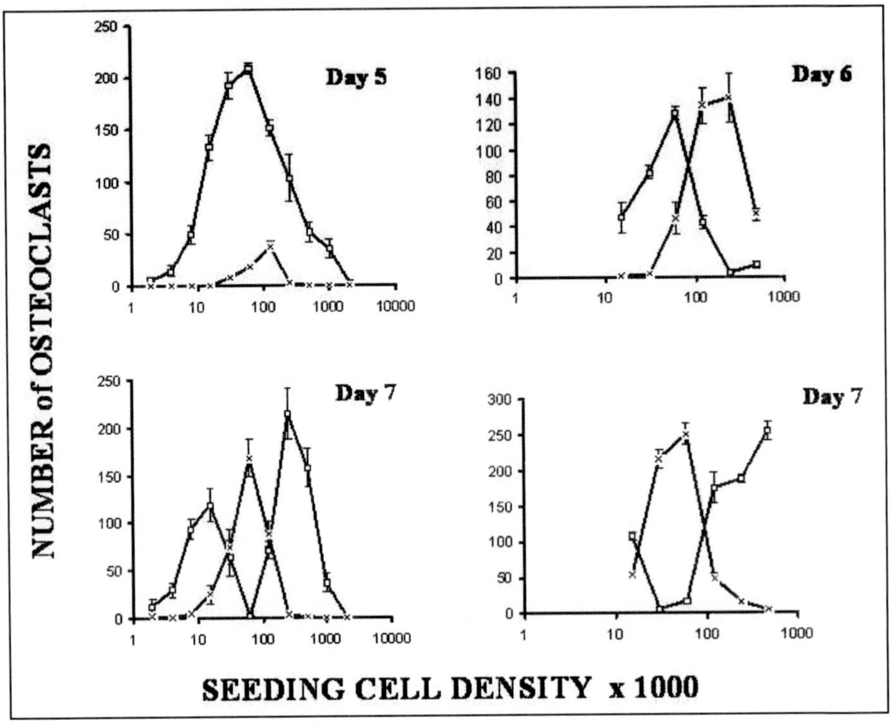

Figure 7. The effects of seeding cell density on the apoptosis of spread osteoclasts on days 5-7 of mouse BMC cultures in 96-well plates. Live spread cells and phantoms were separately identified and counted using Histolab software on microscopic images at 10 times magnification, displayed via a colour camera on a computer screen. Triplicate wells were counted for each group in all experiments and the variability is represented by the standard error of the mean (SEM). Osteoclastogenesis was induced by RANKL (15 ng/ml). When a wide range of seeding cell densities was studied on the same plate (graphs on the left), there was a clear peak for spread cells (squares) at 60 thousand cells/well on day 5 when there were few phantoms (crosses). The phantoms present were in wells having a higher seeding cell density (120 thousand cells /well). On day 7, at the density which was optimum for spread cell development on day 5, there were virtually no live spread cells present, but there were many phantoms. At slightly lower or higher densities there were large numbers of live spread cells and few phantoms. It appears that, when there are large numbers of spread cells developed, the great majority undergo apoptosis. These experiments were repeated, concentrating on the cell densities around this central region (graphs on the right). Day 6 was studied as there are more phantoms in the wells and the numbers of spread cells and phantoms are more equilibrated. The results were confirmed. The peak of spread cells was again at 60 thousand cells/well but the phantom peak was at 120-240 thousand cells. Again, on day 7 virtually all the spread cells at 60 thousand cells/well had died.

This pro-apoptosis signal may be stronger than the survival stimuli provided by RANKL and M-CSF. However, as the cells are strongly adhered in the wells there must be strong fixation processes other than integrins on the surface of osteoclasts.

EXPERIMENTAL FACTORS INFLUENCING OSTEOCLAST APOPTOSIS

Cell Density

It had been shown that the number of cells put in the well at the beginning of a culture (the seeding cell density) was an important factor determining the number of osteoclasts that develop [4]. At lower densities osteoclastogenesis is probably limited by the low numbers of precursors, but high cell densities appear to actively suppress osteoclastogenesis. Studies with a range of seeding cell densities in both spleen cell and BMC cultures indicated that the first phantoms observed in the wells were present at higher cell densities than the population of spread cells (Figs. 2,7). This suggests that, in addition to suppression of osteoclast development, higher surrounding cell density may trigger apoptosis in spread osteoclasts. This effect was no longer evident later in cultures when most of the spread cells were dead, such as on day 7 of BMC cultures (Fig. 7). Increases in cell density induce apoptosis in other cell types such as primary cultures of rat hepatocytes [19]. It was concluded that the viability of these cells is dependent on optimal culture density and that the cell population is regulated, at least in part, by apoptosis. It is interesting that matrigel abolished hepatocyte apoptosis induced by increased cell density, suggesting that matrix binding may be limited when there are many cells in close proximity, and cannot be replaced by cell-cell contacts. This could also be the case for developing osteoclasts in culture.

The real cell density in the well increased with the duration of culture. The number of non-osteoclastic cells increased progressively. This occurred even if RANKL was not added to the wells and there was no osteoclastogenesis. It is difficult to organise the ideal density for osteoclastogenesis as it depends on the duration of the cultures. This can be observed in BMC cultures with several plates cultured together but stopped at days 3,4,5,6 and 7. Independent of the stage of osteoclastogenesis, the peak in osteoclast numbers was at a progressively lower original seeding density the longer the culture continued. The data suggest that there is an ideal real cell density which is the same for all stages of osteoclastogenesis, and as the cell density increases with culture time, the ideal density will be found in wells which had progressively lower seeding densities (Fig. 8). The analysis of BMC cultures in 96-well plates with an estimate of overall cell density progression, suggests that the ideal cell density for osteoclast development is between 400 and 600 thousand cells per well (Fig. 8).

This is also true for apoptosis. As the density in the well increases, if it is above the optimal (ideal) when the spread cells develop, they are more likely to go into apoptosis. It seems possible that the more spread cells are present, the more likely they are to enter into apoptosis. Osteoclastic cells themselves contribute to the increased cell density involving increased surrounding cell contacts. Cell density also has a strong influence on the development of spread cells, and could also in this way influence the number of apoptotic cells.

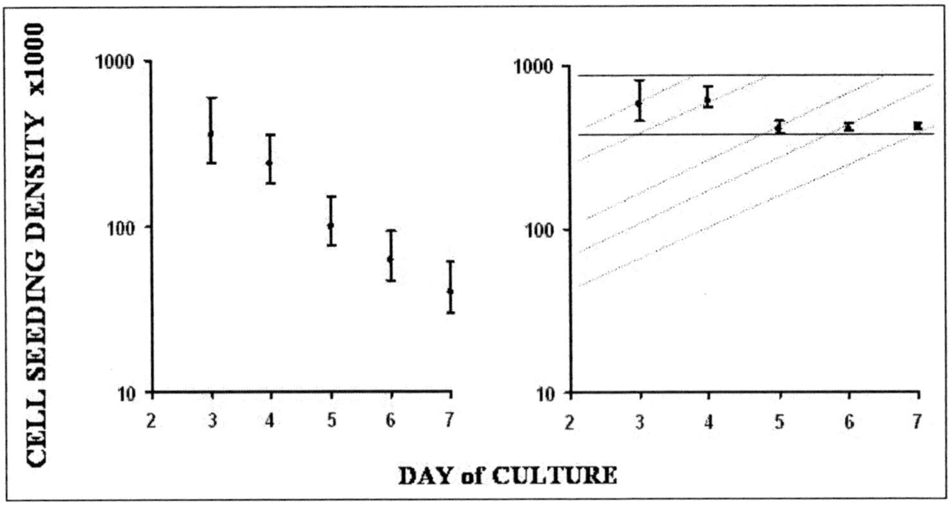

Figure. 8. Effects of culture duration on the observed optimal seeding cell density for osteoclastogenesis. Osteoclastogenesis was induced from mouse BMC cultures in 96-well plates by RANKL (15 ng/ml) in the presence of M-CSF (25 ng/ml). Parallel cultures were stopped daily from day 3 to day 7. As responses can vary from experiment to experiment, two successive experiments were performed with overlapping cell density ranges, to verify that the level of osteoclastogenesis was similar in the two experiments. The different morphological categories of osteoclast were separately identified and counted using Histolab software on microscopic images at 10 times magnification, displayed via a colour camera on a computer screen. 2-4 replicate wells were counted for each group. In each experiment the seeding cell density giving the highest levels of osteoclastogenesis (small cells in shorter duration experiments, spread cells on day 5 and the total of spread cells and phantoms on days 6 and 7) was identified. The variability bars represent the range in which the real optimal density could lie based on the intervals between cell densities used in the experiments, and how distinct was the peak of osteoclastogenesis. Optimal osteoclastogenesis occurred at progressively lower original seeding densities the longer the duration of the experiment (left graph). The actual cell density in the wells increases with the duration of cultures, and this was estimated in control wells from which RANKL had been omitted. There was a fairly regular increase from day to day of in the order of 1.6 fold per day. This estimate was used to calculate the approximate real cell densities in the wells on each day. It can be seen (right graph) that the estimated optimal actual cell density was relatively constant throughout the cultures, in the order of 500 thousand cells/well.

It has been observed that the role of ERK in RANKL-induced osteoclastogenesis from RAW264.7 cells is dependent on the seeding density [20]. When the cells were plated at the standard cell density for the method used, inhibition of the upstream ERK-activating kinase MEK with U0126 or PD98059 resulted in a block of osteoclastogenesis. This suggests that ERK activation had a permissive effect on osteoclastogenesis under these conditions. However, when a higher seeding cell density was used, the inhibitors induced a marked enhancement and acceleration of osteoclastogenesis, suggesting that under high density conditions ERK had an inhibitory effect. This might be one mechanism underlying the inhibition of osteoclastogenesis observed at high seeding densities in spleen cell and BMC cultures. As RAW264.7 is a cell line, there were no other cell types in these cultures except osteoclasts and their precursors, indicating that density can affect osteoclasts alone. As ERK is considered to play a role in osteoclast survival, such a mechanism could also play a part in the control of osteoclast apoptosis induced by high cell density.

In conclusion, cell density in the wells is a variable factor which has strong influences on osteoclastogenesis and apoptosis.

Well Size

It is interesting to compare the numbers of spread cells and phantoms in spleen cell cultures performed in either 24-well plates or in 96-well plates where the surface area of the wells is about 4 times smaller. Even accounting for the different numbers of cells which must be added to the wells to obtain the equivalent seeding cell density per surface area, spread cells developed in 24-well plates died more rapidly, so that there were practically always more phantoms than live spread cells. In contrast, in 96-well plates there were consistently more live spread cells than phantoms (Fig. 2)

Plates with different well sizes may not be made from identical plastic material, even when obtained from the same supplier, and this could be a factor influencing these observations. However, it seems possible that the size of the well is itself a factor in determining the rate of apoptosis of the osteoclasts. It may be related to the different surface area of the medium in the wells which allows gas exchange, but the reasons for this difference are not clear.

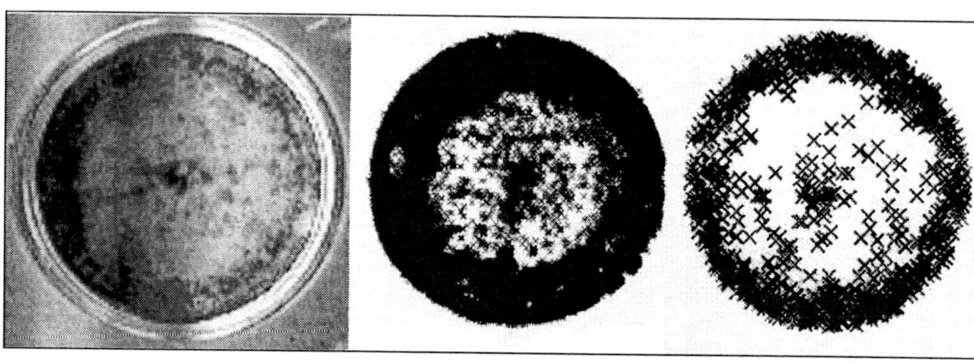

Figure 9. Cellular distribution on day 7 of spleen cell cultures (1 million cells per well in 24-well plates), stimulated with a high concentration of RANKL (100 ng/ml) in the presence of M-CSF (25 ng/ml). The left image is a low magnification image of a culture well. The regions with stronger TRAP staining (violet) shows darker in the black and white image as it was superposed on a light yellow background cell layer; The areas of TRAP staining give an indication of the distribution of osteoclasts in the well. The middle image shows a well with all individual elements (cells or particles) identified using Histolab software and each marked with a cross. The right image shows the plan of a well with the position of all osteoclasts present, irrespective of their stage of development (including phantoms), marked with a cross. All images demonstrate that there is a concentration of cells, including all forms of osteoclasts, around the periphery of the well and also at a point in the centre of the well.

Another possibility is that there is non-homogeneous distribution of cells in the wells. The bottom of the wells used were not completely flat. For example, the 24-well plates used had a dip in the centre and were lower around the edge, leaving a slightly raised ring in between. Cells tended to locate preferentially to the dip in the centre and to the periphery, particularly when larger numbers of osteoclasts developed (Fig. 9). It is possible that the migratory intermediate cells tend to locate in the lower flatter regions. There would be localised changes in cell density, and the increased cell density in the lower regions of the floor of the well would tend to increase the rate of spread cell apoptosis.

This effect was most evident in spleen cell cultures where, despite a higher seeding density, after the first medium change there were less adherent cells to establish a cell layer. It was less often observed in BMC cultures in 24-well plates where a relatively thick adherent cell layer formed.

This effect was also present in 96-well plates but was much less marked and seldom had a strong influence on the overall cell distribution within the plates, which generally remained more homogeneous across the plate.

Effects of Endogenous Mediators

RANKL as a Survival Factor

Both M-CSF and RANKL are considered to be survival factors for osteoclasts, in addition to being necessary for osteoclastogenesis [6,21]. The survival effect of RANKL was studied by either withdrawing it (in the continued presence of M-CSF) or blocking its actions with soluble receptor inhibitors (sRANK or osteoprotegerin (OPG)) in the last 1-2-days of 6 or 7- day BMC cultures. Concentrations of the soluble receptors which totally blocked the osteoclastogenic action of RANKL were chosen, to avoid the difficulties of interpretation of partial block of both the osteoclastogenic and survival actions of RANKL. In the absence of RANKL from day 5, whether withdrawn or blocked, no further osteoclasts developed. Furthermore, the spread cells that were already present rapidly entered into apoptosis, increasing the number of phantoms on day 6, with an equivalent decrease in spread cells. In cultures where RANKL was still present from days 5-6, there was only a small increase in the number of phantoms associated with the increasing number of spread cells. This indicates that the lack of RANKL caused the death of the great majority of spread cells that died by day 6. By day 7 there were large numbers of phantoms in the wells and very few smaller live spread cells (Fig. 10). The absolute numbers of spread cells and phantoms varied from experiment to experiment, but based on data from a series of experiments a general scheme was developed to represent these changes (Fig. 11).

Rapid apoptosis of osteoclasts occurs after removal of survival factors, and the intracellular mechanisms involved have been studied. The withdrawal of M-CSF results in an increase in the pro-apoptotic Bcl-2 family member Bim. This molecule is a member of the sub-family which possesses only a Bcl-2 homology domain 3 (BH3). These molecules transmit the pro-apoptotic signals to the mitochondria from other parts of the cell [14]. M-CSF can regulate Bim expression through activation of the mTOR protein [11], but can also induce ubiquitinylation and proteosomal degradation of Bim which is mediated via the protein c-Cbl [9].

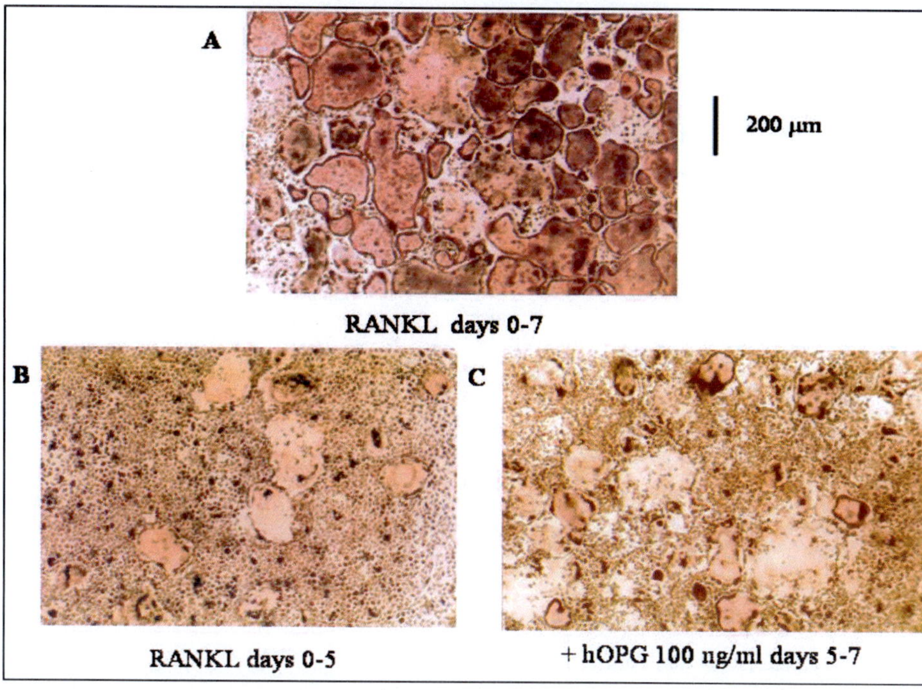

Figure 10. Effects of a lack, or block of RANKL on the survival of spread osteoclasts. Osteoclastogenesis was induced from mouse BMC cultures in 96-well plates (60 thousand cells/well) by RANKL (40 ng/ml) in the presence of M-CSF (25 ng/ml). Representative light microscopic images at 4 times magnification were recorded on day 7 from the central region of the culture well, using Histolab software. A.Cultures were stopped at day 7 when there were still a majority of spread cells with respect to phantoms in this experiment. B. RANKL was omitted from the medium from day 5, resulting in no further osteoclastogenesis and the death of existing spread cells in the last two days of culture. C. Human recombinant osteoprotegerin (hOPG), the endogenous decoy receptor for RANKL, was included in the cultures from day 5 in the continued presence of RANKL. A concentration of osteoprotegerin was used that caused near maximal inhibition of RANKL-induced osteoclastogenesis when it was added at the beginning of cultures. There was again a marked increase in phantoms although some live spread cells remained.

However, the precise mechanisms involved in the survival action of RANKL have not been clearly elucidated. Its signalling receptor RANK recruits TNF receptor associated factors (TRAFs) which act as adaptor molecules. They bind to the receptor and also to other cytoplasmic molecules, forming complexes which further activate intracellular signalling mechanisms. The intracellular part of RANK has three motifs which can bind TRAFs. Motifs 2 and 3 appear to bind TRAFs other than TRAF6, possibly TRAFs 3 and 1,2,5 respectively, but the molecular interactions and functional responses linked to these motifs are less well understood. TRAF binding to either motif leads to the activation of NF-κB and, in addition, p38 kinase can be activated via motif 2 [22]. Motif 1 binds TRAF6, which is considered to be the major TRAF molecule in the control of osteoclast functions. This leads to the activation of several signalling pathways, including the activation of the transcription factor, nuclear factor kappa B (NF-κB). NF-κB activation contributes to RANKL-induced osteoclast differentiation. RANKL can also inhibit osteoclast apoptosis by activating NF-κB, which in turn inhibits the core apoptotic mechanisms, in part via the increased expression of members of the family of inhibitor of apoptosis (IAP) proteins [6,22]. M-CSF can also maintain the stimulation of NF-κB, and this may also contribute to its anti-apoptotic activity [7].

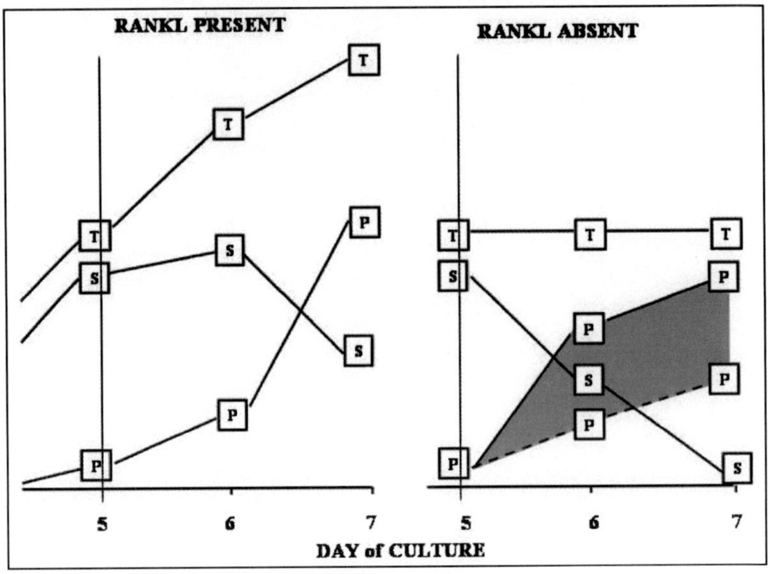

Figure 11. Schematic representation of the ratio of live spread cells and phantoms on days 5-7 of mouse BMC cultures, in the presence or absence of RANKL. This scheme is based on the results of a series of experiments. On day 5 spread cells [S] have begun to develop but there are very few phantoms [P]. In the continued presence of RANKL (left graph), on day 6 there is a clear increase in the total numbers of spread cells (live+phantoms) [T] but with a relatively small increase in phantoms. However, on day 7 as the total spread cells continue to increase the majority undergo apoptosis, despite the continued presence of RANKL and M-CSF. If RANKL is withdrawn or blocked from day 5 (right graph), osteoclastogenesis stops and total spread cells do not increase further. In addition, the removal of the survival effect of RANKL increases the rate of apoptosis, such that the majority of spread cells are dead on day 6, and virtually all are dead on day 7. The broken line represents the estimated increase in apoptosis that would have occurred if RANKL was still present. The grey area would then represent the result of the removal of the survival effect of RANKL.

TRAF6 bound to motif 1 can also recruit and increase the kinase activity of the enzyme c-Src which in turn activates phosphatidylinositol-3-kinase (PI3-K) and several downstream signalling pathways. PI3-K can also be activated via small GTPase binding proteins, particularly Rac-1 [23]. M-CSF activates Rac-1, and this is a critical factor in the activation of PI3-K and the survival effect of M-CSF. RANKL also activates Rac1, but not Rac2, via TRAF6 and this leads to activation of NF-κB [24]. However, RANKL induction of a receptor complex involving TRAF6, Rac1 and NADPH oxidase (Nox)1 in bone marrow monocyte/macrophage cells led to the transient production of reactive oxygen species (ROS) [25]. Inactivation of components of this complex inhibited responses to RANKL, including ROS production and activation of the mitogen-activated protein kinases (MAPKs) extracellular signal-regulated kinase (ERK), jun N-terminal kinase (JNK) and p38 kinase, as well as osteoclast differentiation. Such a mechanism could also be involved in osteoclast survival. The Ras/Raf system also leads to the activation of ERK, and blocking ERK activition with dominant negative Ras induced apoptosis in osteoclasts [26].

Signalling pathways downstream of PI3-K include activation of Akt/protein kinase B (PKB) which may contribute to the differentiation and the survival of osteoclasts, in part via activation of NF-κB [27]. There are conflicting data regarding the role of PKB. The use of small interfering RNAs to selectively inhibit the transcription of specific molecules indicated that PKB plays an important role in osteoclast differentiation via NF-κB activation, but does not play a role in osteoclast survival [11]. In these experiments, the survival effect mediated by PI3-K was mediated via phosphorylation of the mTOR protein (independent of PKB) and subsequent inactivation of the pro-apoptotic Bcl-2 family protein Bim. M-CSF can induce survival of osteoclasts via activation of mTOR and subsequent Bim inactivation [11]. However, selective block of individual signal pathways suggested that the PKB/NF-κB system is one of two PI3-K-activated pathways essential for osteoclast survival, the other being activation of the MAPK ERK. ERK has been shown to be a survival factor for osteoclasts, without affecting their bone resorbing capacity [26]. These experiments caused some doubt about the survival role of NF-κB. Inhibition of NF-κB formation inhibited bone resorption and this was reversed by reactivation of NF-κB, but osteoclast survival was not affected [26]. The influence of ERK on apoptosis appears to depend on the duration of its activation. 17beta-estradiol has been shown to have a pro-apoptotic effect on osteoclasts but an anti-apoptotic effect on osteocytes. These responses are associated with sustained (at least 24 h) or a transient (less than 30 min) activation of ERK, respectively. Experimental manipulation of the duration of ERK activation indicated that a transient ERK activation is associated with a pro-apoptotic effect and a sustained activation is associated with an anti-apoptotic (survival) effect in both cell types [28].

Other MAPKs can also be activated via TRAF6-receptor complexes or via PI3-K. These include the stress-activated enzymes JNK and p38 kinase, both of which may be involved in osteoclast differentiation [22]. In addition, TRAF6 bound to the receptor can increase intracellular calcium which, activates the enzyme calcineurin and, in concert with other activated signalling pathways such as NF-κB, leads to the activation of nuclear factor of activated T cells (NFAT) which has been proposed as the master transcription factor for osteoclastogenesis [29]. Osteoclast production of the enzyme cathepsin K is induced by RANKL and involves coordinated signalling through parallel pathways [30]. RANKL induces NFATc1 and p38 kinase which phosphorylates the NFAT, facilitating its entry into the nucleus. In the nucleus NFAT acts synergistically with other transcription factors (such as PU.1 and microphthalmia transcription factor) to enhance the transcription of a set of genes including the cathepsin K gene. This represents a mechanism of osteoclast maturation, and it remains to be determined whether these signalling pathways are also involved in osteoclast survival.

There is less strong evidence for the role of these stress-activated kinases in osteoclast survival, although it has been suggested that p38 kinase is a survival factor for developing multinuclear osteoclasts but not for their mononuclear precursors [31].

Signal pathways involved in osteoclast differentiation have become more complicated with the discovery of essential co-stimulation via immunoglobulin-like receptors (Fc receptor common gamma chain or DNAX-activated protein 12) and their associate receptor chains [32]. This is analogous with co-stimulatory signalling in immune responses. In cooperation with RANKL, these receptors activate phospholipase Cgamma (PLCgamma) and calcium signalling which are essential for the induction of NFATc1. The contribution of these mechanisms to the survival of the osteoclasts which develop, remains to be elucidated.

Calcitonin

Calcitonin is a major endogenous inhibitor of osteoclast functions. It causes developing osteoclasts to retract and become immobile and it inhibits bone resorption [33,34,35]. Calcitonin has also been shown to be a survival factor for multinuclear osteoclasts [18,36].

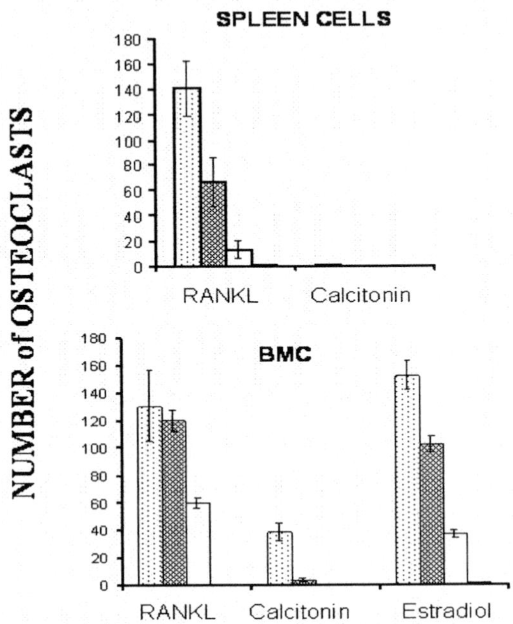

Figure 12. Effects of calcitonin on osteoclastogenesis from mouse precursor cells. Osteoclastogenesis was induced by RANKL in the presence of M-CSF (25 ng/ml) from mouse spleen cells cultured in 24-well plates (RANKL 100 ng/ml) and from mouse BMC cultures in 96-well plates (RANKL 40 ng/ml). The different morphological categories of osteoclast were separately identified and counted using Histolab software on microscopic images at 10 times magnification, displayed via a colour camera on a computer screen. 4 or 3 replicate wells were counted for each group in spleen cell or BMC cultures, respectively. The variability is represented by the standard error of the mean (SEM). Calcitonin (10^{-9}M), added from the beginning of 7 day spleen cell cultures (upper graph), completely inhibited the RANKL-induced development of multinuclear osteoclasts. No osteoclasts were observed in the wells treated with RANKL and calcitonin (Calcitonin) in comparison with the RANKL control wells (RANKL) where numerous small osteoclasts (3-6 nuclei, stippled bar) were present, as well as some intermediate osteoclasts (more than 6 nuclei, cross hatched bar) and a few spread cells (white bar). There were only occasional phantoms (grey bars). In 5-day BMC cultures (lower graph) calcitonin markedly inhibited RANKL-induced osteoclastogenesis although some smaller osteoclasts were still present. In comparison, 10^{-8}M 17β-estradiol (Estradiol) had no marked effect on RANKL-induced osteoclastogenesis in neighbouring wells on the same plate.

In the cultures reported in this chapter, calcitonin (salmon calcitonin 10^{-9} M) markedly inhibited osteoclast development if added at the beginning of cultures. This was particularly evident in spleen cell cultures, where virtually no multinuclear cells developed despite 7 days culture (Fig. 12). In BMC cultures in 96-well plates, calcitonin, added at the beginning of the culture, markedly inhibited osteoclast development on days 3,4 or 5 of culture, but mononuclear TRAP-positive cells and some small osteoclasts, mostly with 3-6 nuclei, were still present (Fig.12).

In the absence of calcitonin, by day 5 of culture a population of osteoclasts, including spread cells, had developed. When calcitonin was added on day 5, there was a decrease in intermediate (>6 nuclei) and spread cells observed on day 7. The decrease in spread cells was particularly marked and accompanied by an increase in phantoms, suggesting increased spread cell apoptosis (Fig. 13). In the absence of RANKL (days 5-7) calcitonin did not have any apparent effects.

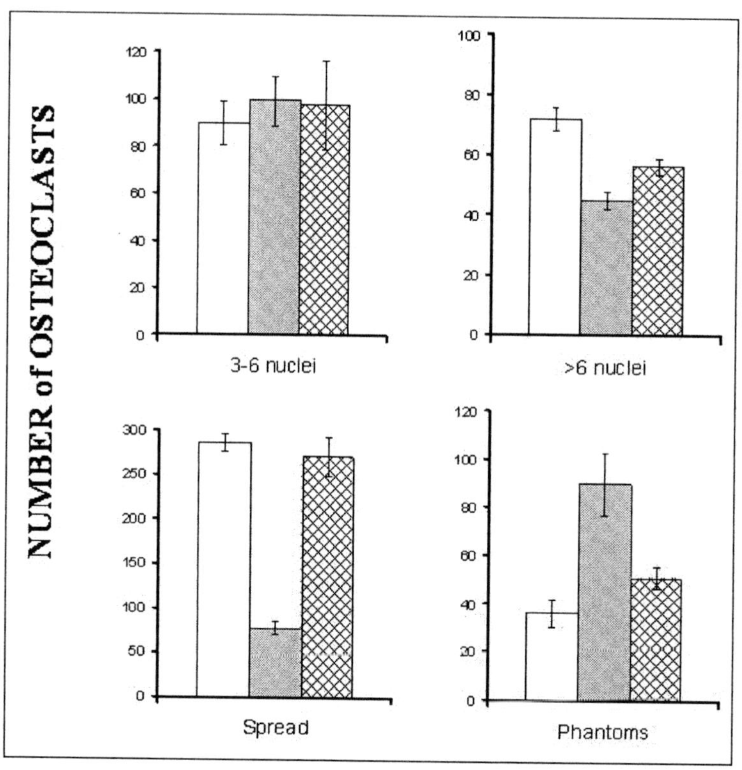

Figure 13. Effects of calcitonin (10^{-9} M, grey bars n=3 wells) or estradiol (10^{-8} M crosshatched bars, n=4) on the numbers of different morphological forms of osteoclasts when administered on only the last two days of 7-day BMC cultures (60 thousand cells/well), after a significant number of spread cells had already been induced by RANKL (40 ng/ml) in the presence of M-CSF (25 ng/ml). The different morphological categories of osteoclast were separately identified and counted using Histolab software on microscopic images at 10 times magnification, displayed via a colour camera on a computer screen. The variability of the counts from the replicate wells is represented by the SEM. In comparison with the RANKL control wells (white bars, n=6), the addition of calcitonin decreased intermediate (>6 nuclei) and live spread cells, and this was associated with an increase in phantoms. Estradiol did not have any marked effect when added at the end of these cultures.

Calcitonin is clearly capable of inhibiting osteoclast development when administered at the beginning of cultures. In BMC cultures this may particularly affect the fusion of small multinuclear cells, perhaps by inhibiting their motility and causing them to retract [33]. Any multinuclear cells that had already developed would remain alive in the wells [18,36]. In clinical use, there is an escape phenomenon from the therapeutic effect of calcitonin (inhibition of bone resorption) and this is due to downregulation of the calcitonin receptor on osteoclasts with continued exposure to calcitonin [37]. It would appear that there is no "escape" from calcitonin, at least in spleen cell cultures, as no osteoclasts began to develop within the 7-day culture period. When a population of osteoclasts had already established, it would be expected that the addition of calcitonin would prevent any further development stimulated by the continued presence of RANKL. This may particularly affect the fusion stage, preventing particularly the development of larger multinuclear cells and their subsequent spreading, that would occur in control wells. The observed reduction of intermediate cells (>6 nuclei) and spread cells, without a decrease in small cells (3-6 nuclei), is consistent with this hypothesis. This analysis does not allow any conclusions regarding the survival of the remaining live cells. Calcitonin has been shown to be a survival factor, at least for non-spread multinuclear osteoclasts [18,36] and it is possible that calcitonin maintains the survival of these cells while inhibiting their development. However, there was a clear increase in phantoms, suggesting that spread cell apoptosis was increased. This probably contributed to the particularly large decrease in spread cells, in addition to the inhibition of their development.

It might be that calcitonin activates the mechanism which causes the retraction of the motile intermediate cells, but this is an inappropriate stimulus in these spread and strongly adhered cells, and it results in apoptosis. Part of the retraction process may involve de-adhering the peripheral area of the cell with disruption of the actin ring structure [33,34,36], and in spread cells, such changes could result in apoptosis. The mechanisms underlying the anti-apoptotic effect in intermediate cells are not fully elucidated, but if they occur in the spread cells, the survival stimulus is not strong enough to prevent their apoptosis. It is interesting that, while calcitonin had a clear survival effect on adhered intermediate cells, like other survival factors (RANKL or M-CSF) it was unable to prevent apoptosis in cells that were not adhered [18]. There may be a balance of conflicting stimuli which control the induction of apoptosis in osteoclasts, as in other cell types [14], which is particularly related to the functions of spread cells.

In the absence of RANKL from day 5 of culture, stimuli for both osteoclast development and survival are reduced or removed. Remaining live cells die rapidly and while the phantoms are a good indication of spread cell death, smaller osteoclasts also enter into apoptosis (see CASPASE ACTIVATION section). Calcitonin induced only a small increase in the numbers of live osteoclasts on day 6, there being little difference from controls on day 7 when few live cells remained. This does not provide strong evidence for a survival effect of calcitonin after RANKL withdrawal. Furthermore, there was little evidence that calcitonin increased the rapid rate of apoptosis of spread cells under these conditions. It is likely that the absence of RANKL is a far stronger stimulus for spread cell apoptosis than calcitonin. By day 7 virtually all spread cells were dead, even in the absence of calcitonin.

Estradiol

Estradiol is a major hormonal control of bone resorption and the lack of estrogens after menopause is considered to be the major cause of increased osteoclastogenesis and bone resorption which occurs in post-menopausal osteoporosis [2,15].

17-β Estradiol (10^{-7}-10^{-8}M) had no apparent effect in the majority of the culture experiments which we performed. A typical result on day 5 of BMC cultures is shown in figure 12, in comparison with a marked inhibitory effect of osteocalcin which was observed in neighbouring wells. A consistent effect across all culture wells was observed only when estradiol was included on days 5-7 of 7 day BMC cultures. A small decrease in intermediate cells together with a small increase in phantoms was observed (Fig. 13). This might indicate a decrease in multinuclear cell development together with an increase in spread cell apoptosis. However, this was not a marked response, and in general terms, estradiol had no major effects in these culture systems.

The cell populations cultured in these experiments include few if any osteoblastic or stromal cells. In this way it was hoped to characterise the direct effects of exogenously-added agents on osteoclastic cells. Direct effects of estrogens on osteoclasts have been demonstrated under certain conditions [6,28], although it has also been suggested that estrogens do not have direct effects on osteoclasts [38] even though androgens do. Direct effects of estrogens (inhibition of development and functions as well as induction of apoptosis) can be demonstrated if the expression of estrogen receptors (particularly ERα) is induced [39,40]. It seems likely that the responsiveness of any given osteoclast population depends on the level of expression of estrogen receptors.

Human primary BMC mononuclear cells can develop into osteoclasts. After 5 days of culture, the expression of ERα, but not ERβ, was detected [41]. It was no longer detected at the end of the cultures (15 days), suggesting that any effects of estrogens occurred at an early stage. 17β-estradiol or the estrogen receptor selective agonist, raloxifen, decreased the numbers of multinucleated osteoclasts which developed in these cultures. However, neither agent affected osteoclast apoptosis, which occurred late in the cultures. It is interesting that RANKL, which is routinely used to induce osteoclastogenesis, decreased estrogen receptor expression [42]. It is therefore possible that, even if estrogen receptors are expressed on the osteoclast precursor cells, RANKL may downregulate the receptors sufficiently rapidly to prevent responses to estrogens later in the cultures.

The RAW264.7 cell line can be induced to differentiate into osteoclasts. These cells were shown to express ERα, but not ERβ, under the experimental conditions used, and responded to estradiol and the phytoestrogens diadzein and genistein [42]. These agonists reduced RANKL stimulation of osteoclast differentiation markers, although never more than 50%. They also inhibited RANKL-induced NF-κB activation and nuclear localisation.

However, other experiments using RAW264.7 cells as osteoclast precursors have used artificially-increased receptor expression to demonstrate the actions of estrogens (decreased osteoclastogenesis, induction of osteoclast apoptosis and decreased RANK signalling as indicated by phosphorylation of ERK) [39,40].

It was originally shown that estrogen downregulates its own receptors [43,131] and that estrogen only directly inhibits osteoclast activity in cell populations taken from donors with low circulating estrogen levels. In the experiments reported in this chapter, the osteoclast precursor cells were obtained from young female mice which presumeably have circulating estrogen levels, and this could be a reason for low estrogen receptor expression and thus the observed lack of clear response to estrogen. However, in RAW264.7 cells, estrogen and phytoestrogens increased ERα expression [42]. It remains to be determined what are the precise effects of RANKL and estrogens on estrogen receptor expression, and thus what is the degree of receptor expression in the *in vitro* cell culture models of osteoclast development and apoptosis. This must be determined in order to know how relevant these models are to the *in vivo* and clinical situations.

It seems most likely that the lack of response to estradiol in the experiments reported in this chapter is due to insufficient expression of estrogen receptors under the experimental conditions used.

TNF Superfamily Apoptosis-Inducing Ligands

RANKL is a member of the TNF ligand superfamily (TNFSF11) and RANK is a member of the TNF receptor superfamily. The ligands all share a homologous homotrimeric structure, with the specificity of interaction with their specific receptors being mostly determined by sequence differences in surface loops [5]. Binding of the ligand to receptors dimerises or trimerises the receptor chains, bringing the intracellular sections into proximity and instigating signalling cascades. TNFα is one of the original members of this family (TNFSF2) and is often involved in inflammatory responses. It has the capability of inducing apoptosis in a wide range of cell types under appropriate conditions when the appropriate receptor is expressed. TNFR-1 is a so-called "death receptor" as the intracellular sequence contains motifs called "death domains" which recruit molecules that induce apoptosis mechanisms within the cell. However, signalling through TNFR1 only seems to induce apoptosis in osteoclasts in limited situations. Apoptosis induced by isolation of osteoclasts from their support cells was reported to be accelerated by TNFα via activation of upstream, initiator caspase-8 and then effector caspases-3 and 6 [44]. However, TNFα has been more frequently shown to induce osteoclast survival [45,46,47,48]. TNFα, in the presence of M-CSF, has also been shown to induce osteoclastogenesis from different precursor cell populations, independent of the RANKL-RANK-TRAF6 system [4,5,49,50,51]. Precursors from spleen cells appear to be particularly responsive to TNFα (threshold concentration of recombinant TNFα of 5-10 ng/ml) [4,5] while it is ineffective at the same lower concentrations (maximum concentration tested 50 ng/ml) in BMC cultures [4,52]. However, higher concentrations of TNFα were shown to induce RANKL-independent osteoclastogenesis in selected M-CSF-dependent bone marrow macrophages [49]. The developing osteoclasts tended to be less numerous and smaller osteoclasts predominated [4] (see Fig. 20). There were relatively few spread cells and they tended to show weaker TRAP staining than cells developed in the presence of RANKL. Consistent with this, TNFα-induced osteoclasts did not efficiently resorb bone, although the addition of IL-1 reinstated bone resorption [49,50] without necessarily increasing osteoclast numbers [50]. This action of TNFα appeared to involve TNFR1 and, to a lesser extent, TNFR2 which contains no death domains [49,50]. A first step in this induction of osteoclastogenesis by TNFα could be the increased expression of c-Fms, the receptor for M-CSF [53].

As TNF receptors and RANK are in the same structural superfamily and tend to activate the same types of intracellular signalling mechanism, it is likely that there are common signalling mechanisms involved in the osteoclastogenesis induced by the two ligands, although differences in the degrees of their activation may explain the qualitative differences between the populations of osteoclasts which develop. For example, TNFα mainly activates TRAF2 rather than TRAF6 in inducing osteoclastogenesis [54]. A lack of TRAF2 markedly reduced TNFα-induced osteoclastogenesis but only caused a minor reduction (20%) in that induced by RANKL.

It has been shown that RANKL stimulation of the receptor RANK leads to an increase in the transcription of TNFα in osteoclasts [55], leaving it possible that, if the protein is released, it could play a role in osteoclastogenesis induced by RANKL. Toll-like receptors (TLR) are a family with structural homology in their intracellular sequences, which includes receptors activated by IL-1 and lipopolysaccharides (LPS). Activation of TLR 9 by CpG-containing oligonucleotides induces osteoclastogenesis in RANKL-primed precursors. This was associated with increased transcription and translation of TNFα [56]. This raises the possibility that RANKL and TNFα co-operate in induction of osteoclastogenesis in some situations *in vivo*.

In order to study the effects of TNFα on spread osteoclast survival, experiments were performed in BMC cultures where TNFα has no effect alone. Osteoclasts were developed in the presence of RANKL and M-CSF until day 5, when many spread cells had developed and some phantoms had begun to appear. Then TNFα (20-50 ng/ml) was added to the cultures in the continued presence or absence of RANKL, and cultures were stopped on days 6 or 7. In all conditions there was a marked increase in the number of live spread cells with a decrease in phantoms. The increase in spread cells may be in part due to accelerated differentiation of osteoclasts as there are less non-spread osteoclasts, which may have fused to form spread cells. By reference to concentration-response curves of RANKL, it can be seen that, if a higher concentration of RANKL had been used under these conditions to induce a similar increase in spread cells, the majority of these cells would die and be observed as phantoms. This limited the numbers of live spread cells. TNFα did not act in the same way as an increased concentration of RANKL, as the large numbers of spread cells which develop do not die, and there are less phantoms in TNFα-treated wells than in control wells, indicating that, in addition, some of the spread cells which would have died in the absence of TNFα have also survived. This is evidence of a strong survival effect of TNFα on spread osteoclasts. This effect was particularly striking in the absence of RANKL. On day 7, in the absence of the survival effect of RANKL, control wells contained virtually no live spread cells but many phantoms. TNFα again induced a marked increase in spread cells but a small decrease in phantoms (Fig.14). It appeared that TNFα had replaced the effect of RANKL on osteoclast differentiation, and perhaps on survival, but that it had superimposed an additional strong survival effect.

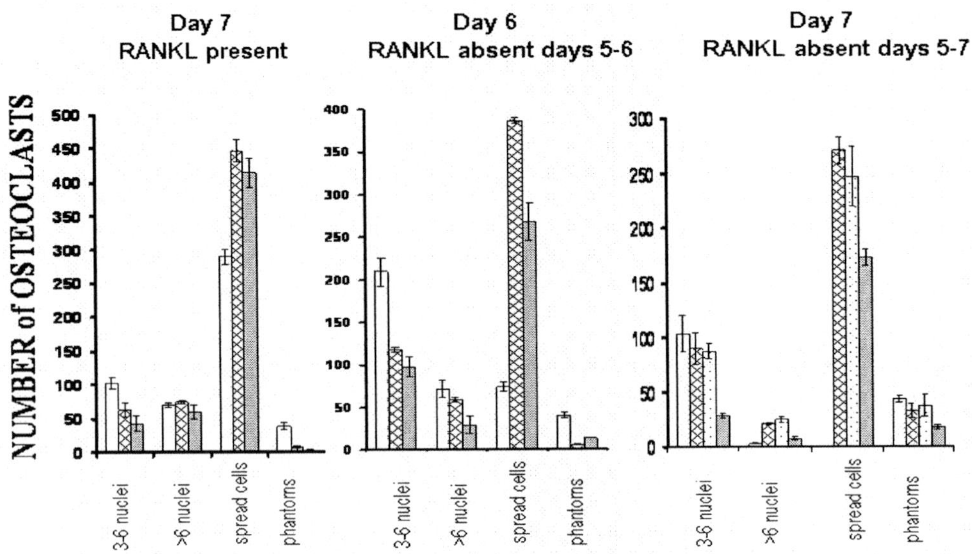

Figure 14. Effects of TNFα or IL-1β on the numbers of different morphological forms of osteoclasts (indicated below the groups of bars) when administered from day 5 in 7-day BMC cultures (60 thousand cells/well) in the continued presence of RANKL (left graph), or in 6 or 7-day cultures after the withdrawal of RANKL on day 5 (centre and right graphs, respectively). By day 5, a significant number of spread cells had already been induced by RANKL (40 ng/ml) in the presence of M-CSF (25 ng/ml). The different morphological states were separately identified and marked on microscopic images displayed on a computer screen and automatically counted using Histolab software. The variability of the counts from the replicate wells is represented by the SEM. Responses in the presence of TNFα (50 ng/ml crosshatched bars, or 20 ng/ml stippled bars) or IL-1β (10 ng/ml grey bars) are grouped with the control data (white bars) for each category of osteoclast, on each graph. Note that both cytokines induced a marked increase in spread cells in the presence of RANKL on day 7 (left graph control n=6, TNFα n=3, IL-1β n=3) or in the absence of RANKL on day 6 (centre graph, control n=6, TNFα n=3, IL-1β n=3) or on day 7 (right graph, control n=6, TNFα 50 ng/ml n=4, 20 ng/ml n=4, IL-1β n=4), when there were virtually no live spread cells in the control wells. This was not associated with an increase in phantoms, which were, in most cases, decreased in numbers.

These data suggest that TNFα is not simply mimicking RANKL and that it must be acting via different mechanisms. Although, survival effects of TNFα on osteoclasts have been observed under a variety of conditions [45,46,47,54], it has not been universally observed. Despite an increase in osteoclast differentiation from mouse BMC, it did not stimulate survival [52]. It was suggested that TNFα acted only at an early stage of osteoclastogenesis. However, other studies have demonstrated survival effects of TNFα on differentiated osteoclasts, and it could be that it only occurs at later stages of differentiation. It has been proposed that the anti-apoptotic effects of RANKL, M-CSF and TNFα may be initiated by different signalling pathways, but all converge on the mTOR pathway as a major component of their anti-apoptotic effects [45]. It has also been proposed that TNF-α induces osteoclast survival via the production of nitric oxide (NO) [46]. As neither RANKL nor IL-1 induced the production of NO, this may be a mechanism involved in the additional strong anti-apoptotic effect observed in the experiments reported above. Other studies have involved the active induction of osteoclast apoptosis by a bisphosphonate and the demonstration of a protective action of TNFα in TNF-transgenic mice. The anti-apoptotic action appeared to be related to an increase in the expression of the Bcl-2 family anti-apoptotic protein Bcl-xL [47].

It remains to be determined which mechanisms underlie the striking anti-apoptotic effect of TNFα on spread osteoclasts in BMC cultures. Release of NO would be a possible candidate. However, most published experiments were performed with non-spread osteoclasts and it is also possible that different mechanisms become operative when the osteoclasts undergo this particular morphological change.

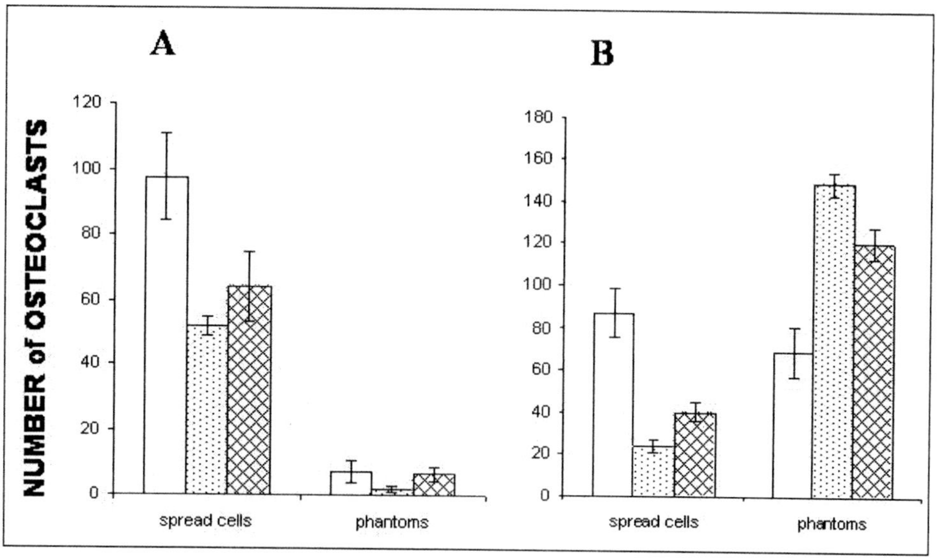

Figure 15. Effects of recombinant FasL (100 ng/ml stippled bars) or TRAIL (100 ng/ml crosshatched bars) on the numbers of live spread osteoclasts and phantoms on day 5 (A) or day 7 (B) of mouse BMC cultures (60 thousand cells /well in 96-well plates). Osteoclastogenesis was induced by RANKL (40 ng/ml) in the presence of M-CSF (25 ng/ml), and FasL and TRAIL were added from the beginning of cultures. The different morphological categories of osteoclast were separately identified and counted using Histolab software on microscopic images at 10 times magnification, displayed via a colour camera on a computer screen. Triplicate wells were counted for all groups and the variability is represented by the SEM. Note that spread cells were decreased by both cytokines and this was associated with an increase in phantoms on day 7 (B).

Other members of the TNF superfamily, Fas ligand (FasL, TNFSF6) or TNF-related apoptosis-inducing ligand (TRAIL, TNFSF10) can also induce apoptosis in many cell types via activation of their own death receptors which are members of the TNF receptor structural superfamily. FasL is well established in the induction of apoptosis, but it is possible that the more-recently discovered TRAIL represents the primary physiological system of tumour surveillance and cell killing [5]. However, there is conflicting evidence regarding the induction of apoptosis by these molecules in osteoclasts. FasL has been shown to induce apoptosis in human, avian and RAW264.7-derived osteoclasts [57]. Similarly, a Fas-activating antibody induced apoptosis in osteoclasts derived from human cord blood monocytes [58]. However, it has been shown that osteoclast precursors and mature osteoclasts express FasL as well as the receptor Fas (CD95), and endogenous FasL plays a permissive role in RANKL-induced osteoclastogenesis from bone marrow-derived macrophages [59]. As an anti-FasL neutralising antibody did not modify the survival of the osteoclasts it was proposed that endogenous FasL does not play a role in the apoptosis of osteoclasts. The roles of FasL in osteoclast functions *in vivo* may be complex. As FasL (like RANKL) is mostly expressed on the cell membrane, rather than being released in soluble form [5], the proximity of FasL-expressing cells is probably necessary for it to play its physiological role *in vivo.*

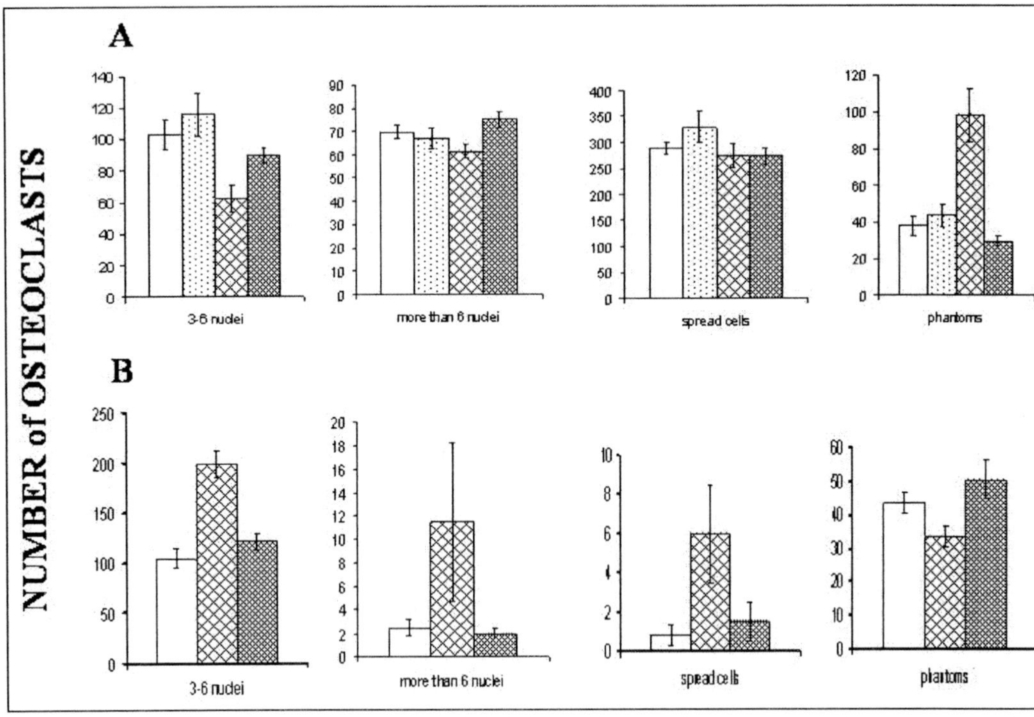

Figure 16. Effects of recombinant FasL (100 ng/ml stippled bars) or TRAIL (100 ng/ml large crosshatched bars or 20 ng/ml small crosshatched bars) on the numbers of different morphological categories of osteoclast (indicated below the groups of bars) when added on days 5-7 of 7-day mouse BMC cultures (60 thousand cells /well in 96-well plates), in the continued presence of RANKL (A) or in the absence of RANKL from day 5 (B). Osteoclastogenesis was induced by RANKL (40 ng/ml) in the presence of M-CSF (25 ng/ml). The different morphological categories of osteoclast were separately identified and counted using Histolab software on microscopic images at 10 times magnification, displayed via a colour camera on a computer screen. 6 control wells were counted in comparison with 4 wells for all groups treated with the cytokines, and the variability is represented by the SEM. Note that TRAIL 100 ng/ml increased phantoms in the presence of RANKL (A, right graph), an effect not observed with a lower concentration of TRAIL or with FasL. This effect was not present in the absence of RANKL (B) but a stimulatory effect of TRAIL on osteoclastogenesis (increase in small cells and in the low numbers of intermediate and spread cells) was revealed.

TRAIL receptors were shown to be expressed on osteoclasts derived from human cord blood monocytes and TRAIL increased the rate of apoptosis [58]. TRAIL, administered early in cultures of either human peripheral blood mononuclear cells or RAW264.7 cells, inhibited RANKL-induced osteoclastogenesis [60]. It is not clear if this was due to induction of cell death in precursor cells or early osteoclasts, but it was associated with a profound reduction of RANKL-induced activition of p38 kinase (but not ERK). However, TRAIL has not been shown to promote osteoclast development at any stage.

These ligands (soluble recombinant proteins) were tested in the BMC model reported in this chapter. When included from the beginning of cultures, both FasL and TRAIL (100 ng/ml) induced a small decrease in osteoclastogenesis, resulting in a small decrease in spread cells on day 5. There was no increase in the low level of phantoms at that stage (Fig. 15). However, when cultures were continued until day 7, a more striking decrease in spread cells was associated with clear increases in phantoms with both ligands (Fig. 15). This suggests that the apoptosis of spread cells which occurs late in cultures is potentiated by these ligands. When the ligands were added for only the last two days of 7-day cultures, TRAIL induced a decrease in small osteoclasts but not in intermediate and spread osteoclasts, but still increased the number of phantoms. This suggests that the great part of the decrease in spread cells occurred early in the cultures, but that apoptosis was still increased by a direct action at the end of the cultures. A lower concentration of TRAIL (20 ng/ml) did not decrease small cells or increase phantoms. FasL-treated cultures showed only a trend to increase small and spread cells, but there was no clear increase in phantoms (Fig. 16). It appears that the pro-apoptotic effect observed on day 7 requires the presence of FasL in the cultures for a short period before the day 5 change of medium.

When TRAIL was included on day 5 but RANKL was withdrawn, TRAIL induced an unexpected, but concentration-related effect. In the absence of RANKL no further osteoclasts developed after day 5 and the existing spread cells entered into apoptosis. Therefore, control wells contained virtually no intermediate or spread osteoclasts but there were many phantoms. However, under these circumstances there are still large numbers of small osteoclasts (3-6 nuclei) which appear to be less affected by the absence of RANKL. TRAIL induced a clear increase in the number of small cells. There were also small increases in the very low numbers of intermediate and spread osteoclasts. However, there was no increase in phantoms. A lower concentration of TRAIL (50 ng/ml) did not induce this response (Fig. 16). It appears that in the presence of rapid apoptosis induced by the absence of RANKL, no potentiation of apoptosis by TRAIL was observed. However, an increase in the early phase of osteoclastogenesis was uncovered.

Early effects of FasL on osteoclastogenesis may be related to the expression of Fas on the precursors. During the first day of osteoclastogenesis from either RAW264.7 cells or mouse bone marrow monocyte/macrophage progenitors, RANKL increased the expression of Fas [61]. In the absence of RANKL these cells showed very low levels of Fas expression and were relatively resistant to FasL-induced apoptosis [59]. In this way, when stimulated by RANKL, the precursor cells become sensitive to FasL and may undergo apoptosis when stimulated at an early stage in cultures. This may explain the small decrease in early osteoclastogenesis observed on days 3 and 4 of BMC cultures (data not shown) which resulted in the decrease in spread cells observed on day 5. During this period the osteoclasts which had developed did not appear to be affected by FasL. Osteoclasts maturing in the presence of a high concentration of RANKL showed downregulation of Fas, leading to a lack of responsiveness to FasL [61]. However, in these studies, FasL was able to increase apoptosis in mature osteoclasts, although it was a much weaker response than that induced by growth factor withdrawal [61]. Therefore, Fas may be sufficiently expressed on spread cells to induce the increase in apoptosis of these cells observed on day 7. However, FasL did not increase phantoms when it was present only on days 5-7. This suggests that FasL exposure was required for a short period before day 5, to initiate the response seen on day 7. In contrast, there was a trend to increased live cells suggesting that there may have been a small increase in osteoclastogenesis. Under certain circumstances FasL ligation of Fas can lead to increased osteoclastogenesis, as was indicated by the role of endogenous FasL expressed on osteoclasts [59]. Increased intracellular calcium plays an important role in RANKL-induced osteoclastogenesis, activating calmodulin and a downstream pathway that leads to the expression of the transcription factor NFAT [61]. Calmodulin binds to the intracellular component of Fas [63] and this may be a signalling mechanism involved in the effects of FasL on osteoclasts. It is possible that responses of osteoclasts to FasL are determined by a changing balance of conflicting effects which depend upon the environmental conditions.

Osteoclast precursors express TRAIL receptors (R1-4) [58]. However, not all these receptors result in intracellular signalling. TRAILR3 has no intracellular domain and is considered to be a decoy receptor. TRAILR1 and 2 are signalling receptors which contain intracellular death domains, however, TRAILR4 has a truncated, non-functional death domain, but it can still lead to the activation of NF-κB [5]. The balance of expression of these receptors is important in determining the response. However, TRAIL receptor expression was not detected in osteoclasts extracted from normal bone, although osteoclasts from the bones of myeloma patients did express TRAIL receptors to a variable extent [58]. Thus it is not clear how results from *in vitro* experiments are relevant to healthy patients. However, early stages of osteoclastogenesis in cellular models were inhibited by TRAIL [60], and this may correspond to the small decrease in osteoclastogenesis and spread cell development observed in the experiments reported above. It is not established whether this effect is due to increased apoptosis of early osteoclasts, and it is possible that it is directly due to a reduction in RANKL-induced intracellular signalling, as it was related to a profound inhibition of RANKL-induced p38 kinase activity [60]. However, similar to FasL, TRAIL increased the number of phantoms on day 7 of culture, indicating an increase in apoptosis of spread cells. This effect was also induced when TRAIL was added to cultures only on days 5-7. This makes it likely that it is a result of a direct effect of TRAIL on spread cells, perhaps by potentiation of the low level of apoptosis which occurred despite the continued presence of RANKL. However, in the absence of RANKL, TRAIL had a surprising effect. There was a clear increase in live cells, particularly small osteoclasts with 3-6 nuclei. There was no increase in phantoms in the presence of a relatively high rate of apoptosis due to the absence of RANKL. The results from these experiments suggest that the pro-apoptotic effects of FasL and TRAIL are far weaker than that of RANKL withdrawal. It has also been observed that apoptosis induced by FasL in osteoclasts was weaker than that induced by withdrawal of trophic factors [59]. It is possible that any pro-apoptotic effect of TRAIL is insignificant in conjunction with that induced by RANKL withdrawal or indeed, that it cannot take place because the mechanisms involved have already been maximally activated by RANKL withdrawal. However, a stimulatory effect was uncovered, either on the process of osteoclastogenesis or on the survival of smaller osteoclasts, which was not apparent in the presence of RANKL. The absence of RANKL may have induced conditions in which TRAIL increases the numbers of small osteoclasts, in opposition to the tendency for these cells to die following the removal of the survival effect of RANKL. It is possible that TRAIL can at least partially replace RANKL by activating some common signalling pathways.

IL-1 Cytokine Family

IL-1 is a potent stimulator of osteoclastogenesis in the presence of stromal/osteoblastic cells via stimulation of the expression of RANKL on the surface of these cells. However, IL-1 alone does not induce osteoclastogenesis in cultures of cells containing osteoclast precursors but no stromal/osteoblastic cells, even in the presence of M-CSF [64]. This is true for the spleen cell and BMC cultures used in the studies reported in this chapter. IL-1 does potentiate osteoclastogenesis in these cultures induced by either RANKL of TNFα [65,66]. This potentiation of RANKL in murine bone marrow monocyte cultures was blocked by IL-1 receptor antagonist (IL-1RA), a ligand homologue which binds to the receptors but has no intrinsic activity [66]. IL-1 did not show this effect in precursors isolated from IL-1 type I receptor-deficient mice, but interestingly, the effect of RANKL was reduced in the absence of IL-1 receptors. RANKL-induced osteoclastogenesis was also reduced by IL-1RA, suggesting that, in part, the action of RANKL was mediated via IL-1. This hypothesis was supported by an observed increase in IL-1α protein production induced by RANKL [66]. This raises the possibility that RANKL stimulates the production of other cytokines such as IL-1 and TNFα during osteoclastogenesis to maintain or amplify the response. IL-1α and IL-1β appear to act on the same receptors, and both ligands are active on osteoclasts. IL-1 can also increase the resorptive activity of osteoclasts [67]. It has also been established that IL-1 increases the survival of osteoclasts [26,36,68,69].

The effects of IL-1β (1-10 ng/ml) on RANKL-induced osteoclastogenesis were studied in both spleen cell and BMC culture models. When included from the beginning of 7-day spleen cell cultures, IL-1β (10 ng/ml) markedly changed the shape of the concentration-response curves of RANKL for the induction of the different morphological stages of osteoclast (Fig. 17). IL-1β particularly induced a marked increase in spread cells at all RANKL concentrations, associated with an equivalent decrease in phantoms. Small osteoclasts (3-6 nuclei) were increased at the lowest concentration of RANKL (6.25 ng/ml) but were decreased at all other concentrations. This suggests that IL-1β stimulated osteoclastogenesis, increasing the number of small osteoclasts, but that these cells fuse to form intermediate then spread cells at higher RANKL concentrations. Observed increases in the numbers of intermediate cells are consistent with this hypothesis and a decrease in numbers with respect to the RANKL control at the highest concentration (50 ng/ml) could reflect the increased spreading of these cells. In these spleen cell cultures in 24-well plates, the majority of live spread cells which developed at higher RANKL concentrations, were dead. Despite a marked, IL-1β -induced increase in the number if spread cells, there was no increase in phantoms. To the contrary, the number of phantoms was clearly reduced. This suggests that IL-1β had a strong survival effect on the newly-developed spread cells and also protected against the apoptosis which occurs in the majority of spread cells in the presence of RANKL in the last 2 days of these cultures.

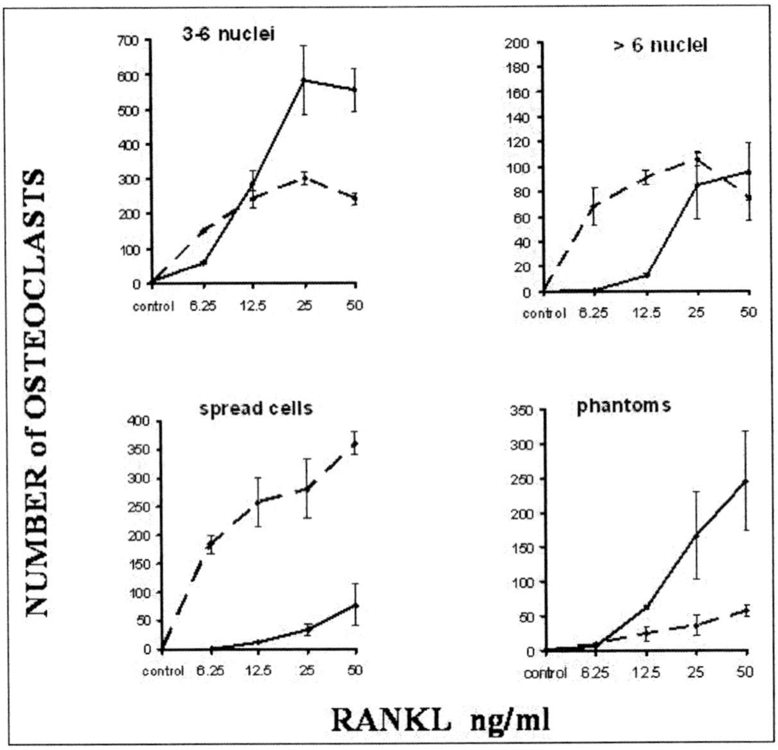

Figure 17. Effects of IL-1β (10 ng/ml) on the concentration-effect curves for RANKL-induced development of different morphological categories of osteoclast (indicated above the graphs) in 7-day spleen cell cultures (24-well plates, 1 million cells/well), in the presence of M-CSF (25 ng/ml). The different morphological categories of osteoclast were separately identified and counted using Histolab software on microscopic images at 10 times magnification, displayed via a colour camera on a computer screen. Duplicate wells were counted for all groups, and the variability is represented by the SEM. RANKL alone (solid lines) induced concentration-related increases in small and intermediate osteoclasts at 6.25-25 ng/ml, which did not further increase at 50 ng/ml, probably due to the fusion and spreading of cells. The majority of spread cells that developed at higher concentrations underwent apoptosis such that there were more phantoms than live spread cells. IL-1β (broken lines) increased non-spread osteoclasts in combination with low concentrations of RANKL, but there was a decrease, particularly in small osteoclasts (3-6 nuclei), with higher concentrations of RANKL.This was associated with a marked increase in spread cells and may result from the fusion and spreading of smaller cells. Despite the increase in spread cells there was a clear decrease in phantoms, suggesting an anti-apoptotic effect of IL-1β.

These changes were particularly clear in an experiment with a constant concentration of RANKL (30 ng/ml) combined with a range of lower concentrations of IL-1β (Fig. 18). Again, spread cells were increased by IL-1β and phantoms were decreased, both in a concentration-related manner. The threshold concentration for IL-1β was below the lowest concentration used (1 ng/ml). Small and intermediate cells were increased at lowest IL-1β concentrations, this being particularly clear at 4 ng/ml. However, at 8 ng/ml there were less cells, particularly small osteoclasts, and this correlated with a large increase in the number of spread cells. Again, it seems likely that this is due to rapid fusion and differentiation to form spread cells. This confirms that IL-1β stimulates osteoclastogenesis, particularly the formation of spread cells, and has a strong survival effect on spread osteoclasts, even at low concentrations (1-4 ng/ml).

However, there may be another factor influencing osteoclasts in the presence of IL-1β.

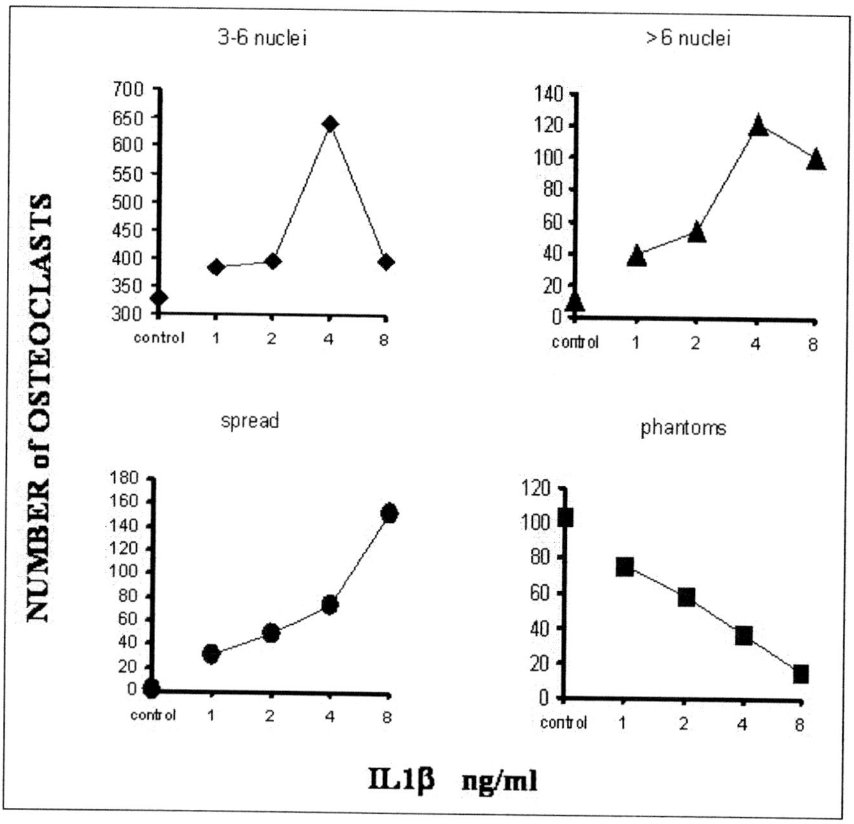

Figure 18. Concentration-effect curves for the effects of low concentrations of IL-1β on development of different morphological categories of osteoclast (indicated above the graphs, and by different symbols) induced by RANKL (30 ng/ml) in spleen cell cultures (24-well plates, 1 million cells/well). The different morphological categories of osteoclast were separately identified and counted using Histolab software on microscopic images at 10 times magnification, displayed via a colour camera on a computer screen. A representative single line of wells was counted. Lower concentrations of IL-1β (1-4 ng/ml) induced a concentration-related increase in non-spread and spread cells, but at 8 ng/ml, small (3-6 nuclei, diamonds) and intermediate (>6 nuclei, triangles) osteoclasts decreased, associated with a large increase in spread cells (circles). IL-1β induced a clear concentration-related decrease in phantoms (squares), indicating an anti-apoptotic effect.

The non-osteoclastic cell density in cultures was increased by IL-1β treatment. This was particularly evident in BMC cultures where the cell layer density was increased several fold by IL-1β, although this was not systematically quantified. This increase in cell density was also observed in wells which were not stimulated by RANKL when there was consequently no osteoclastogenesis. The difference was striking, resulting in a change in colour under the microscope from light yellow to dark yellow and to brown depending on the duration of the culture and thus on the increase in density. If the cell density in the control well was below optimal, the IL-1β -induced increase in density would tend to increase osteoclastogenesis.

Although this factor will certainly influence the observed effects of IL-1β, it cannot explain all of the observed responses. Most cultures were performed at seeding cell densities considered to give near the optimal cell density in the wells at the end of the cultures. Under these conditions, in the presence of IL-1β, the cell density in the wells at the end of the cultures would be well above optimal. This would tend to have the effect of inhibiting osteoclastogenesis and increasing spread cell apoptosis. In the light of this additional factor, the survival effect of IL-1β is even more impressive, because it occurs despite this additional pro-apoptotic stimulus.

IL-1β induced similar responses when included from the beginning of BMC cultures, inducing a marked increase in spread cells with a decrease in small osteoclasts on day 4 when no phantoms were present. A similar response was observed on day 5, accompanied by a decrease in the phantoms that had now developed in the control wells (data not shown).

In order to study the survival effect of IL-1β, osteoclasts were developed in the presence of RANKL and M-CSF until day 5, when many spread cells had developed and phantoms had begun to appear. Then IL-1β (7.5-10 ng/ml) was added to the cultures in the continued presence or absence of RANKL, and cultures were stopped on days 6 or 7. In all experiments there was a marked increase in the number of live spread cells, irrespective of whether there were large numbers (day 7 in the presence of RANKL) or very few (day 7 in the absence of RANKL) already present. There was a decrease in phantoms in all experiments. The increase in spread cells may be in part due to accelerated differentiation of osteoclasts, as there were less non-spread osteoclasts, which may have fused and formed spread cells (Fig. 14). The effect of IL-1β is not equivalent to an increased concentration of RANKL, as the large numbers of spread cells which developed did not die, and there were less phantoms in IL-1β-treated wells than in control wells, indicating that, in addition, some of the spread cells which would have died in the absence of IL-1β have also survived. The effect of IL-1β is qualitatively similar to that of TNFα (Fig. 14).

There was a morphological difference in a proportion of the population of spread cells developed in the presence of IL-1β, which was particularly seen in isolated cells or cells on the periphery of large groups. This was observed in both spleen and BMC cultures and appeared to be specific to IL-1β. If there were large numbers of spread cells, up to 40% of them showed some degree of morphological change. The cells began to lose TRAP staining around the periphery. This could result in a clearly-defined TRAP-negative area around the periphery of the cell (Fig. 19). This was accompanied by a loss of the definite ring structure around the periphery of the cell and a re-grouping of nuclei in the centre of the cell. In some cells the loss of staining was observed over the majority of the cell area, to the extent that a minority of cells were TRAP-negative except for a small area around the nuclei in the centre of the cell. The periphery became less structured and regular, and the cells became more round in profile. If these changes were particularly marked, the cells, although still being multinuclear, otherwise began to morphologically resemble TRAP-negative cells of other lineages that develop from the common precursors, such as macrophages or dendritic cells (Fig. 19).

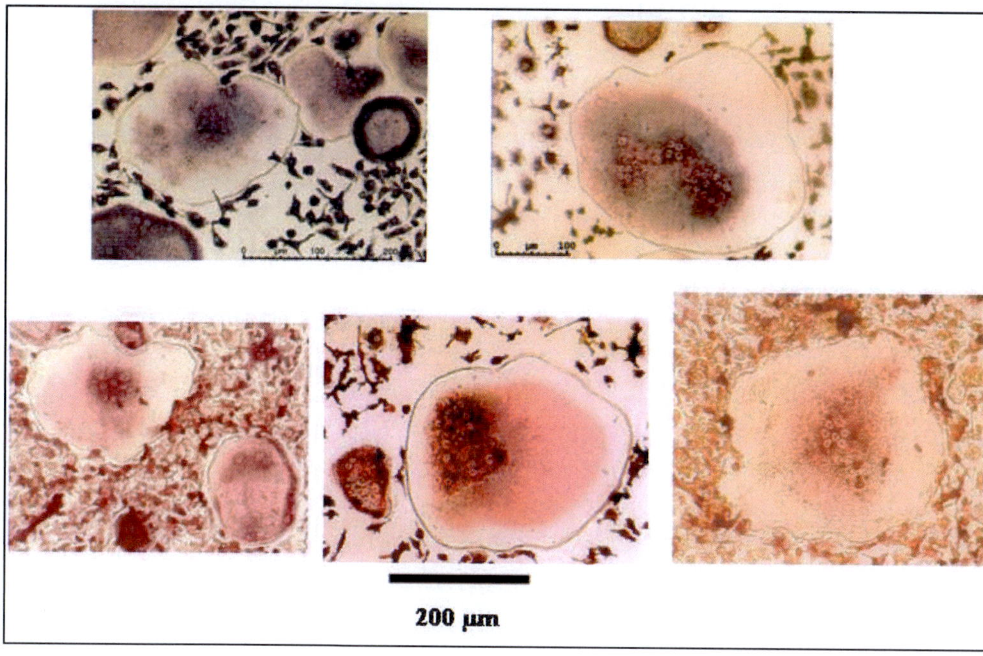

Figure 19. Examples of spread osteoclasts showing IL-1β-induced changes in morphology. Microscopic images at 10 times magnification were displayed via a colour camera on a computer screen and were selected and recorded using Histolab software. Osteoclastogenesis was induced by RANKL (40 ng/ml) in the presence of M-CSF (25 ng/ml) and IL-1β (10 ng/ml) in 7-day BMC cultures (96-well plates, 60 thousand cells/well). Note the loss of TRAP staining around the periphery of the cells and the grouping of nuclei in the centre of the cell, associated with the loss of the raised peripheral ring structure. Not all spread cells showed the IL-1β-induced changes in morphology and the classical morphology can be seen in nearby small spread cells in the two images on the left. TRAP staining was lost throughout the cytoplasm of some cells, except around the central group of nuclei, and the periphery of the cells became less regular (bottom right image).

It has been suggested that part of the action of RANKL in initiating the differentiation of osteoclasts is the suppression of the macrophage phenotype of the precursors. It has been considered that, in the absence of RANK expression the precursor cells would develop "by default" into macrophages. It is less common for well-differentiated cells like these spread osteoclasts to change phenotype. Further evidence, such as macrophage- or dendritic cell-specific markers are required to verify such a phenotypic change, but the morphological changes are quite striking. IL-1β is capable of stimulating the development and activation of macrophages and it is possible that such stimuli predominate in these cells which are protected from apoptosis. It may be that the common signalling mechanisms induced by IL-1 and RANKL are not sufficient to maintain the osteoclast phenotype, but the cells remain alive due to the strong survival effect of IL-1β, which may be a different survival mechanism from that of RANKL. Based on this hypothesis, as TNFα did not apparently induce these morphological changes, it may activate sufficient common signalling mechanism to RANKL to maintain the phenotype. This underlines the possibility that neither IL-1 nor TNFα nor RANKL induce identical patterns of intracellular signalling, although they do activate some common pathways.

In other studies, the osteoclast survival effect of IL-1 appeared to be less strong than that of RANKL [36]. It is necessary to determine the mechanisms underlying the observed powerful survival effects of both TNFα and IL-1β on spread cells. As the actions of these two cytokines on the morphology of the spread cells were not identical, it is possible that they induce spread cell survival by mechanisms which are not identical. Some mechanisms may be common (ERK and other MAP kinases, PI3-kinase/PKB, NF-κB and calcium-NFAT) while others may be different (induction of NO by TNF). However, it would be interesting to identify the mechanisms which appear to make both these cytokines more effective than RANKL as survival stimuli under these circumstances. IL-1α was shown to activate NF-κB and ERK in osteoclasts, but its survival effect was mediated through ERK while NF-κB was important for resorbing activity [26]. The survival effects of IL-1 and LPS depended on the presence of myeloid differentiation factor 88 (MyD88) which is a key adaptor molecule which associates with the intracellular domains of IL-1 receptors and other TLRs such as those activated by LPS [68]. MyD88 can lead to the activation of ERK, but it is not clear which downstream pathways were involved in this survival effect of IL-1α. IL-1 receptor-associated kinase M (IRAK-M) is an endogenous inhibitor of IL-1 receptor signalling. Mice which lack IRAK-M develop severe osteoporosis which is associated with increased osteoclast development, activation and survival. IRAK-M prevents the dissociation of the receptor-associated complex containing MyD88, IRAKs and TRAF6 and blocks downstream signalling. In IRAK-M-deficient osteoclasts, IL-1 induced hyperphosphorylation of I-kB and the resulting activation of NF-kB as well as hyperphosphorylation of the MAP kinases ERK1/2 and JNK. p38 Kinase was constitutively active in IRAK-M+ and IRAK-M- osteoclasts [69]. Therefore, osteoporosis in mice lacking IRAK-M, and the associated increase in osteoclast survival, may not be related to the activity of p38 kinase. However, it is not possible to distinguish which of the other signalling pathways is associated specifically with the survival effect of IL-1, as RANKL also activates TRAF6 and all the other signalling pathways. This may depend on the balance of activity in several pathways. TNFα-induced survival could be selectively associated with NO production, as IL-1 and RANKL do not activate this mechanism [46].

IL-18 is a structural homologue of the IL-1 ligand family. Its specific receptors activate some of the intracellular signalling mechanisms activated by IL-1 receptors [70,71]. However, in co-culture models it inhibits osteoclastogenesis in contrast to IL-1 [72]. IL-18-transgenic mice had fewer osteoclasts and this may be related to increased secretion of interferon (IFN) gamma [73]. The inhibitory effect in co-cultures may be in part explained by a selective increase in expression of osteoprotegerin, the decoy receptor for RANKL, without increase in RANKL expression [70]. In immune systems IL-18 is a strong inducer of the release of IFNγ, but this cytokine does not seem to mediate the inhibitory effects of IL-18 on osteoclasts. However, it was shown that IL-18 inhibits osteoclastogenesis in co-cultures via release of GM-CSF from T cells [74,75]. This may explain why IL-18 did not inhibit bone resorption in a neonatal mouse calvaria models, as in these preparations there are few, if any T cells present [72,76].

In spleen cell cultures, in the absence of osteoblastic/stromal cells, IL-18 was shown to inhibit osteoclastogenesis [77, see Fig. 20], but this is likely to involve the release of GM-CSF from T cells [77]. IL-18 did not affect osteoclastogenesis from RAW264.7 cells, suggesting that it does not act directly on osteoclast precursors [77]. Similarly, the bone resorbing activity of isolated matured osteoclasts was not affected by IL-18 [76] or was only inhibited weakly at an extremely high concentration of IL-18 (1000 ng/ml) [71]. There is a complex relationship between IL-18 and IL-12 which have synergistic inhibitory effects on osteoclasts [71,77]. Thus, a combination of IL-18 and IL-12, both of which were inactive alone, clearly inhibited bone resorption by isolated mature osteoclasts [71]. This response seemed to involve IFNγ as it was blocked by a neutralising antibody, although IFNγ by itself did not inhibit osteoclastic bone resorption.

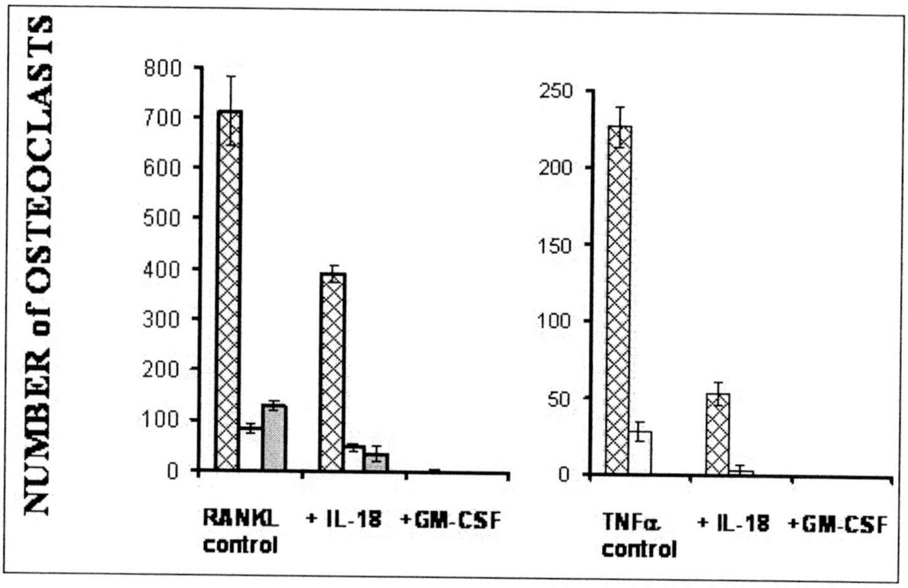

Figure 20. Effects of recombinant IL-18 or GM-CSF on osteoclastogenesis in 7-day mouse spleen cell cultures in 24-well plates (1 million cells/well). Osteoclastogenesis was induced in mouse spleen cells, by either RANKL (50 ng/ml) or recombinant mouse TNFα (40 ng/ml) in the presence of recombinant human osteoprotegerin (400 ng/ml), to exclude the involvement of RANKL in the response. All cultures were performed in the presence of M-CSF (25 ng/ml). Note that TNFα was a weaker stimulus for osteoclastogenesis, inducing far fewer osteoclasts. The different morphological categories of osteoclast were separately identified and counted using Histolab software on microscopic images at 10 times magnification, displayed via a colour camera on a computer screen. 2-4 replicate wells were counted for each group (n=2 for RANKL-stimulated culture, n=4 for TNFα-stimulated cultures) and the variability is represented by the SEM. IL-18 (50 ng/ml), added from the beginning of cultures (+IL-18), clearly reduced osteoclastogenesis induced by either RANKL (left graph) or TNFα (right graph), but there were some remaining non-spread (crosshatched bars) and spread (white bars) cells. There was also a reduced number of phantoms (grey bars) still present in IL-18-treated, RANKL-stimulated cultures, indicating that osteoclastogenesis was slowed, but not completely blocked at this concentration of IL-18. GM-CSF (100 ng/ml), added from the beginning of cultures, completely inhibited osteoclastogenesis induced by either RANKL or TNFα. No osteoclasts were observed in the wells treated with both GM-CSF and either RANKL or TNFα (+GM-CSF), in comparison with the RANKL or TNFα control wells where numerous small non-spread osteoclasts were present as well as a few spread cells and phantoms.

A similar synergistic effect on osteoclastogenesis in spleen cell cultures also involved the release of inhibitory factors from T cells, although these factors did not include GM-CSF or

IFNγ [77]. It is interesting that the IL-18 gene was amongst a group of genes which were downregulated during RANKL-induced osteoclastogenesis from RAW264.7 cells [78]. Therefore, osteoclasts may be able to produce IL-18 and it may have a functional effect which is not conducive to osteoclast development (an inhibitory effect).

Recombinant IL-18 (50 ng/ml) was tested in spleen cell cultures in comparison with GM-CSF, a potent inhibitor of osteoclastogenesis thought to be involved in the actions of IL-18 [74,75]. IL-18 inhibited osteoclastogenesis induced by either RANKL of TNFα (Fig. 20). Inhibition was not total at this concentration of IL-18 but 100 ng/ml GM-CSF totally inhibited the development of osteoclasts (Fig. 20).

GM-CSF has been shown to increase osteoclastogenesis early in cultures, an effect caused by an increase in M-CSF [79,80]. In human precursors, treatment with GM-CSF transiently increased M-CSF mRNA expression at 3h, but suppressed expression later at 7-14 days [81]. Inhibition of TNF-induced osteoclastogenesis from mouse precursors by GM-CSF involved the downregulation of TNF receptor types I and 2 [82]. However, a major mechanism of action of GM-CSF may be to divert the common myeloid precursors into other lineages such as dendritic cells and macrophages. Later in cultures of human precursors, GM-CSF inhibited osteoclast development, associated with the formation of extensive dendritic cell clusters [81]. In the experiments shown in figure 20 there was a visible increase in the cell density in the GM-CSF-treated culture wells although virtually no osteoclasts had developed. Clearly there had been proliferation of some non-osteoclastic cells.

IL-1 can also induce the production of GM-CSF in various cell types and GM-CSF was shown to be a key mediator of IL-1-induced, T cell-dependent arthritis [83]. GM-CSF is a survival factor for osteoclast precursors. Mouse long-term BMC cultures generate osteoclast precursor cells completely free from marrow stromal cells. The survival of substrate-attached cells was clearly increased by GM-CSF [84]. The apoptosis of serum-deprived RAW264.7 cells was reduced by GM-CSF but not M-CSF [85]. GM-CSF could potentially be a survival factor for mature cells in the myeloid lineages, including osteoclasts. If IL-1 induced GM-CSF in combination with a strong osteoclastogenic signal, this could result in the observed marked survival effect of IL-1β. It might also contribute to the increase in total cell density induced by IL-1β, as GM-CSF induced a similar effect in spleen cell cultures. Blocking or removing GM-CSF did not uncover a stimulatory effect of IL-18 on osteoclast development [74,75,77], so clearly IL-1β and IL-18 do not share the stimulatory action on osteoclast development. This effect of IL-1β on RANKL-induced osteoclast development was particularly clear on the development of spread cells and at later stages, when large numbers of spread cells developed, there was a marked decrease in small osteoclasts. This may result from the fusion of these cells to acheive the critical size for spreading, with more cells used than produced from precursors. However, TNFα consistently induced larger numbers of spread cells with less decrease in small cells (see Fig.14). It is possible that IL-1β has an additional inhibitory effect on the early stages of osteoclastogenesis. When administered from the beginning of cultures, IL-1β did stimulate a small increase in RANKL-induced osteoclastogenesis. However, this effect was not the same as with an increased concentration of RANKL. There was a precocious increase in small rounded spread cells, often accompanied by a marked decrease in small osteoclasts (3-6 nuclei) (data not shown).

Whether GM-CSF release and/or other mechanisms are involved in this particular response to IL-1β remains to be elucidated.

Release of GM-CSF may also provide a mechanism for the phenotype change in a proportion of the IL-1-induced spread cells. GM-CSF primarily induces differentiation into dendritic cells in combination with other cytokines, and it is possible that, despite the multinuclear state of the developing osteoclasts, it may induce changes which give them some characteristics more like dendritic or macrophagic cells. Differentiation of monocyte/macrophage precursors into osteoclasts induced by M-CSF and RANKL was completely inhibited by GM-CSF at early stages of differentiation. Cells treated with GM-CSF alone expressed RANK and c-Fms but also expressed markers detected on dendritic cells. At this stage, the cells appeared to have a mixed phenotype. M-CSF and GM-CSF reciprocally regulate the differentiation of osteoclast and dendritic cell lineages [86]. Further experiments are necessary with surface markers to characterise the exact phenotype of these cells with modified morphology in the presence of IL-1β.

However, IL-18 did not appear to release GM-CSF from human PHA-stimulated T cells or from synovial T cells from patients with rheumatoid arthritis [87]. IL-18, like IL-1β and TNFα, induced the expression of RANKL and when these T cells were incubated with human osteoclast precursors they differentiated into osteoclasts. The actions of these cytokines may also depend upon the state of activation of the cells. The effects of IL-18 as well as those of GM-CSF on the survival of mature osteoclasts, and in particular spread cells, remain to be determined. However, it is likely that they will not be identical those of IL-1β. Clues to which signalling mechanisms are important in osteoclast development, activity and survival might come from the differences in mechanisms activated by these cytokines.

Transforming Growth Factor Beta (TGFβ)

TGFβ has been shown to modulate osteoclast development, activity and apoptosis, and it was initially suggested that estradiol acts by increasing the production of TGFβ [88,89]. 17β-Estradiol can release TGFβ from osteoblasts, but can also release it by direct action on osteoclasts, where it induced an isotype switch from TGFβ2 to TGFβ3 [89]. Estrogen receptor activation increased the transcription of the TGFβ3 gene and TGFβ3 protein was shown to inhibit the differentiation and bone resorbing activities of both murine and avian osteoclasts [88].

However, lower concentrations of TGFβ have been shown to increase osteoclastogenesis. In a co-culture model of support cells in combination with different populations of osteoclast precursors (spleen or bone marrow), a biphasic effect of TGFβ was seen. Low concentrations stimulated osteoclastogenesis while high concentrations inhibited osteoclast differentiation [90]. The increase was mediated via an increase in the RANKL/OPG expression ratio on osteoblastic cells, while at high concentrations the RANKL/OPG ratio and M-CSF expression were both repressed. TGFβ can also increase osteoclastogenesis by direct actions on osteoclast precursors [90]. It has been shown to enhance RANKL-induced osteoclastogenesis from bone marrow or monocyte precursors, and a model was proposed in which TGFβ in matrix or released from bone lining or other cells, maintains and enhances the osteoclast-forming potential of precursors as they migrate towards sites of cell-bound RANKL [91]. However, a biphasic action has also been observed directly on osteoclasts. In cultures of highly-purified CD14+ osteoclast precursors TGFβ increased TRAP activity and bone resorption when it was included in the initial culture period (days 1-7 of 21 day cultures). This was associated with a strong activation of p38 kinase in the precursors, but not in the mature osteoclasts [92]. If TGFβ was present throughout the 21 days of culture, TRAP activity, cell fusion and markers of mature osteoclasts were almost completely abrogated, and this was associated with a downregulation of RANK expression.

TGFβ appears to be an important cofactor for osteoclast development. Experiments with mice deficient in components of the RANKL-RANK-TRAF system confirmed that TNFα can induce osteoclastogenesis independent of RANKL [51]. However, TNFα was only effective in bone marrow precursors if they had been pre-incubated with TGFβ as well as M-CSF. This effect correlated with TNFα-induced NFATc1 activation which also only occurred if the precursor cells were pre-incubated in the presence of TGFβ. RANKL was also able to stimulate osteoclast development in TRAF6-deficient mice if the precursors were pre-incubated with M-CSF and TGFβ. TRAF6 had previously been considered essential for RANKL-induced osteoclastogenesis, but it appears that it is not required in TGFβ-pretreated cells. However, TRAF6 does appear to be essential for the functioning of differentiated osteoclasts. Neither mature osteoclasts developed by RANKL from TRAF6-deficient cells nor those developed by TNFα (which does not activate TRAF6) from RANKL-deficient cells, resorbed bone. Addition of IL-1, which activates TRAF6, was able to induce bone resorption by the TNFα-induced osteoclasts, but not in the RANKL-induced osteoclasts which were TRAF6 deficient [51]. Furthermore, it has been shown that TGFβ can activate the early stages of osteoclastogenesis in the absence of either RANKL or TNFα, and this action is not mediated by the release of either of these mediators [93]. TRAP+ cells appear and they also express the osteoclast marker, vitronectin receptors. However, many cells were mononuclear and multinuclear cells were small. It appears that TGFβ alone (in the presence of M-CSF) may be able to initiate osteoclast development, but not to support their further differentiation.

TGFβ can also modulate osteoclast apoptosis. In the presence of osteoblastic/stromal cells TGFβ enhances osteoclast apoptosis, an effect which is, at least in part, mediated by an increase in the expression of OPG [6,94]. Force-induced osteoclast apoptosis *in vivo* is associated with a marked increase in TGFβ and in OPG [95]. More recently, TGFβ has been shown to directly increase the apoptosis of osteoclasts developed from human cord blood monocytes [58].

When TGFβ (100 ng/ml) was included from the beginning of BMC cultures in 96-well plates, when small osteoclasts began to appear on days 3 and 4, there was a clear increase in the number of these cells (data not shown). By day 5, the larger species (intermediate cells with more than 6 nuclei and spread cells) were clearly increased but the numbers of small cells was not changed (Fig. 21). It appears that TGFβ accelerated osteoclastogenesis, and even if it had increased the generation of small cells, they had quickly fused to reach the optimal size for spreading. In this experiment there was a particularly high rate of spread cell apoptosis, there being more phantoms than live spread cells in the RANKL control wells. There was no change in phantoms in TGFβ-treated cultures on day 5. Clearly spread cell apoptosis was not enhanced, and in the presence of a clear increase in live spread cells when an increase in phantoms would be expected in the presence of RANKL alone, the lack of increase in phantoms may reflect a weak anti-apoptotic effect of TGFβ under these conditions.

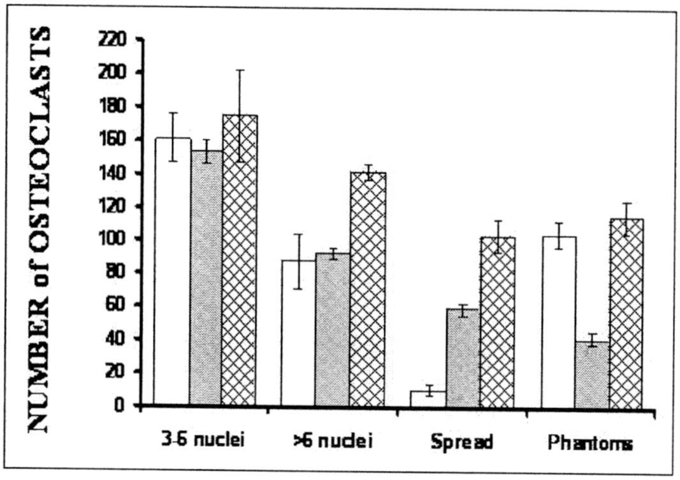

Figure 21. Effects of TGFβ (100 ng/ml, crosshatched bars) or PGE2 (30 ng/ml, grey bars) on RANKL-induced osteoclastogenesis and apoptosis in 5-day mouse BMC cultures (96-well plates, 120 thousand cells/well). The different morphological categories of osteoclast (indicated below the groups of bars) were separately identified and counted using Histolab software on microscopic images at 10 times magnification, displayed via a colour camera on a computer screen. Triplicate wells were counted for each group and the variability is represented by the SEM. In comparison with RANKL controls (white bars), TGFβ increased intermediate (>6 nuclei) and spread osteoclasts, whereas PGE2 increased spread cells and decreased phantoms.

The effects of TGFβ (20-100 ng/ml) on the apoptosis of spread cells were studied at the later stage (days 5-7) of BMC cultures, when sufficient numbers of spread cells had developed. When TGFβ was included for only the last two days of RANKL-stimulated 7-day BMC cultures, there was a clear increase in small and intermediate osteoclasts which tended to be greater at a lower concentration (20 ng/ml) (Fig. 22). There was a small increase in spread cells only at 20 ng/ml. This suggests that, even when administered for only two days late in cultures, TGFβ was enhancing the early phases of osteoclastogenesis. In the presence of little increase in spread cells there was a clear decrease in the relatively small number of phantoms at both concentrations of TGFβ (Fig. 22), suggesting a decrease in spread cell apoptosis. When administered from day 5 in the absence of RANKL, TGFβ (50 ng/ml) was still able to stimulate the early phases of osteoclastogenesis on both day 6 and on day 7, with clear increases in non-spread and spread cells (Fig. 23). This was particularly striking on day 7 as there were very few live cells present in control wells, due to the absence of RANKL. Phantoms tended to decrease on day 6, despite a marked increase in live spread cells, suggesting that TGFβ was still able to exert an anti-apoptotic effect, even in the presence of an accelerated rate of apoptosis due to the absence of RANKL. However, phantoms tended to increase on day 7 in association with a smaller increase in spread cells. It may be that TGFβ could not counter the now strong pro-apoptotic effect of RANKL withdrawal, despite its ability to induce some increase in the numbers of live osteoclasts. The small increase in phantoms does not necessarily reflect a direct enhancement of apoptosis by TGFβ as a proportion of the spread cells that it induced could have entered into apoptosis, thus increasing the numbers of phantoms. It is worthy of note that there were more spread cells than non-spread cells induced on day 6 but there were relatively less live spread cells on day 7 (Fig. 23). This is consistent with the hypothesis that more of the induced spread cells had died on day 7. In conclusion, TGFβ had relatively weak effects on spread cell apoptosis, but it tended to have an anti-apoptotic effect which was particularly evident when it was administered on the last two days of 7-day cultures in the continued presence of RANKL.

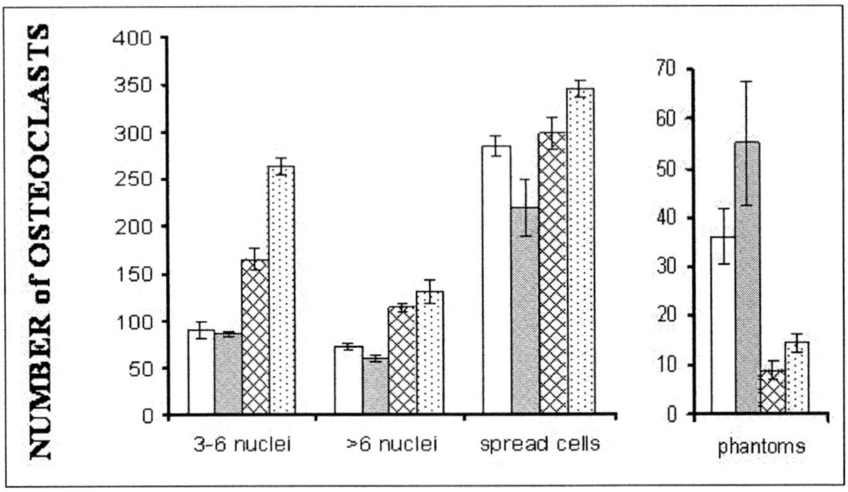

Figure 22. Effects of TGFβ (100 ng/ml crosshatched bars, 20 ng/ml stippled bars) or PGE2 (30 ng/ml, grey bars) on the numbers of different morphological categories of osteoclast (indicated below the groups of bars) when added on days 5-7 of 7-day mouse BMC cultures (60 thousand cells /well in 96-well plates), in the continued presence of RANKL (40 ng/ml) and M-CSF (25 ng/ml). The different morphological categories of osteoclast were separately identified and counted using Histolab software on microscopic images at 10 times magnification, displayed via a colour camera on a computer screen. Quadruplicate wells were counted for each group and the variability is represented by the SEM. In comparison with RANKL controls (white bars), TGFβ increased small (3-6 nuclei), intermediate (>6 nuclei) and spread osteoclasts (only at 20 ng/ml), and clearly decreased phantoms. In contrast, PGE2 had no consistent effect when administered at the end of these cultures.

It is not clear why TGFβ induced an anti-apoptotic effect under these circumstances when it has been shown that it enhances osteoclast apoptosis in other models. Several studies used co-cultures with osteoclast precursors in the presence of osteoblastic/stromal cells [6,94,95]. However, a direct pro-apoptotic action of TGFβ has been demonstrated on osteoclasts developed from human cord blood precursors [58]. Most other studies of TGFβ effects on osteoclast apoptosis have not specifically studied the spread cell form. In this state of differentiation the cells may respond differently to TGFβ. Furthermore, in the experiments reported in this chapter, the more marked anti-apoptotic effects of TGFβ were observed when it was administered late in cultures, after many spread cells had already developed. It is possible that TGFβ could have different effects on this particular cell population if it had not been present during their development.

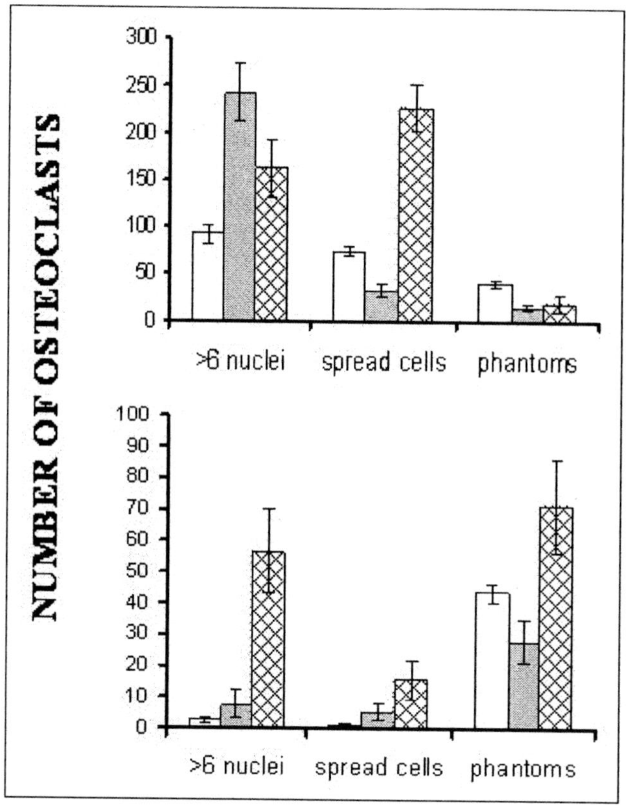

Figure 23. Effects of TGFβ (50 ng/ml crosshatched bars) or PGE2 (30 ng/ml, grey bars) on the numbers of different morphological categories of osteoclast (indicated below the groups of bars) when added from day 5 of 6 or 7-day mouse BMC cultures (60 thousand cells /well in 96-well plates), in the absence of RANKL from day 5. Osteoclastogenesis was induced for 5 days by RANKL (40 ng/ml) in the presence of M-CSF (25 ng/ml), then RANKL was withdrawn from the cultures to induce osteoclast apoptosis. The different morphological categories of osteoclast were separately identified and counted using Histolab software on microscopic images at 10 times magnification, displayed via a colour camera on a computer screen. 3-4 wells were counted for each mediator-treated group, in comparison with 6 control wells, and the variability is represented by the SEM. In comparison with RANKL controls (white bars), on day 6 (upper histogram) TGFβ increased earlier phases of osteogenesis, represented by intermediate (>6 nuclei) and spread osteoclasts, while PGE2 increased only intermediate cells. On day 7 (lower histogram), TGFβ still increased intermediate and spread osteoclasts and tended to increase phantoms, while PGE2 had no consistent effect.

It is interesting to consider which signalling mechanisms might be involved in the the effects of TGFβ on osteoclasts. The canonical signalling pathway for TGFβ and its superfamily of ligand-receptor pairs, is via Smad proteins. There are receptor-associated signalling Smads (1,2,3,5) and a common dimerising Smad4. The Smad-Smad4 dimers enter into the nucleus and function as transcription factors and activate target genes. There are also Smads (6,7) which inhibit signalling. There is little evidence that this pathway has direct influences on osteoclast development or functions, although it is involved in bone formation, as bone morphogenic proteins (BMPs) are members of this superfamily. Activin A is a member of this ligand family and is produced by osteoblasts. It was shown to stimulate osteoclastogenesis in a co-culture model of mouse BMC and primary osteoblasts. In M-CSF-dependent bone marrow macrophages (osteoclast precursors), activin A stimulated the phosphorylation and activation of ERK1/2 and p38 kinase as well as the phosphorylation of Smad2 [96], but it is not clear which of these signals was involved in enhanced osteoclastogenesis. TGFβ also signals via TGFβ-activated kinase (TAK)1, an upstream MAPK kinase kinase, to activate MAPKs, primarily the stress-activated protein kinases JNK and p38 kinase, as well as NF-κB [97,98]. This signalling pathway is important in TGFβ-induced cardiomyocyte hypertrophy with TAK1 sequentially activating MAPK kinases 3 and 6 and p38 kinase [99,100]. TAK1 binds directly to TGFβ receptor type II and also interacts indirectly with the type I receptor ALK5, in cardiomyocytes. TAK1 is rapidly activated by TGFβ within one minute [100]. Inactivation of the TAK1 gene induces marked defects in embryonic cardiovascular development, similar to those observed with loss-of-function mutations of TGFβ receptors, suggesting that TAK1 is an important downstream component of the TGFβ signalling pathway that regulates vertebrate vascular development [101].

TAK1 plays an important role in multiple signalling pathways, including signalling via receptors which modify osteoclast activity such as TNF receptors, IL-1 receptors and other TLRs [98]. TAK1 is therefore a potential intermediate in TGFβ-induced modifications of osteoclast development and functions via activation of MAPKs and NF-κB.

There is also crosstalk between the two TGFβ signalling pathways and between TGFβ signalling and other receptor signalling pathways. TAK1 can interact directly with most Smads. TAK1 dramatically interfered with receptor-associated Smad transactivation in reporter assays and affects subcellular distribution of Smad proteins. Both synergistic and interfering influences of TAK1 on BMP-mediated Smad signalling have been reported [102]. Inhibition of TGFβ signalling by IL-1β was shown to be due to IL-1β-induced activation of TAK1 and its binding to and phosphorylation of Smad3. IL-1β alleviated the inhibitory effect of TGFβ on heamopoietic myeloid colony formation [103]. TGFβ-induced apoptosis in malignant hepatocytes is limited by its parallel activation of NF-κB and the subsequent production of anti-apoptotic proteins such as XIAP and Bcl-xL. XIAP binds to TAK1 resulting in its polyubiquitination and degradation by the proteosome. In this way, NF-κB activation also inhibits TGFβ-induced activation of JNK, which is dependent on TAK1 [104]. TGFβ can also up-regulate suppressor of cytokine signalling (SOCS)3 in osteoclasts induced by either RANKL or TNF, and this plays a role in TGFβ-induced enhancement of osteoclastogenesis [105]. SOCS proteins inhibit signalling in the Janus kinase (JAK)/signal transducer and activator of transcription (STAT) pathway which is notably activated by IL-6 family cytokines and IFNs. As SOCS3 overexpression reduced the inhibition of osteoclastogenesis by IFNβ, it was suggested that TGFβ may in part increase osteoclast development by antagonising inhibitory actions of JAK/STAT signals, priming precursors for resorption rather than inflammatory functions.

It is not clear which mechanisms underlie the actions of TGFβ on osteoclasts, particularly on their apoptosis, but it seems likely that TAK1 is an important component.

Prostaglandins

Prostaglandins, particularly the E series, have been implicated as mediators of bone resorption [106]. PGE2 is the primary species in bone and is released from osteoblastic cells stimulated by a range of agents that increase bone resorption, such as IL-1, TNFα, IL-17, IL-6 in the presence of its permissive soluble receptor, and its close homologue IL-11. Bone resorption induced by these mediators is mediated via PGE2, as it is blocked by inhibition of cyclo-oxygenases (COXs), the enzymes which produce prostaglandins, particularly the inducible form, COX-2.

PGE2 stimulates several prostaglandin receptors (EP1-4) which do not all activate the same intracellular signalling pathways. The effects of PGE2 on bone resorption appear to be mediated via EP2 and EP4 receptors which signal via cyclic AMP and cAMP-dependent protein kinase (PKA) [106]. This can be demonstrated using co-cultures of osteoblastic cells and cell populations containing osteoclast precursors. IL-1-induced osteoclastogenesis and PGE2 production were inhibited by the selective COX-2 inhibitor NS398, and the osteoclastogenesis response could be rescued by the addition of PGE2 [107]. It is possible that the PGE2 released from osteoblasts may influence osteoclastogenesis by acting on this same population of osteoblastic cells. IL-6 release stimulated by IL-1, TNFα or IL-6 in the presence of its soluble receptor, is mediated via release of PGE2 and its subsequent autocrine action on the osteoblasts [108]. Thus it is clear that PGE2 can stimulate the osteoblasts in an autocrine manner, and another response to this stimulation is an increase in the RANKL/OPG expression ratio [107,109]. Experiments with prostaglandin receptor-deficient mice suggested that this effect of PGE2 involved EP2 and EP4 receptors [110,111].

However, PGE2 may also have direct actions on osteoclast precursors and differentiated osteoclasts. Initial observations showed an inhibitory effect of prostaglandins on bone resorption induced by otherwise unstimulated osteoclasts isolated from rabbit long bones [112]. This effect was less pronounced than that of calcitonin (which also acts via a cAMP-linked receptor) and was of relatively short duration [113]. Following the discovery of the role of RANKL in osteoclast development, it was shown that PGE2 had a marked synergistic effect on RANKL-induced osteoclastogenesis from stroma-free precursor cell populations (spleen cells and purified bone marrow cells) [114]. PGE2 particularly induced the formation of large, well spread multinuclear cells, and this action appeared to involve a cAMP-dependent signalling mechanism. The oxidised analogue 8-isoPGE2 increased osteoclast development induced by RANKL from bone marrow precursors, and this was inhibited by calcitonin [115]. Receptor antagonist studies suggested that this was mediated by stimulation of EP2/DP receptors, via the cAMP-PKA pathway. Subsequently, biphasic effects of PGE2 on osteoclasts have been reported. PGE2 decreased the numbers of RANKL-induced osteoclasts on days 5 and 6 of spleen cell cultures, but increased osteoclast numbers on days 8 or 9 [116]. PGE2 also increased osteoclast size on days 7-8, decreased apoptosis measured on day 7, and increased bone resorption at day 8. Studies with selective agonists suggested that EP2 receptors were involved in the increased osteoclastogenesis and this did not occur in cells from EP2-deficient mice. Removal of T cells from the cultures reduced this stimulatory effect of PGE2. The early inhibitory effect of PGE2 may have been mediated by EP3 and EP2 receptors [116]. A biphasic effect of PGE2 was also observed in cultures of haematopoietic stem cells isolated from murine bone marrow. Low concentrations of PGE2 stimulated osteoclastogenesis but higher concentrations had the opposite effect [117]. EP3 receptors were expressed at high levels in the precursors and the expression decreased during differentiation. The expression of EP1, EP2 and EP4 did not change during osteoclast differentiation. EP3 receptors do not activate cAMP and the stimulatory effect of PGE2 was associated with a decrease in PKA activity. The mechanism underlying these effects of PGE2 remain to be clarified. In isolated rat osteoclasts, PGE2 reduced cell surface area and decreased motility, and these responses appeared to involve EP2, and to a lesser extent EP4 receptors and activation of a cAMP-dependent pathway [118]. This resulted in an outwardly-rectifying chloride current and long-lasting depolarisation of the cell membrane, which might be associated with the morphological changes in the osteoclasts.

There appear to be some differences between osteoclasts derived from animal models and human osteoclasts. Osteoclasts extracted from the long bones of human foetuses expressed only EP3 and EP4 receptors [119]. PGE2 decreased actin ring formation in these cells, probably via EP4 receptors. An EP3 agonist increased the number of lamellipodia in these cells. RANKL-induced osteoclastogenesis from human peripheral blood mononuclear cells was inhibited by PGE2 [120]. This effect appeared to involve EP2 and EP4 receptors. The conditioned medium from osteoclast precursor cells pre-treated with PGE2 inhibited RANKL-induced osteoclast formation, not only in human cell cultures but also in mouse macrophage cultures, suggesting that PGE2 inhibits osteoclast formation through the production of an inhibitory factor [120].

PGE2 may act on different receptors on osteoclasts at different stages of their differentiation. Bone marrow osteoclast precursors expressed EP1, EP2, EP3beta and EP4 receptors, and expression of EP2 and EP4 receptors was down-regulated during RANKL-induced osteoclastogenesis [121]. In contrast, functional EP1 was continuously expressed in mature osteoclasts. Artificial expression of EP4 permitted an inhibitory effect of PGE2 with inhibition of actin ring formation and of resorption. These effects were similar to those of calcitonin, which also activates a cAMP-linked receptor. Activation of EP1 with PGE2 or a selective agonist did not result in osteoclast inhibition.

It was suggested that EP2 and EP4 receptors are down-regulated during osteoclastogenesis, to avoid the inhibitory effects which result from their activation in mature osteoclasts.

COX-2 was identified as a gene induced in RAW 264.7 cells by RANKL, via activation of the Rac1-NF-κB pathway, and shows that these cells can produce PGE2 [122]. Selective inhibition of COX-2 inhibited RANKL-induced osteoclastogenesis, which could be rescued by the addition of exogenous PGE2. Therefore, PGE2 may be released from osteoclasts on stimulation with RANKL and could play an autocrine role in osteoclast development. Osteoclastogenesis from BMC or bone marrow macrophages stimulated with RANKL and PGE2 was blocked by the putative specific EP1 antagonist SC-19220. This antagonist reduced RANK and c-Fms (M-CSF receptor) expression in osteoclast precursors, together with downstream signals c-Src and NFAT [123]. This leaves it possible that RANKL induces the release of PGE2 in osteoclast precursors which then stimulates EP1 and suppresses the expression of RANK and inhibits further osteoclast development.

It appears that the responses of osteoclast to PGE2 depend on the origin of the cells and their expression of the different EP receptor types, as well as on the state of differentiation and of activation of the cells.

Although there is much data on the effects of PGE2 on osteoclast development and on the functions of mature osteoclasts, there is less information directly addressing the mechanisms of osteoclast apoptosis. In RANKL-stimulated spleen cell cultures PGE2 decreased osteoclast apoptosis at day 7, in association with increased osteoclast numbers and increased bone resorption [116]. At this stage the osteoclasts are more mature and these observations may be more relevant to the studies of spread cell apoptosis.

The effects of PGE2 (10-80 ng/ml) were initially tested in 7 day RANKL-stimulated spleen cell cultures, where the majority of spread cells die rapidly, such that there are more phantoms than live spread cells. Lower concentrations of PGE2 (10-20 ng/ml) induced an increase in both non-spread and spread osteoclasts, accompanied by a decrease in phantoms. At 40 ng/ml there was an evident decrease in overall cell density in the wells but there was a further increase in non-spread, and particularly in spread cells. Despite this increase in spread cells there was a further decrease in phantoms (Fig. 24), suggesting an anti-apoptotic effect of PGE2 on spread cells.

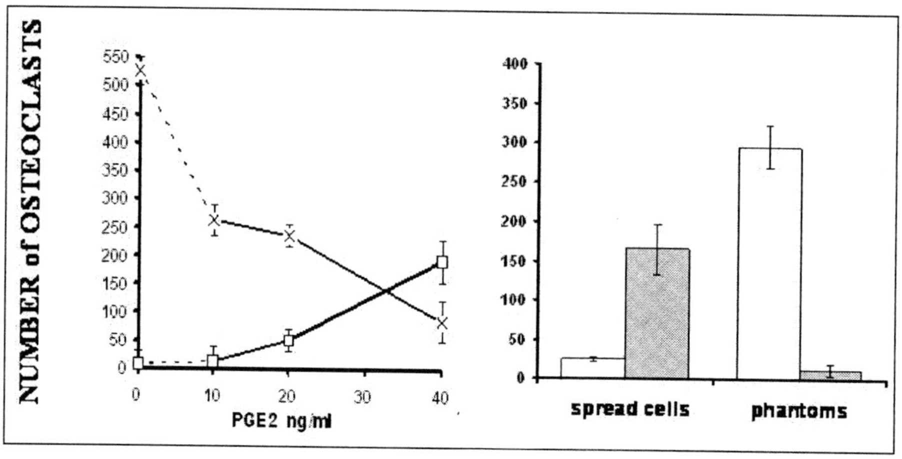

Figure 24. Effects of PGE2 on the number of spread osteoclasts and their phantoms induced by RANKL (30 ng/ml) in spleen cell cultures (24-well plates, 1 million cells/well). The different morphological categories of osteoclast were separately identified and counted using Histolab software on microscopic images at 10 times magnification, displayed via a colour camera on a computer screen. 2-4 replicate wells were counted for each group (n=2 for the PGE2 concentration-effect curve on the left, n=3 for the single concentration effect on the right) and the variability is represented by the SEM. Left graph. Concentration-effect curve for the effects of PGE2 on development of spread osteoclasts (squares) and phantoms (crosses). In these cultures there were more phantoms than live spread cells and PGE2 induced a striking decrease in the number of phantoms, associated with an increase in spread cells. Right histogram. This experiment was performed in collagen-coated plates and there was no clear difference from responses in the concentration effect curve in uncoated plates, for either RANKL controls (white bars) or PGE2-treated cultures (grey bars). PGE2 (30 ng/ml) markedly decreased the number of phantoms, with an associated increase in spread cells.

No direct inhibitory effects on osteoclastogenesis were seen, although there was a marked reduction in total cell numbers at 40-80 ng/ml. This effective decrease in cell density in the wells was associated with lower numbers of all classes of osteoclast counted in the wells. For this reason, the cell counts at 80 ng/ml PGE2 are not included on the concentration-effect curve shown in figure 24. In subsequent experiments this effect limited the maximum concentration of PGE2 used to 30 ng/ml. At this concentration, PGE2 induced a clear increase in RANKL-induced spread cells together with a marked decrease in phantoms (Fig. 24).

On days 3 and 4 of RANKL-stimulated BMC cultures the early phases of osteoclastogenesis were enhanced by PGE2 if the seeding cell density was at or just above that required to give optimal real cell density in the wells at the end of the cultures. However, with seeding densities below optimal, with 30 ng/ml PGE2, a reduction the total number of cells was particularly evident in the wells and this was associated with limited increases or net decreases in the total numbers of small osteoclasts counted in the wells. Such a decrease in total cell numbers in osteoclast cell cultures has been previously observed and quantified [114]. Total cell counts were decreased by 40-50% in the presence of PGE2, depending on the day of culture. It is possible that differences in effective cell density in different experiments may in part determine whether there is a net observed increase or decrease in osteoclast development and functions.

On day 5 there was no longer an increase in small or intermediate cells but a clear increase in spread cells remained, associated with a decrease in phantoms (Fig. 21). This suggests that spread cell apoptosis was reduced, and this alone could explain the increase in spread cells. The lack of increase in non-spread osteoclasts suggests that there was no longer such marked stimulation of the early phases of osteoclastogenesis, although there could be a balance between the number of cells developed and the number which fused and subsequently became spread cells.

When PGE2 was administered for only the last 2 days of 7 day BMC cultures, using the predicted optimal seeding cell density, there was no increase in spread or non-spread osteoclasts but there was a trend to a decrease in spread cells with an increase in phantoms. Clearly PGE2 was not able to increase osteoclastogenesis when administered at this late stage of the cultures, in contrast to TGFβ in neighbouring lines of wells on the plate (Fig. 22). In addition there was no anti-apoptotic effect, suggesting that PGE2 had little effect on RANKL-stimulated BMC cultures when administered on days 5-7. In the absence of RANKL after day 5, PGE2 induced a clear increase in non-spread osteoclasts on day 6, but a trend to a decrease in spread cells. This may suggest that, when added for this short period late in the cultures in the absence of RANKL, PGE2 still stimulated the early phases of osteoclastogenesis, or reduced the RANKL withdrawal-induced apoptosis of these cells. This increase in non-spread cells was insufficient to result in the generation of more spread cells. Despite a small decrease in phantoms, there was a small decrease in spread cells, suggesting that PGE2 was not able to protect spread cells from the accelerated rate of apoptosis induced by the absence of RANKL. On day 7, when virtually all the spread cells in control wells were dead, PGE2 was not able to significantly increase the numbers of live osteoclasts (non-spread or spread) although there was still a trend to decreased phantoms (Fig. 23). This suggests that PGE2 has limited, if any capability of preventing spread cell apoptosis in the absence of RANKL.

The reduction of the degree of stimulation of osteoclastogenesis in BMC cultures on day 5, with respect to days 3-4, and the lack of marked effects of PGE2 when administered late in the cultures, suggests that the cells become less responsive to PGE2. It is still possible to increase osteoclastogenesis at this stage of BMC cultures as other mediators could do so, including TGFβ in nearby wells on the same plate. It is possible that this lack of responsiveness results from a decrease in the expression of the receptors which mediate these effects. Expression of EP2 and EP4 receptors was down-regulated during RANKL-induced osteoclastogenesis [121]. However, artificial expression of EP4 permitted an inhibitory effect of PGE2 on actin ring formation and resorption in matured osteoclasts. A potential explanation is that there is a change in signalling from these receptors from early stimulation to later inhibition and that they are down-regulated later in cultures to avoid the inhibitory consequences of their stimulation [121]. Another possibility is that EP1 receptors are expressed at a later stage in cultures in the absence of EP2 and EP4 [121] and their stimulation suppresses the expression of RANK and inhibits further osteoclast development [123]. This would also be expected to result in an increase in apoptosis of osteoclasts, due to the reduction in the RANKL survival effect.

It has been suggested that PGE2 increases osteoclastogenesis by reducing the production of GM-CSF [109,124,110]. Another factor which could influence the observed effect of PGE2 in culture models is therefore the availability of cells which can release GM-CSF. In spleen cell cultures, T cells are a source of large quantities of GM-CSF, and it is thought that IL-18 inhibits osteoclastogenesis by releasing GM-CSF from T cells [77]. For a consistent activation of osteoclastogenesis by PGE2 via inhibition of GM-CSF release, there must be a tonic release of GM-CSF. In cultures where there are few, if any osteoblastic/stromal cells, the effects of PGE2 are qualitatively similar in spleen cell or BMC cultures although there are far fewer T cells in BMC cultures. It is not clear how a comparable tonic level of GM-CSF release could be obtained in these two culture types unless other cells contribute to tonic the release of GM-CSF.

It would be interesting to determine how EP signalling pathways impinge on the pathways known to be activated by RANKL. PGE2 enhanced osteoclast differentiation of RAW 264.7 cells in the presence of RANKL through EP2 and EP4 receptors [125]. RANKL-induced degradation of IκBα and phosphorylation of p38 kinase and JNK were enhanced by PGE2, in a PKA-dependent manner, suggesting cross-talk between EP-activated PKA and RANK signals. TAK1, an upstream MAPK kinase kinase involved in several cytokine signalling pathways including the RANK receptor-associated complex, was directly phosphorylated by PKA in RAW 264.7 cells. TAK1 was shown to be a direct link between PKA and the RANK-activated pathways. In a similar way, TNFα-induced osteoclastogenesis and associated activation of IκBα degradation and p38 kinase in RAW 264.7 cells were all potentiated by the PKA activator, forskolin. A TAK1 dominant negative molecule blocked this enhancement by forskolin [125]. Therefore, TAK1 appears to serve as a link between receptors which activate PKA, such as prostaglandin receptors EP2 and EP4, and the downstream pathways activated by RANK stimulation. If PGE2 signals via other receptors which do not activate PKA, it is less clear how the resulting intracellular signals impinge on the RANK-activated pathways.

General Discussion

The morphological changes observed in spread osteoclast cell death suggest that the cells died from a process of apoptosis. Observed changes included cytoplasmic condensation and enparcelment, followed by fragmentation, and groups of pyknotic nuclei surrounded by disintegrating cytoplasm in isolated regions of the cell. The *in situ* fluorescence induced by cleavage of a fluorescent caspase substrate confirms the involvement of members of the caspase family of enzymes in this cell death. As the fluorescent substrate is particularly cleaved by caspase-3, this effector caspase which induces the destruction of intracellular structures and functions, is probably involved. The involvement of caspases confirms that the cells die from apoptosis. The great majority of processes of apoptotic cell death involve caspases as one of the core apoptotic mechanisms [14]. Caspase activation has been detected in other studies of osteoclast apoptosis and caspase-3 plays a major role [7,8].

The fact that both the spread cells and their phantoms remain strongly adhered to the plates under these experimental conditions allows the counting of the live and dead cells as an index of apoptosis. The phantoms do not disintegrate rapidly and remain visible for at least two days. Thus reliable counts of the numbers of live and dead cells at the time of fixation of the cultures, can be obtained. However, the rate of osteoclastogenesis is also a factor controlling the number of spread cells, in balance with the rate of apoptosis. The rate of osteoclastogenesis must be taken into account, as an index of the rate of production of spread cells.

The manipulation of the numbers of live and dead cells by mediators confirms the relevance of these measures as a model of spread osteoclast apoptosis. It is well established that RANKL has a survival effect on mature osteoclasts, in addition to its essential role in stimulating the differentiation of osteoclasts [6,21]. The lack of RANKL for the last two days of 7-day BMC cultures, either due to its withdrawal or its block with specific antagonists, resulted not only in the block of further osteoclastogenesis, but also in the apoptosis of existing spread cells. The number of live spread cells decreased with an associated increase in phantoms. These data further suggested that the counting of spread cells and their phantoms is a reliable index of the rate of apoptosis of spread osteoclasts.

It is interesting that spread cell apoptosis occurs in spleen cell or BMC cultures despite the continued presence of RANKL. There is probably a strong pro-apoptotic stimulus which overcomes the survival effect of RANKL, and this may be related to unbound integrins on the surface of these cells which give strong intracellular signals for apoptosis [12,16,17]. This spread osteoclast apoptosis is modulated by the overall cell density in the culture wells, as is the rate of osteoclastogenesis. Changes in cell density have marked effects on osteoclast development and apoptosis. The initial population of spread cells that enter into apoptosis are larger cells and are more frequent at higher cell densities. The larger the number of spread cells which develop, the greater the number of phantoms. This has practical implications in the design of experiments with osteoclast cultures. The cell density in the wells increases progressively throughout the experiment with the proliferation of non-osteoclastic cells. It is the cell density in the well at the time of observation that determines the observed rate of development of osteoclasts and their rate of apoptosis, and there is an optimum cell density for both spleen cell and BMC cultures. The seeding cell density must be determined by the duration of the cultures and the desired outcome. For the study of apoptosis it is better to use a seeding cell density which will result in an effective cell density slightly higher than optimal at the end of the experiment when spread cell have developed and begin to die. A slightly higher cell density increases the rate of spread cell apoptosis. A further consideration in the study of the effects of mediators in this model, is that several mediators modify the overall cell density in the wells by actions on non-osteoclastic cells. IL-1β and GM-CSF both induce marked increases in total cell density while PGE2 induces a decrease. It is likely that these changes in cell density will modify the direct effects of these mediators on osteoclasts and this must be taken into account when interpreting the responses.

There are technical difficulties in performing mechanistic studies on this population of spread cells. Although the spread morphology is distinct and allows definite identification and the possibility of mathematical analysis of frequency and size, they have certain characteristics which hinder molecular biology studies. Although it may be possible to separate them from other cells in the culture on the basis of their strong adhesion, this same characteristic poses problems for further manipulation. It is difficult to then totally de-adhere the purified population in a reproducible manner, even with quite drastic procedures. Furthermore, it is not possible to distinguish between live and dead cells, although it is assumed that virtually all proteins involved in signalling pathways would be destroyed in cells in which the process of apoptosis was complete. In addition, it seems likely that the spread cells rapidly undergo apoptosis as soon as they are artificially detached. When analysed by flow cytometry, it appeared that the detached cells had folded on themselves or rolled up, making analysis of any fluorescent marking very difficult. In addition, the purified, but still adhered population responded variably to several lysis buffers which were investigated, shedding doubt on the reliability of the analyses of the lysates. *In situ* immunohistofluorescence is a possible approach to the study of intracellular mechanisms in these cells, given that the cell permeabilisation procedures used work more reliably than the lysis buffers. It would be particularly interesting to study the survival of the purified, but still adhered spread osteoclasts. Some information can be obtained from the used of specific inhibitors which exist for many of the different signalling pathways, again, given that they penetrate into these cells in a reproducible manner. Attempts are being made to overcome the technical problems in order to be able to study the intracellular signalling mechanisms active in this specific morphological state of osteoclasts.

It is not established how mediators which activate different classes of receptors with different associated primary signalling mechanisms can all influence the key mechanisms which control osteoclast development, functions and apoptosis. While there is much data on the signalling pathways underlying osteoclast differentiation and the functions of mature osteoclasts (actin ring formation, motility and bone resorption), there are fewer studies which directly measure the mechanisms involved in apoptosis of osteoclasts. The influences of the different mediators on osteoclast apoptosis probably involve the convergence of direct receptor-activated signals onto the essential intermediate signalling mechanisms which control osteoclast apoptosis. RANKL is the principle inducer of osteoclast differentiation and is a survival factor for mature osteoclasts. Although there are a wide range of signal pathways activated by RANKL, the mechanisms most likely to be involved in osteoclast apoptosis are activation of ERK, PI3K and NF-κB [11,22], and the outcome probably depends on a pattern of activation of all these mechanisms. These pathways then influence the core apoptotic mechanisms [14] (Fig. 25). They may modulate the expression, release or activity of the Bcl-2 family members, either the BH3-only members which act as messengers, bringing pro-apoptotic signals to the mitochondria from other cellular locations, or those on the mitochondrial membrane which gate large channels which release pro-apoptotic molecules such as cytochrome c [14]. For example, PI3K inactivates the BH3-only molecule Bim via the mTOR protein [11]. The core apoptosis mechanisms may also be modulated at the level of the internal control of caspase activity mediated by the IAP family proteins. NF-κB activation leads to the expression of members of this family [6,22].

However, it is not clear how all mediators which modulate osteoclast apoptosis contribute to the activation of ERK, PI3K and NF-κB, either in the presence or absence of RANKL. There must be mechanisms activated by all the different primary intracellular receptor complexes which influence the pattern of activity of these key intermediate signalling pathways in a similar way. The influences of TGFβ on osteoclasts were initially difficult to explain, as the canonical Smad signalling pathway does not have direct influences on the signalling pathways involved in osteoclasts. However, the characterisation of TAK1 added an alternative pathway activated by TGFβ which influenced particularly MAPKs and NF-κB [97]. TAK1 appears to have an important role in TGFβ signalling involved in vascular development [101] and cardiac hypertrophy [100]. It is interesting that TAK1 might be a target of the inhibitor of apoptosis protein XIAP in hepatocytes [104]. There is crosstalk between two signalling pathways activated by TGFβ. NF-κB activation induces XIAP production and this inhibits JNK activation which depends on TAK1. XIAP associates with TAK1 and the ring domain of XIAP mediates polyubiquitination of TAK1 which targets this molecule for proteosomal degradation. In this way, XIAP inhibits a potential pro-apoptotic activity of JNK in these cells.

However, it soon became evident that TAK1 plays a role in signalling through many different receptor types, including TNFR1, IL-1R and TLRs 3 and 4 [98]. TAK1 gene deletion in mice results in an early embryonic lethal phenotype. However, in fibroblasts from the live embryos, responses indicated that TAK1 mediates IKK activation in TNFα and IL-1 signalling pathways where it functions downstream of RIP1-TRAF2 and MyD88-IRAK1-TRAF6, respectively. However, TAK1 was not required for activation of NF-κB after stimulation with another TNF superfamily ligand, lymphotoxin (LT)-β. TAK1 depletion impaired TGFβ-induced activation of NF-κB and JNK but had no apparent effect on Smad signalling [98].

Suppression of TAK1 with small interfering RNA suppressed TNFα-induced activation of JNK and p38 kinase in murine colon cancer cells, and inhibited their migration and pulmonary metastasis, indicating that TAK1 plays an essential role in the TNFα-induced metastasis of these cells [126]. In fibroblasts, expression of a kinase-inactive mutant of TAK1 significantly modified the profile of TNF-regulated genes. 48 TNF-induced genes were suppressed by the inactive mutant, including a selected group of inflammation-related genes [127]. The induction of one such gene, ccl2, by TNF or IL-1 was also suppressed by TAK1 antisense RNA or by deficiency in JNK1/2 or p65NF-κB, suggesting that TAK1 regulates this gene via activation of JNK and/or NF-κB. However, TAK1 plays different roles in the signalling of different TNF superfamily receptors. LTβ activates NF-κB independent of

TAK1 in embryonic fibroblasts [98]. CD40 activates IKK and thus NF-κB, and TAK1 is involved in the pathway, but another MAPK kinase kinase (MAP3K8) is recruited to the CD40 complex and functions upstream of TAK1 [128]. Death receptor 3 (DR3), when stimulated by its ligand TL1A, induced IL-8 secretion in human peripheral blood monocyte-derived macrophages. TNFα induced a similar response via TNFR1 [129]. TAK1 was an upstream component of DR3 signalling, leading to the activation of JNK2 and the production of IL-8. However, TAK1 was not essential in the TNFR1 signalling for this response. JNK1, JNK2 and p38 kinase were activated downstream, leading to the secretion of IL-8. In the human chondrosarcoma cell line SW1353, suppression of TAK1 with a small interfering RNA indicated that it was involved in the IL-1-stimulated signalling pathways that lead to the secretion of inflammatory mediators TNFα and matrix metalloproteinase enzymes [130]. TAK1 is involved in the IL-1-induced activation of NF-κB, JNK and p38 kinase and is involved in the MyD88/IRAK1/TRAF6 receptor complex. The suppressor of cytokine signalling, SOCS-3, known as an inhibitor of the JAK-STAT signalling pathway, also inhibited this IL-1 signalling pathway. SOCS-3 inhibits the association of TRAF6 and TAK1, preventing the activation of TAK1 [131]. This indicates that TAK1 activation is essential for IL-1-induced activation of NF-κB, JNK and p38 kinase.

TAK1 also plays an essential role in some immune responses. Studies of mice with a B cell-specific deficit in TAK1 indicated that TAK1 is indispensable for cellular responses to CD40, TLR ligands and B cell receptor crosslinking. In addition, antigen-induced immune responses were impaired in mice with B cell TAK1 deficiency [132]. TAK1-deficient cells did not activate NF-κB or MAPKs in reponse to TNF, IL-1 or TLR ligands, although B cell receptor stimulation was able to activate NF-κB but not JNK. It was not clear how the B cell antigen receptor led to the activation of IKK. This activation is dependent on PKCβ, but the intermediate steps were not known. It has now been shown that PKCβ phosphorylates the adaptor molecule CARMA1 which recruits both TAK1 and IKK to the complex. TAK1 is essential for IKK activation in response to B cell receptor stimulation [133]. Given that TAK1 is involved in B cell receptor activation of NF-κB, there must also be another pathway for its activation which was operative in TAK1-deficient B cells.

TAK1 has also been shown to be involved in signalling pathways in osteoclasts. RANKL-induced activation of NF-κB in osteoclast precursors involved recruitment of TRAF6 to RANK and the activation of Rac1. TAK1 was activated downstream of Rac1, leading to the activation of NF-κB [24]. A dominant negative form of TAK1 completely blocked RANK-induced activation of JNK and AP1 [134]. RANKL-induced osteoclastogenesis from bone marrow precursors was reduced by a dominant negative TAK1. TAK1 was shown to be upstream of p38 kinase in a pathway leading to the activation of p65 and induction of NFATc1 [135]. PGE2 enhanced osteoclast differentiation of RAW 264.7 cells in the presence of RANKL through EP2 and EP4 receptors [125]. RANKL-induced degradation of IκBα and phosphorylation of p38 kinase and JNK were enhanced by PGE2, in a PKA-dependent manner, suggesting cross-talk between EP-activated PKA and RANK signals. TAK1 was directly phosphorylated by PKA in RAW 264.7 cells. TAK1 was shown to be a direct link between PKA and the RANK-activated pathways.

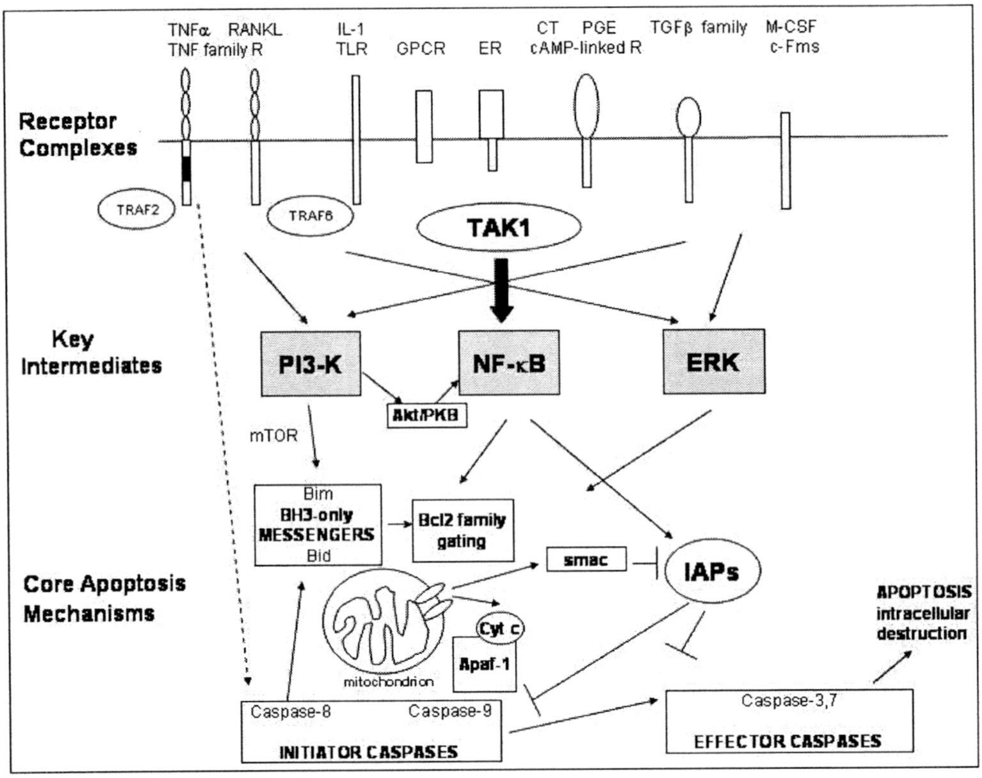

Figure 25. Intracellular signalling mechanisms involved in the control of osteoclast apoptosis. Signalling through diverse families of receptors (R) and their associated, recruited signalling complexes converge on key intermediate signals, which then influence the core apoptosis mechanisms. The modulation of apoptosis mechanisms can be via modulation of Bcl-2 family proteins, either the molecules with only a BH3 domain which bring pro-apoptotic signals to the mitochondrion from other locations in the cell (such as the modulation of Bim by PI3-K via the protein mTOR), or via modulating the expression of family members bound to mitochondrial membranes which gate large mitochondrial channels. This modulation can also be via inhibitor of apoptosis proteins (IAPs) which control the activation of caspases. Second mitochondrial activator of caspases (smac) is an endogenous pro-apoptotic molecule released through mitochondrial channels together with cytochrome c (Cyt c), the primary instigator of mitochondrial apoptosis. Smac neutralises the controlling effects of IAPs on caspase activation and amplifies the cytochrome c-mediated pro-apoptotic signal. TNFR1 (left) can recruit and activate the upstream initiator caspase-8 via the death domain (black), and this can activate apoptosis, either via the BH3-only molecule Bid and opening of Bcl-2 family-gated mitochondrial channels, or by direct activation of effector caspases. This mechanism operates in many cell types but may not be activated under most circumstances in osteoclasts. RANKL is a member of the TNF ligand superfamily and the adaptor molecule TRAF6 is recruited to its activated receptor RANK. TRAF6 plays an important, although not indispensible role in RANK signalling controlling osteoclast development and functions as well as in signalling through IL-1 receptors and other Toll-like receptors (TLR). Signalling through diverse other receptors such as G-protein-coupled receptors (GPCR), estrogen receptors (ER), cAMP-linked receptors, TGFβ family receptors and M-CSF receptors (c-Fms), can influence osteoclast apoptosis. With the discovery of TGFβ-activated kinase1 (TAK1), a signalling mechanisms was discovered which is involved in the signalling complexes associated with many of these receptors and this provides a mechanism for convergence of signalling through these diverse receptors to the key intermediates, at least to NF-κB. In addition, there is potential crosstalk between the key intermediates. Activation of NF-κB by PI3-K via PKB is an example.

There is no direct evidence that TAK1 is involved in osteoclast apoptosis. However, it plays a role in the activation of some of the key mechanisms that are thought to be involved. The activation of TAK1 by a range of different receptors makes it a potential common signalling mechanism for the effects of diverse mediators (RANKL, TNFα, IL-1β, TGFβ, PGE2) on osteoclast development and functions and potentially on osteoclast apoptosis. However, although TAK1 is a MAPK kinase kinase which can lead to the activation of the stress-activated MAPKs JNK and p38 kinase, as well as NF-κB, there is less evidence that TAK1 leads to the activation of ERK. In addition, there is little evidence that TAK1 can lead to the activation of PI3-K. In this way, TAK1 may be a common mechanism linking many receptor types to signalling pathways involved in osteoclast apoptosis, but for only some of the signalling mechanisms (Fig. 25).

The control of osteoclast apoptosis probably involves a pattern of activation involving all these pathways. However, the simple activation of a pathway may not be the appropriate contributing signal. The outcome may depend on the duration of activation of a pathway, as with ERK activation in osteoclast apoptosis [28]. It also may depend upon the intensity of activation of pathways. There may be threshold levels of activation, below which there is no change in the overall outcome, even though the pathway may be clearly activated. This was observed for TRAF6 activation in the control of osteoclastogenesis [136]. There may also be crosstalk between the essential intermediate mechanisms. PI3-K can activate NF-κB via activation of PKB [27]. It is interesting that PKB, which is activated by PI3-K, has been established as an anti-apoptotic molecule in many other cellular systems and acts on several pathways which impinge on the core apoptosis mechanisms [22]. It was considered that it may play the same role in osteoclasts [22], but recent evidence suggested that PKB is not an essential component of the signal pathways activated by PI3-K which have an anti-apoptotic effect [11]. It is possible that there are other crosstalk mechanisms between the PI3-K, NF-κB and ERK systems which add further subtlety to the control of the combination of signals that are essential for determining whether osteoclasts enter into apoptosis.

Further studies are required to elucidate the exact nature of the activation of each contributing pathway and their combination and interaction, that controls the overall outcome, which in this case, is the survival or apoptosis of mature osteoclasts.

REFERENCES

[1] Lean, J.M., Matsuo, K., Fox, S.W., Fuller, K., Gibson, F.M., Draycott, G., Wani, M.R., Bayley, K.E., Wong, B.R., Choi, Y., Wagner E.F. & Chambers T.J. (2000) Osteoclast lineage commitment of bone marrow precursors through expression of membrane-bound TRANCE. *Bone 27*, 29-40.

[2] Reddy, S.V. (2004) Regulatory mechanisms operative in osteoclasts. *Crit Rev Eukaryot Gene Expr 14*, 255-270.

[3] Tanaka, S., Nakamura, K., Takahasi, N. & Suda, T. (2005) Role of RANKL in physiological and pathological bone resorption and therapeutics targeting the RANKL-RANK signaling system. *Immunol Rev 208*, 30-49.

[4] Blanqué, R. & Gardner, C.R. (2000) TNF induces osteoclastogenesis in mouse spleen cell cultures independent of T cells or of RANK ligand. *J Bone Miner Res 15*, suppl. 1, S389.

[5] Gardnerova, M., Blanqué, R. & Gardner, C.R. (2000) The use of TNF family ligands and receptors and agents which modify their interaction as therapeutic agents. *Curr Drug Targets 1*, 327-364.

[6] Greenfield, E.M., Bi, Y. & Miyauchi A. (1999) Regulation of osteoclast activity. *Life Sci 65*, 1087-1102.

[7] Lacey, D.L., Tan, H.L., Lu, J., Kaufman, S., Van, G., Qiu, W., Rattan, A., Scully, S., Fletcher, F., Juan, T., Kelley, M., Burgess, T.L., Boyle, W.J. & Polverino, A.J. (2000) Osteoprotegerin ligand modulates murine osteoclast survival in vitro and in vivo. *Am J Pathol 157*, 435-448.

[8] Okahashi, N., Koide, M., Jimi, E., Suda, T. & Nishihara, T. (1998) Caspases (interleukin-1β-converting enzyme family proteases) are involved in the regulation of the survival of osteoclasts. *Bone 23*, 33-41.

[9] Akiyama, T., Bouillet, P., Miyazaki, T., Kadono, Y., Chikuda, H., Chung, U.I., Fukuda, A., Hikita, A., Seto, H., Okada, T., Inaba, T., Sanjay, A., Baron, R., Kawaguchi, H., Oda, H., Nakamura, K., Strasser, A. & Tanaka, S. (2003) Regulation of osteoclast apoptosis by ubiquitylation of proapoptotic BH3-only Bcl-2 family member Bim. *EMBO J, 22*, 6653-6664.

[10] Oursler, M.J., Bradley, E.W., Elfering, S.L. & Giulivi, C. (2005) Native, not nitrated, cytochrome c and mitochondria-derived hydrogen peroxide drive osteoclast apoptosis. *Am J Physiol Cell Physiol 288*, C156-168.

[11] Sugatani, T. & Hruska, K.A. (2005) Akt1/Akt2 and mammalian target of rapamycin/Bim play critical roles in osteoclast differentiation and survival, respectively, whereas Akt is dispensable for cell survival in isolated osteoclast precursors. *J Biol Chem, 280*, 3583-3589.

[12] Zhao, H., Ross, F.P. & Teitelbaum , S.L. (2005) Unoccupied alpha(v)beta3 integrin regulates osteoclast apoptosis by transmitting a positive death signal. *Mol Endocrinol 19*, 771-780.

[13] Shaw, A.K., Pascoe, D.A., Giulivi, C. & Oursler, M.J. (2001) Mechanisms controlling osteoclast apoptosis. *J Bone Miner Res 16*, suppl, S385.

[14] Gardner, C.R. (2004) Anticancer drug development based on modulation of the Bcl-2 family core apoptosis mechanism. *Expert Rev Anticancer Ther 4*, 1157-1177.

[15] Weinstein, R.S. & Manolagas, S.C. (2000) Apoptosis and osteoporosis. *Am J Med 108*, 153-164.

[16] Villanova, I., Townsend, P.A., Uhlmann, E., Knolle, J., Peyman, A., Amling, M., Baron, R., Horton, M.A. & Teti, A. (1999) Oligonucleotide targeted to the αv gene inhibits αv integrin synthesis, impairs osteoclast function, and activates intracellular signals to apoptosis. *J Bone Miner Res 14*, 1867-1879.

[17] Ross, F.P. & Teitelbaum, S.L. (2005) Alphav beta3 and macrophage colony-stimulating factor: partners in osteoclast biology. *Immunol Rev 208*, 88-105.

[18] Sakai, H., Kobayashi, Y., Sakai, E., Shibata, M. & Kato, Y. (2000) Cell adhesion is a prerequisite for osteoclast survival. *Biochem Biophys Res Commun 270*, 550-556.

[19] Qiao, L. & Farrell, G.C. (1999) The effects of cell density, attachment substratum and dexamethasone on spontaneous apoptosis of rat hepatocytes in primary culture. *In Vitro Cell Dev Biol Anim 35*, 417-424.

[20] Hotokezaka, H., Sakai, E., Kanaoka, K., Saito, K., Matsuo, K., Kitaura, H., Yoshida, N. & Nakayama, K. (2002) U0126 and PD98059, specific inhibitors of MEK, accelerate differentiation of RAW264.7 cells into osteoclast-like cells. *J Biol Chem 277*, 47366-47372.

[21] Fuller, K., Owens, J.M., Jagger, C.J., Wilson, A., Moss, R. & Chambers, T.J. (1993) Macrophage colony-stimulating factor stimulates survival and chemotactic behaviour in isolated osteoclasts. *J Exp Med 178*, 1733-1744.

[22] Feng, X. (2005) Regulatory roles and molecular signaling of TNF family members in osteoclasts. *Gene 350*, 1-13.

[23] Fukuda, A., Hikita, A., Wakeyama, H., Akiyama, T., Oda, H., Nakamura, K. & Tanaka, S. (2005) Regulation of osteoclast apoptosis and motility by small GTPase binding protein Rac1. *J Bone Miner Res 20*, 2245-2253.

[24] Lee, N.K., Choi, H.K., Kim, D.K. & Lee, S.Y. (2006) Rac1 GTPase regulates osteoclast differentiation through TRANCE-induced NF-kappaB activation. *Mol Cell Biochem 281*, 55-61.

[25] Lee, N.K., Choi, Y.G., Baik, J.Y., Han, S.Y., Jeong, D.W., Bae, Y.S., Kim, N. & Lee, S.Y. (2005) A crucial role for reactive oxygen species in RANKL-induced osteoclast differentiation. *Blood 106*, 852-859.

[26] Miyazaki, T., Katagiri, H., Kanegae, Y., Takayanagi, H., Sawada, Y., Yamamoto, A., Pando, M.P., Asano, T., Verma, I.M., Oda, H., Nakamura, K. & Tanaka, S. (2000) Reciprocal role of ERK and NF-κB pathways in survival and activation of osteoclasts. *J Cell Biol 148*, 333-342.

[27] Gingery, A., Bradley, E., Shaw, A. & Oursler, M.J. (2003) Phosphatidylinositol 3-kinase co-ordinately activates the MEK/ERK and AKT/NFkappaB pathways to maintain osteoclast survival. *J Cell Biochem 89*, 165-179.

[28] Chen, J.R., Plotkin, L.I., Aguirre, J.I., Han, L., Jilka, R.L., Kousteni, S., Bellido, T. & Manolagas, S.C. (2005) Transient versus sustained phosphorylation and nuclear accumulation of ERKs underlie anti-versus pro-apoptotic effects of estrogens. *J Biol Chem, 280*, 4632-4638.

[29] Takatsuma, H., Asagiri, M., Kubota, T., Oka, K., Osada, T., Sugiyama, C., Saito, H., Aoki, K., Ohya, K., Takayanagi, H. & Umezawa, K. (2005) Inhibition of RANKL-induced osteoclastogenesis by (-)-DHMEQ, a novel NF-kappaB inhibitor, through downregulation of NFATc1. *J Bone Miner Res 20*, 653-662.

[30] Matsumoto, M., Kogawa, M., Wada, S., Takayanagi, H., Tsujimoto, M., Katayama, S., Hisatake, K. & Nogi, Y. (2004) Essential role of p38 mitogen-activated protein kinase in cathepsin K gene expression during osteoclastogenesis through association of NFATc1 and PU.1. *J Biol Chem 279*, 45969-45979.

[31] Karsdal, M.A., Foged, N.T., Delaissé, J.M. & Lochter, A. (2001) Inhibition of p38 MAP kinase induces apoptosis in mature primary osteoclasts but not in mononucleated precursors. *J Bone Min Res 16*, suppl, S384.

[32] Takayanagi, H. (2005) Mechanistic insight into osteoclast differentiation in osteoimmunology. *J Mol Med 83*, 170-179.

[33] Marzia, M., Chiusaroli, R., Neff, L., Kim, N.Y., Chishti, A.H., Baron, R. & Horne, W.C. (2006) Calpain is required for normal osteoclast function and is down-regulated by calcitonin. *J Biol Chem 281*, 9745-9754.

[34] Nakamura, H., Nagaoka, N., Hirata, A., Inoue, M., Ozawa, H. & Yamamoto, T. (2005) Distribution of actin filaments, non-muscle myosin, M-Ras, and extracellular signal-regulated kinase (ERK) in osteoclasts after calcitonin administration. *Arch Histol Cytol 68*, 143-150.

[35] Fuller, K., Kirstein, B. & Chambers, T.J. (2006) Murine osteoclast formation and function: differential regulation by humoral agents. *Endocrinology 147*, 1979-1985.

[36] Kanaoka, K., Kobayashi, Y., Hashimoto, F., Nakashima, T., Shibata, M., Kobayashi, K., Kato, Y. & Sakai, H. (2000) A common downstream signalling activity of osteoclast survival factors that prevent nitric oxide-promoted osteoclast apoptosis. *Endocrinology 141*, 2995-3005.

[37] Ikegame, M., Ejiri, S. & Ozawa, H. (2004) Calcitonin-induced change in serum calcium levels and its relationship to osteoclast morphology and number of calcitonin receptors. *Bone 35*, 27-33.

[38] Michael, H., Harkonen, P.L., Vaananen, H.K. & Hentunen, T.A. (2005) Estrogen and testosterone use different cellular pathways to inhibit osteoclastogenesis and bone resorption. *J Bone Miner Res 20*, 2224-2232.

[39] Piva, R., Penolazzi, L., Lambertini, E., Giordano, S. & Gambari, R. (2005) Induction of apoptosis of human primary osteoclasts treated with a transcription factor decoy mimicking a promoter region of estrogen receptor alpha. *Apoptosis 10*, 1079-1094.

[40] Kanno, S., Hirano, S. & Kayama, F. (2004) Effects of the phytoestrogen coumestrol on RANK-ligand-induced differentiation of osteoclasts. *Toxicol 203*, 211-220.

[41] Ramalho, A.C., Couttet, P., Baudoin, C., Morieux, C., Graulet, A.M., de Vernejoul, M.C. & Cohen-Solal, M.E. (2002) Estradiol and raloxifene decrease the formation of multinucleate cells in human bone marrow cultures. *Eur Cytokine Netw 13*, 39-45.

[42] Garcia Palacios, V., Robinson, L.J., Borysenko, C.W., Lehmann, T., Kalla, S.E. & Blair, H.C. (2005) Negative regulation of RANKL-induced osteoclastic differentiation in RAW264.7 cells by estrogen and phytoestrogens. *J Biol Chem 280*, 13720-13727.

[43] Pederson, L., Kremer, M., Goged, N.T., Winding, B., Ritchie, C., Fitzpatrick, L.A. & Oursler, M.J. (1997) Evidence of a correlation of estrogen receptor level and avian osteoclast estrogen responsiveness. *J Bone Miner Res 12*, 742-752.

[44] Shaw, A.K. & Oursler, M.J. (2001) TNF-alpha mediated apoptosis in osteoclasts is driven by caspases. *J Bone Miner Res 16*, suppl, S509.

[45] Glantschnig, H., Fisher, J.E., Wesolowski, G., Rodan, G.A. & Reszka, A.A. (2003) M-CSF, TNFalpha and RANK ligand promote osteoclast survival by signaling through mTOR/S6 kinase. *Cell Death Differ, 10,* 1165-1177.

[46] Lee, S.K., Huang, H., Lee, S.W., Kim, K.H., Kim, K.K., Kim, H.M., Lee, Z.H. & Kim, H.H. (2004) Involvement of iNOS-dependent NO production in the stimulation of osteoclast survival by TNF-alpha. *Exp Cell Res 298*, 359-368.

[47] Zhang, Q., Badell, I.R., Schwarz, E.M., Boulukos, K.E., Yao, Z., Boyce, B.F. & Xing, L. (2005) Tumor necrosis factor prevents alendronate-induced osteoclast apoptosis in vivo by stimulating Bcl-xL expression through Ets-2. *Arthritis Rheum 52*, 2708-2718.

[48] Kwak, H.B., Jin, H.M., Ha, H., Kang, M.J., Lee, S.B., Kim, H.H. & Lee, Z.H. (2005) Tumor necrosis factor-alpha induces differentiation of human peripheral blood mononuclear cells into osteoclasts through the induction of p21(WAF1/Cip1). *Biochem Biophys Res Commun 330*, 1080-1086.

[49] Kobayashi, K., Takahashi, N., Jimi, E., Udagawa, N., Takami, M., Kotake, S., Nakagawa, N., Kinosaki, M., Yamaguchi, K., Shima, N., Yasuda, H., Morinaga, T., Higashio, K., Martin, T.J. & Suda, T. (2000) Tumor necrosis factor α stimulates osteoclast differentiation by a mechanism independent of the ODF/RANKL-RANK interaction. *J Exp Med 191*, 275-285.

[50] Azuma, Y., Kaji, K., Katogi, R., Takashita, S. & Kudo, A. (2000) Tumor necrosis factor-α induces differentiation of and bone resorption by osteoclasts. *J Biol Chem 275*, 4858-4864.

[51] Kim, N., Kadono, Y., Takami, M., Lee, J., Okada, F., Kim, J.H., Kobayashi, T., Odgren, P.R., Nakano, H., Yeh, W.C., Lee, S.K., Lorenzo, J.A. & Choi, Y. (2005) Osteoclast differentiation independent of the TRANCE-RANK-TRAF6 axis. *J Exp Med 202*, 589-595.

[52] Zhao, Y.M., Ge, L.H. & Grigoriadis, A.E. (2005) Effect of TNF-alpha on murine osteoclast differentiation. *Beijing Da Xue Xue Bao 37*, 433-435.

[53] Yao, Z., Li, P., Zhang, Q., Schwarz, E.M., Keng, P., Arbini, A., Boyce, B.F.& Xing, L. (2006) TNF increases circulating osteoclast precursor numbers by promoting their proliferation and differentiation in the bone marrow through up-regulation of c-fms expression. *J Biol Chem 281*, 11846-11855.

[54] Kanazawa, K. & Kudo, A. (2005) TRAF2 is essential for TNF-alpha-induced osteoclastogenesis. *J Bone Miner Res 20*, 840-847.

[55] Zou, W., Amcheslavsky, A., Takeshita, S., Drissi, H. & Bar-Shavit, Z. (2005) TNF-alpha expression is transcriptionally regulated by RANK ligand. *J Cell Physiol 202*, 371-378.

[56] Amcheslavsky, A., Zou, W. & Bar-Shavit, Z. (2004) Toll-like receptor 9 regulates tumor necrosis factor-alpha expression by different mechanisms. Implications for osteoclastogenesis. *J Biol Chem 279*, 54039-54045.

[57] Wu, X., McKenna, M.A., Feng, X., Nagy, T.R. & McDonald, J.M. (2003) Osteoclast apoptosis: the role of Fas in vivo and in vitro. *Endocrinology 144*, 5545-5555.

[58] Roux, S., Lambert-Comeau, P., Saint-Pierre, C., Lépine, M., Sawan, B. & Parent, J.L. (2005) Death receptors, Fas and TRAIL receptors, are involved in human osteoclast apoptosis. *Biochem Biophys Res Commun 333*, 42-50.

[59] Park, H., Jung, Y.K., Park, O.J., Lee, Y.J., Choi, J.Y. & Choi, Y. (2005) Interaction of Fas ligand and Fas expressed on osteoclast precursors increases osteoclastogenesis. *J Immunol 175*, 7193-7201.

[60] Zauli, G., Rimondi, E., Nicolin, V., Melloni, E., Celeghini, C. & Secchiero, P. (2004) TNF-related apoptosis-inducing ligand (TRAIL) blocks osteoclastic differentiation induced by RANKL plus M-CSF. *Blood 104*, 2044-2050.

[61] Wu, X., Pan, G., McKenna, M.A., Zayzafoon, M., Xiong, W.C. & McDonald, J.M. (2005) RANKL regulates Fas expression and Fas-mediated apoptosis in osteoclasts. *J Bone Miner Res 20*, 107-116.

[62] Seales, E.C., Micoli, K.J. & McDonald, J.M. (2006) Calmodulin is a critical regulator of osteoclastic differentiation, function, and survival. *J Cell Biochem 280*, 29964-29970.

[63] Wu, X., Ahn, E.Y., McKenna, M.A., Yeo, H. & McDonald, J.M. (2005) Fas binding to calmodulin regulates apoptosis in osteoclasts. *J Biol Chem 280*, 29964-29970.

[64] Ma, T., Miyanishi, K., Suen, A., Epstein, N.J., Tomita, T., Smith, R.L. & Goodman, S.B. (2004) Human interleukin-1-induced murine osteoclastogenesis is dependent on RANKL, but independent of TNF-alpha. *Cytokine 26*, 138-144.

[65] Fox, S.W., Fuller, K. & Chambers, T.J. (2000) Activation of osteoclasts by interleukin-1: divergent responsiveness in osteoclasts formed in vivo and in vitro. *J Cell Physiol, 184*, 334-340.

[66] Lee, S.K., Gardner, A.E., Kalinowski, J.F., Jastrzebski, S.L. & Lorenzo, J.A. (2006) RANKL-stimulated osteoclast-like cell formation in vitro is partially dependent on endogenous interleukin-1 production. *Bone 38*, 678-685.

[67] Kamolmatyakul, S., Chen, W., Yang, S., Abe, Y., Moroi, R., Ashique, A.M. & Li, Y.-P. (2003) IL-1α stimulates cathepsin K expression in osteoclasts via the tyrosine kinase-NF-κB pathway. *J Dent Res 83*, 791-796.

[68] Sato, N., Takahashi, N., Suda, K., Nakamura, M., Yamaki, M., Ninomiya, T., Kobayashi, Y., Takada, H., Shibata, K., Yamamoto, M., Takeda, K., Akira, S., Noguchi, T. & Udagawa, N. (2004) MyD88 but not TRIF is essential for osteoclastogenesis induced by lipopolysaccharide, diacyl lipopeptide, and IL-1α. *J Exp Med 200*, 601-611.

[69] Li, H., Cuartas, E., Cui, W., Choi, Y., Crawford, T.D., Ke, H.Z., Kobayashi, K.S., Flavell, R.A. & Vignery A. (2005) IL-1 receptor-associated kinase M is a central regulator of osteoclast differentiation and activation. *J Exp Med 201*, 1169-1177.

[70] Makiishi-Shimobayashi, C., Tsujimura, T., Iwasaki, T., Yamada, N., Sugihara, A., Okamura, H., Hayashi, S. & Terada, N. (2001) Interleukin-18 up-regulates osteoprotegerin expression in stromal/osteoblastic cells. *Biochem Biophys Res Commun 281*, 361-366.

[71] Yamada, N., Niwa, S., Tsujimura, T., Iwasaki, T., Sugihara, A., Futani, H., Hayashi, S., Okamura, H., Akedo, H. & Terada, N. (2002) Interleukin-18 and interleukin-12 synergistically inhibit osteoclastic bone-resorbing activity. *Bone 30*, 901-908.

[72] Blanqué, R., Battmann, T., Cottereaux, C. & Gardner C.R. (1998) Effects of IL10, IL17 and IL18 on osteoclast development and bone resorption. *Eur Cytokine Netw 9*, 422.

[73] Kawase, Y., Hoshino, T., Yokota, K., Kuzuhara, A., Nakamura, M., Maeda, Y., Nishiwaki, E., Zenmyo, M., Hiraoka, K., Aizawa, H. & Yoshino, K. (2003) Bone malformations in interleukin-18 transgenic mice. *J Bone Miner Res 18*, 975-983.

[74] Udagawa, N., Horwood, N.J., Elliott, J., Mackay, A., Owens, J., Okamura, H., Kurimoto, M., Chambers, T.J., Martin, T.J. & Gillespie, M.T. (1997) Interleukin-18 (interferon-gamma-inducing factor) is produced by osteoblasts and acts via

granulocyte/macrophage colony-stimulating factor and not via interferon-gamma to inhibit osteoclast formation. *J Exp Med 185*, 1005-1012.

[75] Horwood, N.J., Udagawa, N., Elliott, J., Grail, D., Okamura, H., Kurimoto, M., Dunn, A.R., Martin, T. & Gillespie, M.T. (1998) Interleukin 18 inhibits osteoclast formation via T cell production of granulocyte macrophage colony-stimulating factor. *J Clin Invest 101*, 595-603.

[76] Cornish, J., Gillespie, M.T., Callon, K.E., Horwood, N.J., Moseley, J.M. & Reid, I.R. (2003) Interleukin-18 is a novel mitogen of osteogenic and chondrogenic cells. *Endocrinology 144*, 1194-1201.

[77] Horwood, N.J., Elliott, J., Martin, T.J. & Gillespie, M.T. (2001) IL-12 alone and in synergy with IL-18 inhibits osteoclast formation in vitro. *J Immunol 166*, 4915-4921.

[78] Zhu, L.L., Zaidi, S., Moonga, B.S., Troen, B.R. & Sun, L. (2005) RANK-L induces the expression of NFATc1, but not of NFkappaB subunits during osteoclast formation. *Biochem Biophys Res Commun 326*, 131-135.

[79] Hodge, J.M., Kirkland, M.A., Aitken, C.J., Waugh, C.M., Myers, D.E., Lopez, C.M., Adams, B.E. & Nicholson, G.C. (2004) Osteoclastic potential of human CFU-GM: biphasic effect of GM-CSF. *J Bone Miner Res 19*, 190-199.

[80] Fujikawa, Y., Sabokbar, A., Neale, S.D., Itonaga, I., Torisu, T. & Athanasou, N.A. (2001) The effect of macrophage-colony stimulating factor and other humoral factors (interleukin-1, -3, -6, and -11, tumor necrosis factor-alpha, and granulocyte macrophage-colony stimulating factor) on human osteoclast formation from circulating cells. *Bone 28*, 261-267.

[81] Hodge, J.M., Kirkland, M.A. & Nicholson, G.C. (2004) GM-CSF cannot substitute for M-CSF in human osteoclastogenesis. *Biochem Biophys Res Commun 321*, 7-12.

[82] Yogesha, S.D., Khapli, S.M. & Wani, M.R. (2005) Interleukin-3 and granulocyte-macrophage colony-stimulating factor inhibits tumor necrosis factor (TNF)-alpha-induced osteoclast differentiation by down-regulation of expression of TNF receptors 1 and 2. *J Biol Chem 280*, 11759-11769.

[83] Lawlor, K.E., Wong, P.K., Campbell, I.K., van Rooijen, N. & Wicks, I.P. (2005) Acute CD4+ T lymphocyte-dependent interleukin-1-driven arthritis selectively requires interleukin-2 and interleukin-4, joint macrophages, granulocyte-macrophage colony-stimulating factor, interleukin-6, and leukemia inhibitory factor. *Arthritis Rheum 52*, 3749-3754.

[84] Liggett, W.H., Lian, J.B., Greenberger, J.S. & Glowacki, J. (1994) Osteocalcin promotes differentiation of osteoclast progenitors from murine long-term bone marrow cultures. *J Cell Biochem 55*, 190-199.

[85] Wei, J., Sun, Z., Chen, Q. & Gu, J. (2006) Serum deprivation induced apoptosis in macrophage is mediated by autocrine secretion of type I IFNs. *Apoptosis 11*, 545-554.

[86] Miyamoto, T., Ohneda, O., Arai, F., Iwamoto, K., Okada, S., Takagi, K., Anderson, D.M. & Suda, T. (2001) Bifurcation of osteoclasts and dendritic cells from common progenitors. *Blood 98*, 2544-2554.

[87] Dai, S.M., Nishioka, K. & Yudoh, K. (2004) Interleukin (IL) 18 stimulates osteoclast formation through synovial T cells in rheumatoid arthritis: comparison with IL1beta and tumor necrosis factor alpha. *Ann Rheum Dis 63*, 1379-1386.

[88] Yang, N.N., Bryant, H.U., Hardikar, S., Sato, M., Galvin, R.J., Glasebrook, A.L. & Termine, J.D. (1996) Estrogen and raloxifene stimulate transforming growth factor-beta 3 gene expression in rat bone: a potential mechanism for estrogen- or raloxifene-mediated bone maintenance. *Endocrinology 137*, 2075-2084.

[89] Robinson, J.A., Riggs, B.L., Spelsberg, T.C. & Oursler, M.J. (1996) Osteoclasts and transforming growth factor-beta: estrogen-mediated isoform-specific regulation of production. *Endocrinology 137*, 615-621.

[90] Karst, M., Gorny, G., Galvin, R.J. & Oursler, M.J. (2004) Roles of stromal cell RANKL, OPG, and M-CSF expression in biphasic TGF-beta regulation of osteoclast differentiation. *J Cell Physiol 200*, 99-106.

[91] Fuller, K., Lean, J.M., Bayley, K.E., Wani, M.R. & Chambers, T.J. (2000) A role for TGFbeta(1) in osteoclast differentiation and survival. *J Cell Sci 113*, 2445-2453.

[92] Karsdal, M.A., Hjorth, P., Henriksen, K., Kirkegaard, T., Nielsen, K.L., Lou, H., Delaissé, J.M. & Foged, N.T. (2003) Transforming growth factor-beta controls human osteoclastogenesis through the p38 MAPK and regulation of RANK expression. *J Biol Chem 278*, 44975-44987.

[93] Itonaga, I., Sabokbar, A., Sun, S.G., Kudo, O., Danks, L., Ferguson, D., Fujikawa, Y. & Athanasou, N.A. (2004) Transforming growth factor-beta induces osteoclast formation in the absence of RANKL. *Bone 2004* 34, 57-64.

[94] Murakami, T., Yamamoto, M., Yamamoto, M., Ono, K., Nishikawa, M., Nagata, N., Motoyoshi, K. & Akatsu, T. (1998) Transforming growth factor-β1 increases mRNA levels of osteoclastogenesis inhibitory factor in osteoblastic/stromal cells and inhibits the survival of murine osteoclast-like cells. *Biochem Biophys Res Commun 252*, 747-752.

[95] Kobayashi, Y., Hashimoto, F., Miyamoto, H., Kanaoka, K., Miyazaki-Kawashita, Y., Nakashima, T., Shibata, M., Kobayashi, K., Kato, Y. & Sakai, H. (2000) Force-induced osteoclast apoptosis in vivo is accompanied by elevation in transforming growth factor β and osteoprotegerin expression. *J Bone Miner Res 15*, 1924-1934.

[96] Murase, Y., Okahashi, N., Koseki, T., Itoh, K., Udagawa, N., Hashimoto, O., Sugino, H., Noguchi, T. & Nishihara, T. (2001) Possible involvement of protein kinases and Smad2 signaling pathways on osteoclast differentiation enhanced by activin A. *J Cell Physiol 188*, 236-42.

[97] Kaminska, B., Wesolowska, A. & Danilkiewicz, M. (2005) TGF beta signalling and its role in tumour pathogenesis. *Acta Biochim Pol 52*, 329-337.

[98] Shim, J.H., Xiao, C., Paschal, A.E., Bailey, S.T., Rao, P., Hayden, M.S., Lee, K.Y., Bussey, C., Steckel, M., Tanaka, N., Yamada, G., Akira, S., Matsumoto, K. & Ghosh, S. (2005) TAK1, but not TAB1 or TAB2, plays an essential role in multiple signaling pathways in vivo. *Genes Dev 19*, 2668-2681.

[99] Matsumoto-Ida, M., Takimoto, Y., Aoyama, T., Akao, M., Takeda, T. & Kita, T. (2006) Activation of TGF-beta1-TAK1-p38 MAPK pathway in spared cardiomyocytes is involved in left ventricular remodeling after myocardial infarction in rats. *Am J Physiol Heart Circ Physiol 290*, H709-715.

[100] Watkins, S.J., Jonker, L. & Arthur, H.M. (2006) A direct interaction between TGFbeta activated kinase 1 and the TGFbeta type II receptor: implications for TGFbeta signalling and cardiac hypertrophy. *Cardiovasc Res 69*, 432-439.

[101] Jadrich, J.L., O'Connor, M.B. & Coucouvanis, E. (2006) The TGF{beta} activated kinase TAK1 regulates vascular development in vivo. *Development 133*, 1529-1541.

[102] Hoffmann, A., Preobrazhenska, O., Wodarczyk, C., Medler, Y., Winkel, A., Shahab, S., Huylebroeck, D., Gross, G. & Verschueren, K. (2005) Transforming growth factor-beta-activated kinase-1 (TAK1), a MAP3K, interacts with Smad proteins and interferes with osteogenesis in murine mesenchymal progenitors. *J Biol Chem 280*, 27271-27283.

[103] Benus, G.F., Wierenga, A.T., de Gorter, D.J., Schuringa, J.J., van Bennekum, A.M., Drenth-Diephuis, L., Vellenga, E. & Eggen, B.J. (2005) Inhibition of the transforming growth factor beta (TGFbeta) pathway by interleukin-1beta is mediated through TGFbeta-activated kinase 1 phosphorylation of SMAD3. *Mol Biol Cell 16*, 3501-3510.

[104] Kaur, S., Wang, F., Venkatraman, M. & Arsura, M. (2005) X-linked inhibitor of apoptosis (XIAP) inhibits c-Jun N-terminal kinase 1 (JNK1) activation by transforming growth factor beta1 (TGF-beta1) through ubiquitin-mediated proteosomal degradation of the TGF-beta1-activated kinase 1 (TAK1). *J Biol Chem 280*, 38599-38608.

[105] Lovibond, A.C., Haque, S.J., Chambers, T.J. & Fox, S.W. (2003) TGF-beta-induced SOCS3 expression augments TNF-alpha-induced osteoclast formation. *Biochem Biophys Res Commun 309*, 762-767.

[106] Gardner, C.R., Blanqué, R. & Cottereaux, C. (2001) Mechanisms involved in prostaglandin-induced increase in bone resorption in neonatal mouse calvaria. *Prostaglandins Leukot Essent Fatty Acids 64*, 117-125.

[107] Ha, H., Lee, J.H., Kim, H.N., Kim, H.M., Kwak, H.B., Lee, S., Kim, H.H. & Lee, Z.H. (2006) alpha-Lipoic acid inhibits inflammatory bone resorption by suppressing prostaglandin E2 synthesis. *J Immunol 176*, 111-117.

[108] Blanqué, R., Cottereaux, C. & Gardner, C.R. (2000) IL-6 autoregulation during osteoclastogenesis and bone resorption in mouse tissue cultures and in MC3T3 osteoblastic cells. Eur Cytokine Netw 11, special issue ICS/ISICR, 141.

[109] Okada, Y., Pilbeam, C., Raisz, L. & Tanaka, Y. (2003) Role of cyclooxygenase-2 in bone resorption. *J UOEH 25*,185-195.

[110] Li, X., Okada, Y., Pilbeam, C.C., Lorenzo, J.A., Kennedy, C.R., Breyer, R.M. & Raisz, L.G. (2000) Knockout of the murine prostaglandin EP2 receptor impairs osteoclastogenesis in vitro. *Endocrinology 141*, 2054-2061.

[111] Li, X., Pilbeam, C.C., Pan, L., Breyer, R.M. & Raisz, L.G. (2002) Effects of prostaglandin E2 on gene expression in primary osteoblastic cells from prostaglandin receptor knockout mice. *Bone 30*, 567-573.

[112] Chambers, T.J., McSheehy, P.M., Thomson, B.M. & Fuller, K. (1985) The effect of calcium-regulating hormones and prostaglandins on bone resorption by osteoclasts disaggregated from neonatal rabbit bones. *Endocrinology 116*, 234-239.

[113] Fuller, K. & Chambers, T.J. (1989) Effect of arachidonic acid metabolites on bone resorption by isolated rat osteoclasts. *J Bone Miner Res 4,* 209-215.

[114] Wani, M.R., Fuller, K., Kim, N.S., Choi, Y. & Chambers, T. (1999) Prostaglandin E2 cooperates with TRANCE in osteoclast induction from hemopoietic precursors:

synergistic activation of differentiation, cell spreading, and fusion. *Endocrinology 140*, 1927-1935.

[115] Tintut, Y., Parhami, F., Tsingotjidou, A., Tetradis, S., Territo, M. & Demer, L.L. (2002) 8-Isoprostaglandin E2 enhances receptor-activated NFkappa B ligand (RANKL)-dependent osteoclastic potential of marrow hematopoietic precursors via the cAMP pathway. *J Biol Chem 277*, 14221-14226.

[116] Ono, K., Kaneko, H., Choudhary, S., Pilbeam, C.C., Lorenzo, J.A., Akatsu, T., Kugai, N. & Raisz, L.G. (2005) Biphasic effect of prostaglandin E2 on osteoclast formation in spleen cell cultures: role of the EP2 receptor. *J Bone Miner Res 20*, 23-29.

[117] Fujita, D., Yamashita, N., Iita, S., Amano, H., Yamada, S. & Sakamoto, K. (2003) Prostaglandin E2 induced the differentiation of osteoclasts in mouse osteoblast-depleted bone marrow cells. *Prostaglandins Leukot Essent Fatty Acids 68*, 351-358.

[118] Okamoto, F., Kajiya, H., Fukushima, H., Jimi, E. & Okabe, K. (2004) Prostaglandin E2 activates outwardly rectifying Cl(-) channels via a cAMP-dependent pathway and reduces cell motility in rat osteoclasts. *Am J Physiol Cell Physiol 287*, C114-124.

[119] Sarrazin, P., Hackett, J.A., Fortier, I., Gallant, M.A. & de Brum-Fernandes, A. (2004) Role of EP3 and EP4 prostaglandin receptors in reorganization of the cytoskeleton in mature human osteoclasts. *J Rheumatol 31*, 1598-1606.

[120] Take, I., Kobayashi, Y., Yamamoto, Y., Tsuboi, H., Ochi, T., Uematsu, S., Okafuji, N., Kurihara, S., Udagawa, N. & Takahashi, N. (2005) Prostaglandin E2 strongly inhibits human osteoclast formation. *Endocrinology 146*, 5204-5214.

[121] Kobayashi, Y., Take, I., Yamashita, T., Mizoguchi, T., Ninomiya, T., Hattori, T., Kurihara, S., Ozawa, H., Udagawa, N. & Takahashi, N. (2005) Prostaglandin E2 receptors EP2 and EP4 are down-regulated during differentiation of mouse osteoclasts from their precursors. *J Biol Chem 280*, 24035-24042.

[122] Han, S.Y., Lee, N.K., Kim, K.H., Jang, I.W., Yim, M., Kim, J.H., Lee, W.J. & Lee, S.Y. (2005) Transcriptional induction of cyclooxygenase-2 in osteoclast precursors is involved in RANKL-induced osteoclastogenesis. *Blood 106*, 1240-1245.

[123] Tsujisawa, T., Inoue, H. & Nishihara, T. (2005) SC-19220, antagonist of prostaglandin E2 receptor EP1, inhibits osteoclastogenesis by RANKL. *J Bone Miner Res 20*, 15-22.

[124] Morgan, H., Tumber, A. & Hill, P.A. (2004) Breast cancer cells induce osteoclast formation by stimulating host IL-11 production and downregulating granulocyte/macrophage colony-stimulating factor. *Int J Cancer 109*, 653-660.

[125] Kobayashi, Y., Mizoguchi, T., Take, I., Kurihara, S., Udagawa, N. & Takahashi, N. (2005) Prostaglandin E2 enhances osteoclastic differentiation of precursor cells through protein kinase A-dependent phosphorylation of TAK1. *J Biol Chem 25*, 11395-11403.

[126] Choo, M.K., Sakurai, H., Koizumi, K. & Saiki, I. (2006) TAK1-mediated stress signaling pathways are essential for TNF-alpha-promoted pulmonary metastasis of murine colon cancer cells. *Int J Cancer 118*, 2758-2764.

[127] Thiefes, A., Wolter, S., Mushinski, J.F., Hoffmann, E., Dittrich-Breiholz, O., Graue, N., Dorrie, A., Schneider, H., Wirth, D., Luckow, B., Resch, K. & Kracht, M. (2005) Simultaneous blockade of NFkappaB, JNK, and p38 MAPK by a kinase-inactive mutant of the protein kinase TAK1 sensitizes cells to apoptosis and affects a distinct

spectrum of tumor necrosis factor target genes. *J Biol Chem 280*, 27728-27741. Erratum: *J Biol Chem 280,* 32564.

[128] Chan, H. & Reed, J.C. (2005) TRAF-dependent association of protein kinase Tpl2/COT1 (MAP3K8) with CD40. *Biochem Biophys Res Commun 328*, 198-205.

[129] Su, W.B., Chang, Y.H., Lin, W.W. & Hsieh, S.L. (2006) Differential regulation of interleukin-8 gene transcription by death receptor 3 (DR3) and type I TNF receptor (TNFRI). *Exp Cell Res 312*, 266-277.

[130] Klatt, A.R., Klinger, G., Neumuller, O., Eidenmuller, B., Wagner, I., Achenbach, T., Aigner, T. & Bartnik, E. (2006) TAK1 downregulation reduces IL-1beta induced expression of MMP13, MMP1 and TNF-alpha. *Biomed Pharmacother 60*, 55-61.

[131] Frobose, H., Ronn, S.G., Heding, P.E., Mendoza, H., Cohen, P., Mandrup-Poulsen, T. & Billestrup, N. (2006) Suppressor Of Cytokine Signaling (SOCS)-3 inhibits IL-1 signaling by targeting the TRAF-6/TAK1 complex. Mol Endocrinol Mar 16 [Epub ahead of print ID 16543409].

[132] Sato, S., Sanjo, H., Takeda, K., Ninomiya-Tsuji, J., Yamamoto, M., Kawai, T., Matsumoto, K., Takeuchi, O. & Akira, S. (2005) Essential function for the kinase TAK1 in innate and adaptive immune responses. *Nat Immunol 6*, 1087-1095.

[133] Shinohara, H., Yasuda, T., Aiba, Y., Sanjo, H., Hamadate, M., Watarai, H., Sakurai, H. & Kurosaki, T. (2005) PKC beta regulates BCR-mediated IKK activation by facilitating the interaction between TAK1 and CARMA1. J Exp Med 202, 1423-1431.

[134] Lee, S.W., Han, S.I., Kim, H.H. & Lee, Z.H. (2002) TAK1-dependent activation of AP-1 and c-Jun N-terminal kinase by receptor activator of NF-kappaB. *J Biochem Mol Biol 35*, 371-376.

[135] Huang, H., Ryu, J., Ha, J., Chang, E.J., Kim, H.J., Kim, H.M., Kitamura, T., Lee, Z.H. & Kim, H.H. (2006) Osteoclast differentiation requires TAK1 and MKK6 for NFATc1 induction and NF-kappaB transactivation by RANKL. Cell Death Differ Feb 24 [Epub ahead of print ID 16498455].

[136] Gohda, J., Akiyama, T., Koga, T., Takayanagi, H., Tanaka, S. & Inoue, J. (2005) RANK-mediated amplification of TRAF6 signaling leads to NFATc1 induction during osteoclastogenesis. *EMBO J 24*, 790-799.

In: Focus on Cell Apoptosis Research
Editor: Daniel W. Cho, pp. 69-80
ISBN: 1-60021-433-9

Chapter II

LOW MOLECULAR WEIGHT GLYCOSAMINOGLYCANS AND APOPTOSIS: POTENTIAL TREATMENT OF NEURODEGENERATIVE DISORDERS INCLUDING ALZHEIMER'S DISEASE

Bertalan Dudas
Neuroendocrine Organization Laboratory (NEO),
Lake Erie College of Osteopathic Medicine, Erie, PA 16509, USA

ABSTRACT

Previous studies have shown that apoptosis (programmed cell death) plays a crucial role in the pathomechanism of neurodegenerative disorders. It is a general consensus that apoptosis can be triggered by dysregulation of nerve growth factor physiology and/or oxidative stress processes. Since both of these pathological changes are characteristic to AD, it is conceivable that influencing apoptosis can be a valuable tool in the treatment of neurodegenerative disorders.

Glucosaminoglycans (GAGs) exhibit neuroprotective properties in several animal models of AD. Low molecular weight GAG, C3, attenuates Aβ(25-35)-induced tau-2 immunoreactivity, and AF64A-induced cholinergic lesion in rat brain. Recent studies have also revealed that C3 influences p75 nerve growth factor receptor expression. However, the exact mechanism of the neuroprotective attributes of C3 is not elucidated yet. Since apoptosis is believed to play a pivotal role in numerous neurodegenerative disorders, C3 may induce neuroprotection/neurorepair via affecting apoptotic processes.

The neuroprotective effect of glucosaminoglycans including C3 thus may involve multiple mechanisms, including an influence on processes that lead to programmed cell death. Since apoptosis is a crucial event not only in neurodegenerative diseases but also in the general process of aging, C3 may be a valuable therapeutic agent in Alzheimer's disease-related senile dementia and in other age-related malfunctions.

Pathology of Alzheimer's Disease: Introductory Comments

Alzheimer's disease (AD) is a progressive and irreversible brain disorder that is manifested in dementia, motor lesions and behavioral deficits. Since as many as 10% of the population above 65 years of age is affected by AD in the United States, AD is becoming a growing concern in health care. The unequivocal diagnosis of AD is based on histopathological evidence at brain autopsy or biopsy. The major pathological hallmarks of AD include abnormal biochemical alteration of regular proteins that leads to accumulation of the modified molecules in the brain. Intracellular deposition of hyperphosphorylated, polymerized tau protein, forming flame-like tangles disrupts axonal and dendritic transport by destroying the normal microtubular structure. Pathological cleavage of the amyloid precursor protein (APP) normally present in the brain results in beta amyloid fragments that are aggregated to form extracellular build-up of amyloid plaques. These histological hallmarks are associated with extensive loss of cholinergic neurons, which is believed to be the major cause of symptoms in AD. However, the exact patho-mechanism leading to neuronal death is not yet elucidated.

Apoptosis in Neurodegenerative Disorders

Apoptosis (programmed cell death) is a natural process that ensures that damaged cells are packaged and removed by the surrounding cells cleanly, in order to prevent inflammation typical in necrotic processes. Apoptosis (Figure 1) is typically triggered by external stimuli (extrinsic pathway) and internal events (intrinsic pathway). The extrinsic pathway is initiated outside of the cell, when conditions in the extracellular environment determine that a cell must die. Extrinsic pathway is triggered by binding ligands to specific "death receptors" on the surface of the cells, leading to activation of cysteinyl proteases that are called initiator caspases. The intrinsic apoptosis pathway begins when an injury occurs within the cell. Mitochondrial damage, for example, can initiate the intrinsic pathway, when cytochrome-c, released from damaged mitochondria, activates an initiator caspase. The extrinsic and intrinsic pathways merge at caspase-3 that is activated by initiator caspases. Subsequent processes result in multiple events, including degradation of DNA, and packaging of the cell into small units that are easily taken up by neighboring cells.

Several studies indicated that apoptosis is one of the crucial factors that may be responsible for neuronal loss, characteristic for neurodegeneratory disorders including AD. Caspases are believed to be involved in generation of toxic fragments of protein substrates, that can trigger the earlier stages of neuronal dysfunction including protein aggregation in Huntington's and Alzheimer's disease [1]. Thus, caspase inhibition may be a valuable adjunct in the treatment of neurodegenerative disorders. Previous findings also demonstrated dysregulation of apoptotic proteins in AD brain, that may play a crucial role in the neuropathology of AD [2,3]. Since the level of cytochrome c has been reported to be unchanged in the frontal cortex of patients suffering from AD [3], apoptosis may occur via the extrinsic death receptor pathway independent of cytochrome c. However, it is generally believed that multiple factors are responsible for the neuronal cell death in AD. For example, neurofibrillary tangles and amyloid plaques are not associated with caspase-3 induction, in contrast to hippocampal neurons, that exhibit vacuolar degeneration and caspase-3 activation [4]. Thus, caspase-3 may not have a significant role in the widespread neuronal apoptosis that occurs in AD, but may contribute to the specific loss of hippocampal neurons involved in learning and memory.

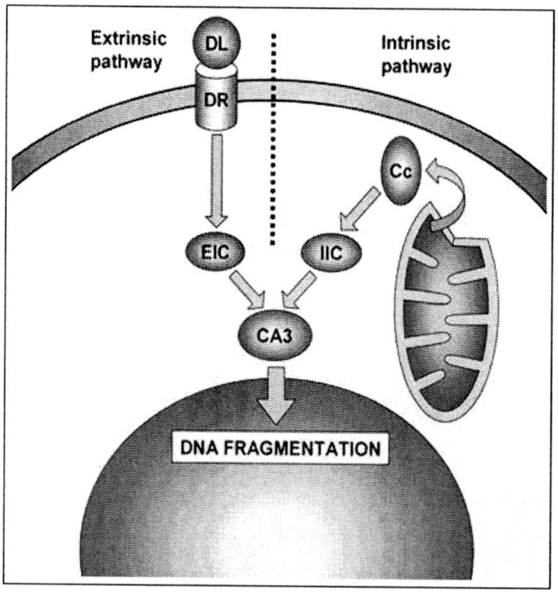

Figure 1. Major processes of the apoptotic pathway. Apoptosis is a natural process that ensures that damaged cells are packaged and removed by the surrounding cells thoroughly, in order to prevent inflammation typical in necrotic processes. Apoptosis is triggered by external stimuli (extrinsic pathway) and internal events (intrinsic pathway). The extrinsic pathway is activated by binding ligands (DL) to specific "death receptors" (DR) on the surface of the cells, leading to activation of initiator caspases of the external pathway (EIC). The intrinsic apoptosis pathway begins with an intracellular lesion that is typically mitochondrial damage, when cytochrome-c (Cc), released from damaged mitochondria, activates initiator caspases of the internal pathway (IIC). The extrinsic and intrinsic pathways merge at caspase-3 (CA3) that is activated by initiator caspases (EIC and IIC). Subsequent processes result in multiple events, including degradation of DNA in the nucleus, and packaging of the cell into small units that are easily taken up by neighboring cells.

It is a common consensus that there are two major pathways playing roles in triggering apoptosis in neurodegenerative disorders: oxidative stress and trophic factor dysregulation.

Oxidative Stress in Alzheimer's Disease

Neuronal cell loss is believed to be responsible for the majority of the clinical symptoms of AD including dementia, motor lesions and behavioral deficits. One of the most important pathways known to trigger apoptosis in the brain is oxidative stress. Oxidative stress refers to the situation of a serious imbalance between production of reactive oxygen species (ROS) and reactive nitrogen species (RNS), and antioxidant defense. ROS/RNS cause cell damage and neuronal death through lipid peroxidation (LPO), oxidation of nucleic acids, and oxidation and nitrosylation of proteins and aminoacids.

It is generally thought that the amyloid deposit (i.e. the senile plaque), which accumulates in the brain of Alzheimer's disease (AD) patients, is the cause of the neurodegenerative changes characteristic of AD, and that oxidative stress induced by amyloid beta peptide (Aβ) is a prominent feature of amyloid-induced neuronal cell death in the AD brain. Indeed, post-mortem studies have shown that 4-hydroxy-nonenal (4-HNE), a lipid peroxidation (LPO) end–product accumulates in the AD brain [5]. Moreover, a recent study has shown that iPF(2α)–IV, another marker of LPO (i.e. oxidative stress) was increased in the urine of patients with mild to moderate dementia and probable AD [6]. Studies have also demonstrated that Aβ can directly cause increases in reactive oxygen species (ROS) and produce 4-HNE [7]. These findings indicate that Aβ-induced oxidative stress is a key feature in AD.

The contention that AD pathology is associated with increased levels of oxidative stress has also been confirmed in an animal model of Alzheimer amyloidosis [8]. Levels of urinary and plasma iPF(2α)–IV were higher in APP-transgenic mice than in littermate wild type animals as early as eight months of age, and remained high for the rest of the study. Furthermore, homogenates from the cerebral cortex and the hippocampus of APP-mice had higher levels of iPF(2α)–IV, and this increase in LPO preceded amyloid plaque formation.

That increased oxidative damage may, indeed, be the earliest event in AD is suggested by findings showing that increases in 8-hydroxyguanosine, an oxidized nucleoside derived from DNA, and nitrotyrosine, an oxidized amino acid, were quantitatively greatest early in the disease [9]. Data confirming that oxidative stress-induced damage precedes the deposition of amyloid in the brain also come from a study of Down's syndrome brains [10]. Oxidized nucleic acid, 8-hydroxy guanosine (8-OHG), oxidized protein and nitrotyrosine were found to be increased significantly in the teens and in the twenties, while Aβ deposits only occurred after age 30. These findings suggest that brain oxidative damage must contribute to AD pathogenesis before Aβ (i.e. senile plaques) accumulates in the AD brain.

Several risk factors may facilitate oxidative stress in the AD brain. One study has showed reduced antioxidant enzyme activity in AD brain [11], and brains of transgenic mice expressing different Alzheimer-linked presenilin-1 gene (PS1) mutations have been shown to have decreased antioxidant enzyme activity [12]. Another risk factor that may facilitate Aβ-induced oxidative brain damage is the epsilon4 allele of apolipoprotein E (APOE), an established risk factor for AD. Levels of LPO were found to be significantly increased in tissues from Alzheimer's cases which were homozygous for the epsilon4 allele of APOE, compared to AD epsilon3/epsilon3 cases and controls [13], suggesting that oxidative stress-induced injury in the frontal cortex of AD cases is related to the APOE genotype. Activated astrocytes (and microglia) are observed in a wide range of CNS pathologies, including brain inflammation, trauma, ischemia, and stroke; neurodegenerative diseases including AD, and normal aging [14,15]. In AD, activated microglia and astrocytes expressing the inducible form of nitric oxide synthase (iNOS) are found in the amyloid plaques surrounded by dead and dystrophic neurites [16,17], and a growing body of evidence suggests that these activated glia contribute to neurotoxicity through the production of reactive nitrogen species (RNS) derived from increased production of iNOS. Moreover, aggregated Aβ peptide can also stimulate iNOS production in rat glial cell cultures [18].

The concentration of 3-nitrotyrosine, a marker of oxidative damage mediated by peroxynitrite, and the 3-nitrotyrosine/tyrosine ratio was also increased six-fold in AD patients when compared with cerebrospinal fluid samples from age matched controls, and increased significantly with decreasing cognitive function [19]. Strong nitrotyrosine (NT) immunoreactivity was also observed in the absence of tangle formation in most terminal deoxynucleotidyl transferase (TdT)-positive neurones in the visual cortex of AD patients, suggesting that neuronal DNA damage may precede tangle formation and is associated with up-regulation of nitrotyrosine in AD brain [20].

These studies indicate that increased oxidative damage must contribute to the pathogenesis of AD before amyloid deposits occur in the brain. Thus, oxidative stress and increased production of iNOS are early events that appear to be central in determining cell death in AD.

Neurotrophic Factors in Alzheimer's Disease

Since nerve growth factor (NGF) induces neuronal growth and regeneration [21-25], it is generally accepted that NGF may play a pivotal role in the repair processes of neurodegenerative disorders [26] and may be a crucial element of the neuronal loss in AD [27-30]. Several studies revealed that deficiency of NGF may directly trigger apoptosis. Neuronal PC12 cells underwent apoptosis within two days following withdrawal of nerve growth factor (NGF) from the culture medium. [31]. Anti-NGF transgenic mice develop neurodegeneratory lesions characteristic for AD including amyloid plaques and neurofibrillary tangles in cortical and hippocampal neurons associated with extensive neuronal loss [32]. Although NGF is commonly accepted as a major factor promoting neuronal regeneration and survival, recent findings indicated that NGF may also play a role in triggering apoptosis. The ambivalent nature of NGF may be explained by the physiology of the NGF receptors, the tropomyosin receptor kinase A (TrkA) and the 75 kDa NGF receptor (p75NTR). TrkA induces survival and regeneration, whereas p75NTR appears to be a death receptor mediating the apoptotic effect of NGF. Indeed, NGF induced apoptosis in p75NTR-expressing human neuroblastoma SK-N-MC cells, but neurotrophin-3 (NT-3) or brain-derived neurotrophic factor (BDNF) lacked apoptotic effect [33]. Moreover, beta-amyloid peptide, commonly present in amyloid plaques in AD, specifically binds the p75NTR in rat cortical neurons, NIH-3T3 cell line and neuronal hybrid cells, and triggers apoptosis [34-36] indicating that neuronal death in Alzheimer's disease may be mediated by the interaction of beta-amyloid with p75NTR. Since p75NTR is up-regulated in neurodegeneratory disorders, p75NTR might be involved in neuronal cell death related to AD [33].

Alzheimer's Disease and Glycosaminoglycans

There is a growing body of evidence that glycosaminoglycans (GAGs) play a pivotal role in the pathogenesis of Alzheimer's disease. Glycosaminoglycans are highly sulfated polysaccharides that are common constituents of the extracellular matrix. GAGs bind to Aβ, and attenuate the proteoglycan-mediated protection of Aβ against proteolysis. It has been also shown that GAGs inhibit the aggregation and toxicity of Aβ itself [37]. Ateroid^{R}, a mixture of low molecular weight GAGs recently used in Europe to reverse the symptoms of AD, improves age-related behavioral deficits in rats. Moreover, human studies revealed that Ateroid^{R} significantly improves dementia [38,39].

C3 is a novel low molecular weight GAG exerting neuroprotective property. C3 is synthesized from unfractionated heparin using a standardized process of γ-irradiation. The chemical and biological properties of C3 have been well characterized including its ability to protect against various lesions in rat animal models. Stereotaxic injection of Aβ fragments (Aβ[25-35]) into the amygdala increases tau-2 immunoreactivity in the rat hippocampus [40]. This phenomenon was blocked by administration of C3 [41]. Moreover, C3 attenuated the cholinergic lesions in the septum and cingulum bundle induced by intraventricular administration of a specific cholinotoxin, ethylcholine aziridinium (AF64A; [42,43]. It has been also shown that C3 increase axonal growth and arborization in the rat hippocampus [44].

Recent studies suggested that GAGs may act via modulating the release of nerve growth factor (NGF) by stimulating the effect of growth factors in cell culture [45,46] and/or reducing the expression of cholinotoxin-stimulated growth factor receptor expression in the rat septum [47]. Glial cell line-derived neurotrophic factor (GDNF), that is vital to the development and maintenance of neural tissues, requires heparan sulphate glycosaminoglycan to promote neuronal survival [48]. However, the exact mechanism of the neuroprotective/neurorepair effect of GAGs is not known.

The bioavailability of various GAGs depends on their potential to traverse the blood bran barrier (BBB). Since C3 has been demonstrated to cross the BBB, it may be a valuable adjunct in the treatment of neurodegenerative disorders including AD. Indeed, C3 is in phase I clinical trials at the time of this writing.

Glycosaminoglycans and Apoptosis: Potential Treatment of Alzheimer's Disease

Previous studies revealed that GAGs express neuroprotective properties against various lesions in cell culture [49] and in vivo [41-43,50]. Since neural damage is often associated with apoptotic processes, it is possible that GAGs may protect against neural lesions via influencing stressor-induced apoptosis. Moreover, C3, the low molecular GAG derived from heparin has been shown to penetrate the blood-brain-barrier [51], thus, it is conceivable that C3 and similar molecules can be used as therapeutic agents to protect against neuronal apoptotic processes, that are major hallmarks of neurodegenerative disorders.

It is a common premise that GAGs play a pivotal role in the molecular processes of programmed cell death in several cell types. However, the data are ambiguous. Numerous studies indicate that GAGs may participate in the inhibition of apoptotic pathways. Hyaluronan, a major glycosaminoglycan component of the extracellular matrix of the mammalian bone marrow, antagonizes dexamethasone-induced apoptosis of malignant multiplex myeloma (MM) cells [52]. Since GAGs are major constituents of the bone marrow, these findings support the idea that GAGs could play a major role in the survival of MM cells in vivo, and could explain why MM cells accumulate in the bone marrow of patients with MM and escape conventional chemotherapy. Apoptosis in fibroblasts was also attenuated by administration of chondroitin sulfate, and heparan sulfate. Platelet-derived growth factor (PDGF) and beta-D-xyloside treatment protected fetal lung fibroblasts against serum starvation-induced apoptosis probably by stimulating the GAG synthesis/secretion of the cells. Indeed, application of sulfated GAGs, chondroitin sulfate, and heparan sulfate, also resulted in diminished apoptosis in fetal lung fibroblasts, possibly via increasing Bcl-associated death promoter phosphorylation and diminished caspase-3 and caspase-7 cleavage [53]

Heparin inhibits mesangial cell apoptosis induced by staurosporine, pyrrolidine dithiocarbamate, and ultraviolet light, and also suppresses oxidant-induced apoptosis of NRK49F fibroblasts and Madin-Darby canine kidney epithelial cells. Furthermore, heparin attenuated spontaneous apoptosis of podocytes in explanted glomeruli. These results indicate the novel potential of heparin as an inhibitor of apoptosis in several cell types, including glomerular cells [54]. Moreover, heparin and aspirin has been shown to inhibit trophoblast apoptosis providing an additional mechanism to explain the clinical benefits of heparin and aspirin on recurrent pregnancy loss [55].

In contrast to these findings, induction of apoptosis by GAGs has been reported by several studies, suggesting that GAGs may be used for eliminating tumor cells. Heparin exerts a significant anti-proliferative and an apoptotic effect on human hepatoma cells in vitro [56], and induces apoptosis of human nasopharyngeal carcinoma CNE2 cells [57]. Internalized heparin induces apoptotic cell death probably by interfering with transcription factor function [58]. Derivatives of heparin and chondroitin sulphate substantially reduce cell viability by induction of apoptosis of myeloma and breast cancer cells in vitro [59]. Heparin also induces apoptosis in peripheral blood neutrophils. This phenomenon may help explain the anti-inflammatory effects resulting from the interaction between vessel wall heparan sulphate and chemoattracted peripheral blood neutrophils [60]. Moreover, heparin is not the only GAG exhibiting apoptotic features. A low molecular weight proteoglycan, lumican induced and/or increased the apoptosis of B16F1 cells, probably via the sulphated GAG component of the molecule [61]. Chondroitin sulphate also induces apoptosis of articular chondrocytes [62].

Discussion and Summary

Apoptosis is believed to play a pivotal role in the pathogenesis of neurodegenerative disorders including AD. Although the exact function of apoptotic processes in brain pathology is not known, it is a common consensus that two major pathways are responsible in triggering apoptosis in neurodegenerative disorders: (1) oxidative stress and (2) trophic factor dysregulation. Oxidative stress appears to be the major pathological hallmark of AD. Moreover, dysregulation in the release of nerve growth factors, and expression of their receptors are characteristic pathological changes attributed to AD.

GAGs are known to exhibit neuroprotective properties in numerous animal models of AD. One possible explanation of the GAG-induced neuroprotection is that GAGs may affect apoptotic processes characteristic in neurodegenerative lesions. The pathomechanism of this phenomenon is not entirely understood; GAGs may affect apoptosis either directly, or via modulating growth factor pathology and/or oxidative stress, that are major hallmarks of AD. Thus, GAGs can be extremely veluable adjunct in the treatment of neurodegenerative disorders. Since a low molecular GAG, C3, is able to cross the blood-brain-barrier, C3 treatment may represent an entirely novel avenue in the therapy of AD. Indeed, C3 has successfully passed the toxicology tests as a low-risk compound, and it is currently in phase I clinical trials.

References

[1] Rideout HJ, Stefanis L. Caspase Inhibition: A Potential Therapeutic Strategy In Neurological Diseases *Histol Histopathol* 2001; 16: 895-908.

[2] Engidawork E, Gulesserian T, Seidl R, Cairns N, Lubec G. Expression Of Apoptosis Related Proteins In Brains Of Patients With Alzheimer's Disease *Neurosci Lett* 2001; 303: 79-82.

[3] Engidawork E, Gulesserian T, Yoo BC, Cairns N, Lubec G. Alteration Of Caspases And Apoptosis-Related Proteins In Brains Of Patients With Alzheimer's Disease *Biochem Biophys Res Commun* 2001; 281: 84-93.

[4] Selznick LA, Holtzman DM, Han BH, Gokden M, Srinivasan AN, Johnson EM, Roth KA. In Situ Immunodetection Of Neuronal Caspase-3 Activation In Alzheimer Disease *J Neuropathol Exp Neurol* 1999; 58: 1020-1026.

[5] Lovell MA, Ehmann WD, Mattson MP, Markesbery WR. Elevated 4-Hydroxynonenal In Ventricular Fluid In Alzheimer's Disease *Neurobiol Aging* 1997; 18: 457-461.

[6] Tuppo EE, Forman LJ, Spur BW, Chan-Ting RE, Chopra A, Cavalieri TA. Sign Of Lipid Peroxidation As Measured In The Urine Of Patients With Probable Alzheimer's Disease *Brain Res Bull* 2001; 54: 565-568.

[7] Mark RJ, Lovell MA, Markesbery WR, Uchida K, Mattson MP. A Role For 4-Hydroxynonenal, An Aldehydic Product Of Lipid Peroxidation, In Disruption Of Ion Homeostasis And Neuronal Death Induced By Amyloid Beta-Peptide *J Neurochem* 1997; 68: 255-264.

[8] Pratico D, Uryu K, Leight S, Trojanoswki JQ, Lee VM. Increased Lipid Peroxidation Precedes Amyloid Plaque Formation In An Animal Model Of Alzheimer Amyloidosis *J Neurosci* 2001; 21: 4183-4187.

[9] Nunomura A, Perry G, Aliev G, Hirai K, Takeda A, Balraj EK, Jones PK, Ghanbari H, Wataya T, Shimohama S, Chiba S, Atwood CS, Petersen RB, Smith MA. Oxidative

Damage Is The Earliest Event In Alzheimer Disease *J Neuropathol Exp Neurol* 2001; 60: 759-767.

[10] Nunomura A, Perry G, Pappolla MA, Friedland RP, Hirai K, Chiba S, Smith MA. Neuronal Oxidative Stress Precedes Amyloid-Beta Deposition In Down Syndrome *J Neuropathol Exp Neurol* 2000; 59: 1011-1017.

[11] Marcus DL, Thomas C, Rodriguez C, Simberkoff K, Tsai JS, Strafaci JA, Freedman ML. Increased Peroxidation And Reduced Antioxidant Enzyme Activity In Alzheimer's Disease *Exp Neurol* 1998; 150: 40-44.

[12] Leutner S, Czech C, Schindowski K, Touchet N, Eckert A, Muller WE. Reduced Antioxidant Enzyme Activity In Brains Of Mice Transgenic For Human Presenilin-1 With Single Or Multiple Mutations *Neurosci Lett* 2000; 292: 87-90.

[13] Ramassamy C, Averill D, Beffert U, Bastianetto S, Theroux L, Lussier-Cacan S, Cohn JS, Christen Y, Davignon J, Quirion R, Poirier J. Oxidative Damage And Protection By Antioxidants In The Frontal Cortex Of Alzheimer's Disease Is Related To The Apolipoprotein E Genotype *Free Radic Biol Med* 1999; 27: 544-553.

[14] Eddleston M, Mucke L. Molecular Profile Of Reactive Astrocytes--Implications For Their Role In Neurologic Disease *Neuroscience* 1993; 54: 15-36.

[15] Kreutzberg GW. Microglia: A Sensor For Pathological Events In The CNS Trends *Neurosci* 1996; 19: 312-318.

[16] Lee SC, Zhao ML, Hirano A, Dickson DW. Inducible Nitric Oxide Synthase Immunoreactivity In The Alzheimer Disease Hippocampus: Association With Hirano Bodies, Neurofibrillary Tangles, And Senile Plaques *J Neuropathol Exp Neurol* 1999; 58: 1163-1169.

[17] Wallace MN, Geddes JG, Farquhar DA, Masson MR. Nitric Oxide Synthase In Reactive Astrocytes Adjacent To Beta-Amyloid Plaques *Exp Neurol* 1997; 144: 266-272.

[18] Akama KT, Van Eldik LJ. Beta-Amyloid Stimulation Of Inducible Nitric-Oxide Synthase In Astrocytes Is Interleukin-1beta- And Tumor Necrosis Factor-Alpha (Tnfalpha)-Dependent, And Involves A Tnfalpha Receptor-Associated Factor- And Nfkappab-Inducing Kinase-Dependent Signaling Mechanism *J Biol Chem* 2000; 275: 7918-7924.

[19] Tohgi H, Abe T, Yamazaki K, Murata T, Ishizaki E, Isobe C. Alterations Of 3-Nitrotyrosine Concentration In The Cerebrospinal Fluid During Aging And In Patients With Alzheimer's Disease *Neurosci Lett* 1999; 269: 52-54.

[20] Su JH, Deng G, Cotman CW. Neuronal DNA Damage Precedes Tangle Formation And Is Associated With Up-Regulation Of Nitrotyrosine In Alzheimer's Disease Brain *Brain Res* 1997; 774: 193-199.

[21] Damon DH, D'Amore PA, Wagner JA. Nerve Growth Factor And Fibroblast Growth Factor Regulate Neurite Outgrowth And Gene Expression In PC12 Cells Via Both Protein Kinase C- And Camp-Independent Mechanisms *J Cell Biol* 1990; 110: 1333-1339.

[22] Gahwiler BH, Enz A, Hefti F. Nerve Growth Factor Promotes Development Of The Rat Septo-Hippocampal Cholinergic Projection In Vitro *Neurosci Lett* 1987; 75: 6-10.

[23] Hartikka J, Hefti F. Development Of Septal Cholinergic Neurons In Culture: Plating Density And Glial Cells Modulate Effects Of NGF On Survival, Fiber Growth, And Expression Of Transmitter-Specific Enzymes *J Neurosci* 1988; 8: 2967-2985.

[24] Hefti F, Knusel B, Lapchak PA. Protective Effects Of Nerve Growth Factor And Brain-Derived Neurotrophic Factor On Basal Forebrain Cholinergic Neurons In Adult Rats With Partial Fimbrial Transections Prog *Brain Res* 1993; 98: 257-263.

[25] Junard EO, Montero CN, Hefti F. Long-Term Administration Of Mouse Nerve Growth Factor To Adult Rats With Partial Lesions Of The Cholinergic Septohippocampal Pathway *Exp Neurol* 1990; 110: 25-38.

[26] Hefti F, Knusel B. Chronic Administration Of Nerve Growth Factor And Other Neurotrophic Factors To The Brain *Neurobiol Aging* 1988; 9: 689-690.

[27] Hefti F, Will B. Nerve Growth Factor Is A Neurotrophic Factor For Forebrain Cholinergic Neurons; Implications For Alzheimer's Disease *J Neural Transm Suppl* 1987; 24: 309-315.

[28] Hefti F, Schneider LS. Nerve Growth Factor And Alzheimer's Disease Clin *Neuropharmacol* 1991; 14 Suppl 1: S62-S76.

[29] Lapchak PA, Hefti F. Emerging Pharmacology Of Nerve Growth Factor *Prog Neuropsychopharmacol Biol Psychiatry* 1991; 15: 851-860.

[30] Phelps CH, Gage FH, Growdon JH, Hefti F, Harbaugh R, Johnston MV, Khachaturian ZS, Mobley WC, Price DL, Raskind M. Potential Use Of Nerve Growth Factor To Treat Alzheimer's Disease *Neurobiol Aging* 1989; 10: 205-207.

[31] Araki W, Wurtman RJ. Increased Expression Of Amyloid Precursor Protein And Amyloid Precursor-Like Protein 2 During Trophic Factor Withdrawal-Induced Death Of Neuronal PC12 Cells *Brain Res Mol Brain Resry* 1998; 56: 169-177.

[32] Capsoni S, Ugolini G, Comparini A, Ruberti F, Berardi N, Cattaneo A. Alzheimer-Like Neurodegeneration In Aged Antinerve Growth Factor Transgenic Mice *Proc Natl Acad Sci U S A* 2000; 97: 6826-6831.

[33] Kuner P, Hertel C. NGF Induces Apoptosis In A Human Neuroblastoma Cell Line Expressing The Neurotrophin Receptor P75ntr *J Neurosci Res* 1998; 54: 465-474.

[34] Yaar M, Zhai S, Pilch PF, Doyle SM, Eisenhauer PB, Fine RE, Gilchrest BA. Binding Of Beta-Amyloid To The P75 Neurotrophin Receptor Induces Apoptosis. A Possible Mechanism For Alzheimer's Disease *J Clin Invest* 1997; 100: 2333-2340.

[35] Tsukamoto E, Hashimoto Y, Kanekura K, Niikura T, Aiso S, Nishimoto I. Characterization Of The Toxic Mechanism Triggered By Alzheimer's Amyloid-Beta Peptides Via P75 Neurotrophin Receptor In Neuronal Hybrid Cells *J Neurosci Res* 2003; 73: 627-636.

[36] Hashimoto Y, Kaneko Y, Tsukamoto E, Frankowski H, Kouyama K, Kita Y, Niikura T, Aiso S, Bredesen DE, Matsuoka M, Nishimoto I. Molecular Characterization Of Neurohybrid Cell Death Induced By Alzheimer's Amyloid-Beta Peptides Via P75ntr/PLAIDD *J Neurochem* 2004; 90: 549-558.

[37] Kisilevsky R, Lemieux LJ, Fraser PE, Kong X, Hultin PG, Szarek WA. Arresting Amyloidosis In Vivo Using Small-Molecule Anionic Sulphonates Or Sulphates: Implications For Alzheimer's Disease *Nat Med* 1995; 1: 143-148.

[38] Conti L, Re F, Lazzerini F, Morey LC, Ban TA, Santini V, Modafferi A, Postiglione A. Glycosaminoglycan Polysulfate (Ateroid) In Old-Age Dementias: Effects Upon Depressive Symptomatology In Geriatric Patients *Prog Neuropsychopharmacol Biol Psychiatry* 1989; 13: 977-981.

[39] Conti L, Placidi GF, Cassano GB. Ateroid In The Treatment Of Dementia: Results Of A Clinical Trial Mod Probl *Pharmacopsychiatry* 1989; 23: 76-84.

[40] Sigurdsson EM, Lorens SA, Hejna MJ, Dong XW, Lee JM. Local And Distant Histopathological Effects Of Unilateral Amyloid-Beta 25-35 Injections Into The Amygdala Of Young F344 Rats *Neurobiol Aging* 1996; 17: 893-901.

[41] Dudas B, Cornelli U, Lee JM, Hejna MJ, Walzer M, Lorens SA, Mervis RF, Fareed J, Hanin I. Oral And Subcutaneous Administration Of The Glycosaminoglycan C3 Attenuates Abeta(25-35)-Induced Abnormal Tau Protein Immunoreactivity In Rat Brain *Neurobiol Aging* 2002; 23: 97-104.

[42] Rose M, Dudas B, Cornelli U, Hanin I. Protective Effect Of The Heparin-Derived Oligosaccharide C3, On AF64A-Induced Cholinergic Lesion In Rats *Neurobiol Aging* 2003; 24: 481-490.

[43] Rose M, Dudas B, Cornelli U, Hanin I. Glycosaminoglycan C3 Protects Against AF64A-Induced Cholinotoxicity In A Dose-Dependent And Time-Dependent Manner *Brain Res* 2004; 1015: 96-102.

[44] Mervis RF, Mckean J, Zats S, Gum A, Reinhard R, Dudas B, Cornelli U, Lee JM, Lorens SA, Fareed J, Hanin I. Neurotrophic Effects Of The Glycosaminoglycan C3 On Dendritic Arborization And Spines In The Adult Rat Hippocampus: A Quantitative Golgi Study *CNS Drug Reviews* 2000; 44-46.

[45] Damon DH, D'Amore PA, Wagner JA. Sulfated Glycosaminoglycans Modify Growth Factor-Induced Neurite Outgrowth In PC12 Cells *J Cell Physiol* 1988; 135: 293-300.

[46] Lesma E, Di Giulio AM, Ferro L, Prino G, Gorio A. Glycosaminoglycans In Nerve Injury: 1. Low Doses Of Glycosaminoglycans Promote Neurite Formation *J Neurosci Res* 1996; 46: 565-571.

[47] Dudas B, Rose M, Cornelli U, Hanin I. Low Molecular Weight Glycosaminoglycan C3 Attenuates AF64A-Stimulated, Low-Affinity Nerve Growth Factor Receptor-Immunoreactive Axonal Varicosities In The Rat Septum *Brain Res* 2005; 1033: 34-40.

[48] Barnett MW, Fisher CE, Perona-Wright G, Davies JA. Signalling By Glial Cell Line-Derived Neurotrophic Factor (GDNF) Requires Heparan Sulphate Glycosaminoglycan *J Cell Sci* 2002; 115: 4495-4503.

[49] Kisilevsky R, Lemieux LJ, Fraser PE, Kong X, Hultin PG, Szarek WA. Arresting Amyloidosis In Vivo Using Small-Molecule Anionic Sulphonates Or Sulphates: Implications For Alzheimer's Disease *Nat Med* 1995; 1: 143-148.

[50] Sigurdsson EM, Lorens SA, Hejna MJ, Dong XW, Lee JM. Local And Distant Histopathological Effects Of Unilateral Amyloid-Beta 25-35 Injections Into The Amygdala Of Young F344 Rats *Neurobiol Aging* 1996; 17: 893-901.

[51] Ma Q, Dudas B, Hejna M, Cornelli U, Lee JM, Lorens S, Mervis R, Hanin I, Fareed J. The Blood-Brain Barrier Accessibility Of A Heparin-Derived Oligosaccharides C3 *Thromb Res* 2002; 105: 447-453.

[52] Vincent T, Molina L, Espert L, Mechti N. Hyaluronan, A Major Non-Protein Glycosaminoglycan Component Of The Extracellular Matrix In Human Bone Marrow, Mediates Dexamethasone Resistance In Multiple Myeloma *Br J Haematol* 2003; 121: 259-269.

[53] Cartel NJ, Post M. Abrogation Of Apoptosis Through PDGF-BB-Induced Sulfated Glycosaminoglycan Synthesis And Secretion *Am J Physiol Lung Cell Mol Physiol* 2005; 288: L285-L293.

[54] Ishikawa Y, Kitamura M. Inhibition Of Glomerular Cell Apoptosis By Heparin *Kidney Int* 1999; 56: 954-963.

[55] Bose P, Black S, Kadyrov M, Weissenborn U, Neulen J, Regan L, Huppertz B. Heparin And Aspirin Attenuate Placental Apoptosis In Vitro: Implications For Early Pregnancy Failure *Am J Obstet Gynecol* 2005; 192: 23-30.

[56] Karti SS, Ovali E, Ozgur O, Yilmaz M, Sonmez M, Ratip S, Ozdemir F. Induction Of Apoptosis And Inhibition Of Growth Of Human Hepatoma Hepg2 Cells By Heparin *Hepatogastroenterology* 2003; 50: 1864-1866.

[57] Li HL, Ye KH, Zhang HW, Luo YR, Ren XD, Xiong AH, Situ R. Effect Of Heparin On Apoptosis In Human Nasopharyngeal Carcinoma CNE2 Cells *Cell Res* 2001; 11: 311-315.

[58] Berry D, Lynn DM, Sasisekharan R, Langer R. Poly(Beta-Amino Ester)S Promote Cellular Uptake Of Heparin And Cancer Cell Death *Chem Biol* 2004; 11: 487-498.

[59] Pumphrey CY, Theus AM, Li S, Parrish RS, Sanderson RD. Neoglycans, Carbodiimide-Modified Glycosaminoglycans: A New Class Of Anticancer Agents That Inhibit Cancer Cell Proliferation And Induce Apoptosis *Cancer Res* 2002; 62: 3722-3728.

[60] Manaster J, Chezar J, Shurtz-Swirski R, Shapiro G, Tendler Y, Kristal B, Shasha SM, Sela S. Heparin Induces Apoptosis In Human Peripheral Blood Neutrophils *Br J Haematol* 1996; 94: 48-52.

[61] Vuillermoz B, Khoruzhenko A, D'Onofrio MF, Ramont L, Venteo L, Perreau C, Antonicelli F, Maquart FX, Wegrowski Y. The Small Leucine-Rich Proteoglycan Lumican Inhibits Melanoma Progression *Exp Cell Res* 2004; 296: 294-306.

[62] Bali JP, Cousse H, Neuzil E. Biochemical Basis Of The Pharmacologic Action Of Chondroitin Sulfates On The Osteoarticular System Semin Arthritis *Rheum* 2001; 31: 58-68.

In: Focus on Cell Apoptosis Research
Editor: Daniel W. Cho, pp. 81-118

ISBN: 1-60021-433-9

Chapter III

LYMPHOCYTES AS CELL MODEL TO STUDY OXIDATIVE STRESS-INDUCED APOPTOSIS AND CYTOPROTECTION IN ALZHEIMER'S AND PARKINSON'S DISEASE

Marlene Jimenez Del Rio* and Carlos Velez-Pardo
Faculty of Medicine, Department of Internal Medicine, Neuroscience Research Program, University of Antioquia (UdeA), SIU- Medellin, Colombia

ABSTRACT

Alzheimer's disease (AD) and Parkinson's disease (PD) are the most common progressive neurodegenerative disorders that not only affect a large population in the world, but also affect a genetically isolated group of individuals known as the "paisa community" living in Antioquia region, Colombia. Both disorders are neuropathologically characterized by specific neuronal and synaptic loss of the hippocampal pyramidal neurons and dopaminergic neurons of the substantia nigra, respectively. Unfortunately, no effective and/or definitive therapeutic treatment aimed to reduce or to retard the clinic and pathologic symptoms induced by both devastating diseases are presently available. Thus, the biochemical mechanism(s) involved in neural loss has mainly been studied in biological models. In this article the authors have proposed lymphocytes as a cellular model to study the molecular signalling mechanism(s) of cell death-apoptosis- induced by oxidative stress and cytoprotection in AD and PD. Remarkably, lymphocytes and neurons are cellularly and biochemically alike. Indeed, lymphocytes express at least six different neurochemical systems

* Correspondence concerning this article should be addressed to Marlene Jimenez Del Rio, Faculty of Medicine, Department of Internal Medicine, Neuroscience Research Programme-University of Antioquia (UdeA), Calle 62 # 52-59, Building 1 Room 412; SIU- Medellin, Colombia. Tel: (574) 210.64.57; Fax: (574) 573.08.65; E-mail address: mdelrio@quimbaya.udea.edu.co.

considered to exclusively pertain to neural cells such as catecholaminergic, serotonergic, acetylcholinergic, glutamatergic, noradrenergic and GABAergic systems. Given that lymphocytes are post-mitotic cells (i.e. G_o cell cycle), possess the synthesis, transport and processing systems of amyloid-beta precursor protein (AβPP), presenilins (PS) and parkin proteins, but most importantly, expresses universal cell death machinery, these cells constitute not only an ideal biomodel to better understanding of the pathological signalling in those neurological entities, but also represent a meaningful tool to design innovative and rational therapeutic strategies focused to arrest the neuronal death in Alzheimer's and Parkinson's patients. Specifically, we present a unified molecular mechanism of cell death induced by dopamine and Aβ, and survival signalling mechanism triggered by insulin-like growth factor and cannabinoids against oxidative stress stimuli.

INTRODUCTION

Alzheimer's disease (AD) and Parkinson's disease (PD) are the most common progressive neurodegenerative disorders that not only affect a large group of individuals in Antioquia, Colombia (Lopera et al., 1997; Sanchez et al. 2004) but also affect other regions in the world (Janca, 2002; Ferri et al., 2005). Little has changed since the first personal description of the clinical and neuropathological properties of AD made by Alois Alzheimer in 1906[1]. AD (OMIM entry 104300) is characterised by memory loss and personality changes, and the pathological hallmarks are senile plaques (SP), neurofibrillary tangles (NFT) deposition in the brain, accompanied by gliosis, neuronal and synaptic loss (Esiri, 2001). NFT has been shown to be composed of hyperphosphorylated microtubule associated protein tau (τ), which assembles in paired helical filaments (PHF) and accumulates in the cytoplasmic compartments in neurons (Wischik et al., 2001). The senile plaques are extracellular deposits in the brain parenchyma, mainly consisting of a 40-43 amino-acid peptide named amyloid beta ($A\beta_{[1\text{-}40/42]}$) (Howlett et al., 2001). This peptide is produced by specific endoproteolytic cleavages of the Aβ amyloid precursor protein (APP), which is a type I cell surface glycoprotein of up to 770 amino-acids, by aspartyl protease β-secretase or BACE-1 (β-site APP-cleaving enzyme) and γ-secretase, which is a membrane protein complex comprised of presenilin (PS), nicastrin (Nct), anterior pharynx-defective phenotype-1 (APH-1), and Presenilin enhancer-2 (PEN-2) proteins (Kimberly et al., 2003). An alternative APP cleavage pathway by α-secretase, identify as a member of the ADAMs (a disintegrin and metalloprotease-like) family of membrane-associated zinc metalloproteinases, and subsequent γ-secretase cleavage, precludes the generation of $A\beta_{[1\text{-}40/42]}$ and leads to the generation of the 16 amino acids shorter not amyloidogenic peptide termed p3 (for a review, see Haass, 2004; Vetrivel and Thinakaran, 2006).

In the last decade, genetic studies have definitively demonstrated that early-onset familial AD is caused by multiple mutations in at least three different genes, i.e. APP (26 pathogenic mutations), presenilin-1 (PS-1; 155 mutations) and PS-2 (10 mutations). Together, these mutations (see htpp://www.molgen.ua.ac.be/ADMutations/ for an up-date review) are

[1] For an English translation of the original Alzheimer's report at 37th Meeting of psychiatrists of South-western Germany in Tubingen on November 3rd and 4th 1906, see Schawab, C. Neuroscience News 2000; 3(4): 6-7.

responsible for 30 to 50% of autosomal dominant AD cases, and about 5% of AD in general. Strikingly, all these mutations and apolipoprotein E (APOE) polymorphisms (as genetic risk factor) have been shown to cause increased formation of total Aβ, or specifically, the more fibrillogenic $A\beta_{[1-42]}$ and eventually resulted in the formulation of the Aβ hypothesis of AD (Hardy and Higgins, 1992)[2]. This theory proposes that progressive cerebral accumulation of amyloid Aβ protein initiates a complex multicellular cascade that includes neuritic dystrophy,

[2] During the last one and a half decade, two contrasting schools of thought have argued about the role of the amyloid-beta peptide (Aβ) and tau protein (τ) in AD. These schools have been nicknamed as "baptists", i.e. proponents of the "amyloid cascade hypothesis", and "tauists", i. e. proponents of the "tau and tangle hypothesis", respectively. On the one hand, the Baptists assert that a dysregulation in APP processing occurs early in the disease process resulting in increased production of the longer amyloid peptide $A\beta_{[1-42]}$, which forms the plaque core. This peptide goes on to induce all the subsequent pathology including tau aggregation, phosphorylation, neuronal erosion and clinical dementia (for example, see Younkin, SG. Evidence that A beta 42 is the real culprit in Alzheimer's disease. Ann Neurol 1995; 37:287-288; Takashima, A; Honda, T; Yasutake, K; Michel, G; Murayama, O; Murayama, M; Ishiguro, K; Yamaguchi, H. Activation of tau protein kinase I/glycogen synthase kinase-3β by amyloid β peptide (25–35) enhances phosphorylation of tau in hippocampal neurons. Neurosci Res 1998; 31: 317–323; Selkoe, DJ. Alzheimer's disease results from the cerebral accumulation and cytotoxicity of amyloid β-protein. J Alzheimer's Disease 2001; 3: 75-81; Tanzi, RE; Bertram, L. Twenty years of the Alzheimer's disease amyloid hypothesis: A genetic perspective. Cell 2005; 120: 545-555). Yet, how $A\beta_{[1-42]}$ induces this neurodegeneration is a matter of intense debate (see Haass, C; Steiner, H. Protofibrils, the unifying toxic molecule of neurodegenerative disorders?. Nat Neurosci 2001; 4: 859-860; Wirths, O; et al.; J Neurochem 2004; 91:513-520; Butterfield, DA; Bush, AI. Alzheimer's amyloid β-peptide (1-42): involvement of methionine residue in the oxidative stress and neurotoxicity properties of this peptide. Neurobiol Aging 2004; 25: 563-568; Klein, WL; Stine, WB; Teplow, DB. Small assemblies of unmodified amyloid beta-protein are the proximate neurotoxin in Alzheimer's disease. Neurobiol Aging 2004; 25: 569-580). On the other hand, the tauists claim that tangles are central to AD. The proponents argue that the normal role of tau in binding to and stabilizing microtubules is impaired by phosphorylation (Lovestone, S; Reynolds, CH. The phosphorylation of tau: a critical stage in neurodevelopmental and neurodegenerative processes. Neuroscience 1997; 78: 309-324), directly by either mutations that alter functions (Brandt, R; Hundelt, M; Shahani, N. Tau alteration and neural degeneration taupathies: mechanism and models. Biochim Biophys Acta 2005; 1739(2-3): 331-354), by mutations that alter isoform expression, or by phosphorylation/ dephosphorylation imbalance in the affected neurons (Iqbal, K; Grundke-Iqbal, I. Metabolic/signal transduction hypothesis of Alzheimer's and other taupathies. Acta Neuropathol (Berl) 2005; 109: 25-31). Decreased tau binding to microtubules might result in increased free tau which will self aggregate to form insoluble paired helical filament, and indeed, gradually replaced by tangles that lead to neurodegeneration and dementia. Several reports support the notion of an "amyloid cascade to tau hypothesis" (see Oddo, S; Caccamo, A; Kitazawa, M; Tseng, BP; LaFerla, FM. Amyloid deposition precedes tangle formation in a triple transgenic model of Alzheimer's disease. Neurobiology of Aging 2003; 241063–1070; Tanzi, RE; Bertram, L. Twenty years of the Alzheimer's disease amyloid hypothesis: A genetic perspective. Cell 2005; 120: 545-555, just to mention a few ones). Moreover, the strongest evidence that tangles cannot cause amyloidosis has been provided by a report of a novel PS mutation (Gly183Val) associated with familial frontotemporal dementia (FTD) that is neuropathologically characterized by Pick-type taupathy, and absence of extracellular Aβ plaques, i.e. there is no parenchymal or cerebrovascular amyloidosis in FTD (Dermaut, B; Kumar-Singh, S; Engelborghs, S; Theuns, J; Rademakers, R; Saerens, J; Pickut, BA; Peeters, K; van den Broeck, C; Vennekens, K; Claes, S; Cruts, M; Cras, P; Martin, JJ; Van Broeckhoven, C; De Deyn, PP. A novel presenilin 1 mutation associated with Pick's disease but not β-amyloid plaques. Ann Neurol 2004; 55: 617–626). Remarkably, PS Gly183Val behaves as a loss of function mutation because of its inability to process APP into Aβ peptides. Although, Aβ plaques were not detected, the possibility remains that this mutation is acting by increasing the levels of toxic Aβ oligomers that induce tangle formation, or that PS1 mutations may contribute to the onset of AD not only by enhancing $A\beta_{[1-42]}$ production but by also accelerating the formation and accumulation of filamentous tau (Tanemura, K; Chui, DH; Fukuda, T; Murayama, M; Park, JM; Akagi, T; Tatebayashi, Y; Miyasaka, T; Kimura, T; Hashikawa, T; Nakano, Y; Kudo, T; Takeda, M; Takashima, A. Formation of tau inclusions in knock-in mice with familial Alzheimer disease (FAD) mutation of presenilin 1 (PS1). J Biol Chem. 2006; 281: 5037-5041). Despite these advances, it is clear that definitive proof of "amyloid cascade to tau hypothesis" validity can only come from successful trials of Aβ- and/or tau/ tangle- therapies in humans.

microgliosis, astrocytosis, neuronal dysfunction and loss, and the synaptic alterations that result in dementia. In fact, support of this hypothesis has come from strong genetic, histopathological, neuropsycological and biochemical evidences. For instance, Clark and collaborators identified in 1995 the largest familial AD (FAD, OMIM entry #104311) kindred so far reported as a result of a missense mutation in the PS-1 gene (g.50024A>C; in chromosome 14, c.839gAa>gCa). Since then, this mutation has been known as PS-1 E280A, i.e., a glutamic acid-to-alanine substitution at codon 280 in PS, or the "paisa mutation" (as reference to the genetic isolate paisa community living in Antioquia region, Colombia, Arcos-Burgos and Muenke, 2002). Several works have enlightened our understanding of this devastating disorder. Shortly after the discovery of the PS-1 E280A mutation, Lemere and co-workers (1996) showed that this mutation produces increased $A\beta_{[1-42]}$ deposition and severe cerebellar pathology[3]. Moreover, Gomez-Isla et al., (1999) established that some PS-1 mutations (M139V, I143F, G209V, R269H) including E280A were associated with faster rates of NFT formation and accelerated neuronal loss in the majority of the patients who harboured them when compared with sporadic Alzheimer's disease patients and that dramatic quantitative differences in clinical and neuropathological features can exist even among family members with the identical PS mutation, in agreement with the description of the clinical features of early-onset Alzheimer's disease reported previously by Lopera and collaborators (1997). Further neuropsychological studies have supported the hypothesis that memory complaints represent the earliest symptom of familial Alzheimer's disease. In addition to the memory difficulties, other minor cognitive impairments were also found, particularly, mild anomie, concentration difficulties and defects in the understanding of complex verbal material (Ardila et al., 2000), and longitudinal studies carried out to investigate the cognitive decline in early onset of FAD have shown these patients to display a decline in verbal memory, language, constructional and abstraction tests according to the Mini-Mental State Exam scores. Patients with sporadic AD (SAD) demonstrated a similar pattern of cognitive decline, but the decline was faster in FAD than in SAD participants (Roselli et al., 2000). One possible explanation to all these cognitive results might come from the fact that PS-1 E280A mutation decreases cerebral perfusion in the posterior parietal and superior frontal cortex in comparison with the normal control subjects. Indeed, regional cerebral perfusion abnormalities based on SPECT are detectable before development of the clinical symptoms of AD in carriers of the PS-1 mutation (Johnson et al., 2001). Alternatively, environmental and genetic factors might also be responsible for variation in the phenotypic expression of the E280A PS-1 mutation. In fact, Pastor and collaborators (2003) have demonstrated, for the first time, that genetic, i.e. APOE 4 allele carriers were more likely to develop AD at an earlier age than subjects without the 4 allele, and environmental factors, i.e. subjects with low education were more likely to develop AD later than those with higher education, influence the age of onset in FAD mutation. Taken together these data suggest that PS-1 mutation affect not only the neurobiology of APP processing, i.e. by accelerating $A\beta_{[1-42]}$ production, but also impairing more subtle brain function such as cognition many years or decades before clinical symptoms are detected. Despite these

[3] Kosik K presents a dramatic personal account of the events associated with the first Familial Alzheimer's diseased "paisa" brain for pathological studies. The Fortune Teller. The Sciences July/ August 1999; 13-17.

advances, how $A\beta_{[1-42]}$ accumulation, once it occurs, initiated and propagates the complex molecular and cellular changes of FAD brain is largely unknown.

Parkinson's disease[4] (OMIM entry #168600) is typified clinically by motor symptoms including bradykinesia, resting tremor, rigidity and gait posture abnormalities followed by postural instability and less frequent non-motor complication such as dementia, depression and autonomic dysfunction (Perkin, 1998). Pathologically, the PD brain is characterised by degeneration of dopaminergic neurons in the substantia nigra pars compacta as well as dopaminergic ventral tegmental area and noradrenergic neurons of the locus coeruleus, the presence of Lewy´s intraneuronal inclusion bodies (Forno, 1996). The cause of all cases of PD remains unknown. However, in the mid-1990s this situation changed with the identification of a mutation in the α-synuclein gene associated with autosomal dominant PD in Italian kindred (Polymeropoulos et al., 1997). In 1998, the first gene that causes autosomal recessive juvenile Parkinsonism (AR-JP) was reported and named *parkin* by Kitada and colleagues. The gene contains 12 exons spanning over 1,5 Mb and encodes a protein of 465 amino acids with a molecular mass of approximately 52.000 M (Kitada et al., 1998). Subsequent studies have shown that *parkin* is a RING-finger-containing protein identified as an E3 protein-ubiquitin ligase (Shimura et al., 2000), which is an integral component of the cytoplasmic ubiquitin/ proteosomal degradation pathway (Pickart, 2001). The reaction promoted by E3 ligases is the addition of a lysine-linked chain four or more ubiquitin molecules to the target protein, which is recognised by the subunits in the proteosome. Since parkin is localized to the microsomal fractions as well as to the cytosol-and Golgi fraction, and endoplasmic protein degradation appears conceivable. Thus, mutation of the *parkin* gene could result in accumulation of misfolded proteins (Tanaka et al., 2001; Imai and Takahashi, 2004). It is, therefore, hypothesized that mutations in *parkin gene*, which result in loss of function, or when the ubiquitin system is inactivated unable to remove enough mutated or misfolded proteins, nigral neurodegeneration might be caused.

Currently, AR-JP (OMIM entry #600116) is consider a distinct genetic entity characterised by early age at onset (<age 45), dystonia with parkinsonism and improvement of symptoms after sleep, slow disease progression, associated signs such as hyperflexia, dysautonomia, peripheral neuropathy; and good response to low doses of L-DOPA (Zhang et al., 2001). One pathological discrepancy between AR-JP and PD is the absence of Lewy bodies[5]. This suggests Lewy bodies formation is not required to evoke nigral cell death.

---

[4] A historical overview of Parkinson's disease is summarised by Goezt, C; Chumura, TA; Lanska, DJ. The history of Parkinson's disease: Part 2 of the MDS-sponsored history of Movement Disorders exhibit, Barcelona, June 2000. Mov Dis 2001; 16:156-161.

[5] It is worth mentioning that either compound heterozygous parkin mutations (e.g. exon 7 R275W substitution + exon 3 deletion, Farrer, M; Chan, P; Chen, R; Tan, L, Lincoln, S; Hernandez, D; Forno, L; Gwinn-Hardy, K; Petrucelli, L; Hussey, J; Singleton, A; Tanner, C; Hardy, J; Langston, JW. Lewy bodies and Parkinsonism in families with parkin mutations. Ann Neurol. 2001; 50: 293-300; delExon 7 + del1072T, Pramstaller, PP; Schlossmacher, MG; Jacques, TS; Scaravilli, F; Eskelson, C; Pepivani, I; Hedrich, K; Adel, S; Gonzales-McNeal, M; Hilker, R; Kramer, PL; Klein, C. Lewy body Parkinson's disease in a large pedigree with 77 parkin mutation carriers. Ann Neurol 2005; 58: 411-422) or homozygous parkin mutations (e. g. exon 3 deletion, Sasaki, S; Shirata, A; Yamane, K; Iwata, M. Parkin-positive autosomal recessive juvenile parkinsonism with alpha-synuclein-positive inclusions. Neurology 2004; 24: 678-682) have been shown neuropathologically to present Lewy body pathology. Therefore, how exactly recessive and/or compound heterozygous mutations in parkin gene lead to Lewy body pathology is at present a mystery.

Additionally, iron deposits are found in PD (Dexter et al., 1989; Sofic et al., 1991; Riederer et al., 1992; Griffiths et al., 1999) as well as in AR-JP (Takanashi et al., 2001). Why dopaminergic neurons in the substantia nigra are particularly vulnerable to the loss of parkin function and iron deposition is yet unknown.

To date, the most common known form of hereditary Parkinsonism, i.e. AR-JP, diagnosed in Antioquia, Colombia is due to the *parkin* C212Y mutation. Pineda-Trujillo et al. identified this parkin mutation (2001) in a genetic isolate community from two paisa family groups (PJF-1, PJF-3). Lastly, three additional genetic defects underlying PD have been identified involving Ubiquitin C-terminal hydrolase-L1 (UCH-L1), DJ-1 and PTEN-induced putative kinase-1 (PINK-1) (for a review, see Cookson, 2005). These mutations have not yet been found in this population. It is noteworthy to mention, however, that several people with a family history of FAD and ARJ-PD are at present living in proximal geographical areas of Antioquia region. Therefore, there is high probability to encounter patients with varying degrees of phenotypic overlap of both disorders in the near future. Unfortunately, this circumstance could aggravate the current socio-economic and health situation of this highly vulnerable paisa community.

Given that there is not yet a definitive cure for AD and PD, it is imperative to answer two basic questions. Firstly, what is (are) the molecular mechanism(s) responsible for the selective neural damage and death that lead to brain dysfunction in both neurological disorders?. Secondly, what are the potential therapeutic approaches that could be implemented to retard, prevent or interrupt the pathobiological impact of FAD and ARJ-PD?. In this chapter, based on our own and others research work, we have attempted to explain those uncertainties by using lymphocytes as a cell model. But, why lymphocytes?

Lymphocytes Resemble Neuronal Cells: A Refreshing Look

Some researches consider lymphocytes as exclusive cells that accomplish only immunological functions neglecting the fact that lymphocytes are a unique group of cell pertaining to the neuroimmunological communication axis, influencing one another's actions (Serafeim and Gordon, 2001). Indeed, the brain and the immune system are involved in functionally relevant cross–talk, whose main function is to maintain homeostasis. Therefore, to play such a role, lymphocytes are equipped with several biochemical systems that display comparable pathways to neural cells. This unusual characteristic makes lymphocytes an excellent in vitro model (defined as a surrogate for human being, or a human biologic system, that can be used to understand normal and abnormal function from gene to phenotype and to provide a basis for preventive or therapeutic intervention in human diseases) to study biochemical and cytopathological mechanisms in AD and PD. As noted by Kriesberg "…this similarity in function is critical because a model is at best a copy, an imitation or representation in some fashion of what we are examining...". Furthermore, "...the reason for [using] in vitro [models] is not just an ethical or humane based concern; often these non whole animal [models] have their own distinct advantages such as the ability to reduce variability, allow for greater control of environment, provide for large numbers of testing strata with less expense, require less chemical materials and thus also produce less toxic waste, and can be quicker and cheaper than in vivo methods. It can also be simpler to isolate the variables with in vitro methods. Yet, if we are to address the challenge of extrapolating to a complex system, the model needs to present enough complexity to be useful. There needs to be some sort of balance between the ability to isolate the phenomenon, to single out variables (the reductionism aspect) and the fidelity to the complexity of the target system..." Eventually, lymphocytes comply with functional, practical and complexity conditions to be used as a neuronal cell model.

In the context of the specific topic of this chapter, lymphocytes appear to be a particularly fascinating cell model for at least five reasons. First, lymphocytes express six homologous neurochemical systems such as (i) catecholaminergic system including tyrosine hydroxylase, monoamine oxidase (Marino et al., 1999 and references within), dopamine transporter (Amenta et al., 2001); dopamine D2-, D3-, D4-, D5-like receptors (Ricci and Amenta, 1994; Ricci et al., 1995, 1997; Amenta et al., 1999; McKenna et al., 2002); (ii) serotonergic system comprising serotonin transporter (SERT, Faraj et al., 1991), serotonin receptors (Stefulj et al, 2000) and tryptophan hydroxylase (Carrillo-Vico et al., 2004); (iii) Cholinergic system encompassing acetylcholine (Ach), muscarinic and nicotinic Ach receptors (mAChRs and nAChRs, respectively), choline acetyltransferase (ChAT), high affinity choline transporter and acetylcholinesterase (for a review, see Kawashima and Fujii, 2004); (iv) glutamatergic system containing ionotrophic glutamate receptors (Lombardi et al., 2001, 2004) and group I metabotropic glutamate receptors (Miglio et al., 2005); (v) Adrenergic system consisting of beta-2 adrenergic receptors (Sanders, 1998); and finally, (vi) gabaergic systems involving γ-aminobutiric acid (GABA) receptors (Tillakaratne et al., 1995). Accordingly, Barbanti et al., (1999, 2000) have shown increased expression of dopamine D1-like and D2-like receptors in PD patients on lymphocytes when compared to either neurologic or healthy control subjects, but reduced density of dopamine D2-like receptors on peripheral blood lymphocytes in AD. In addition, Caronti et al., (1999) have shown that dopamine content and tyrosine hydroxylase immunoreactivity are reduced in peripheral blood lymphocytes in the early stages of PD. Thus, lymphocytes represent a biological tool with which to investigate changes in the neurotransmitter-receptor system in neurodegenerative disorders.

Second, they express similar molecular death machinery leading to typical morphologic and biochemical features of cell death –apoptosis[6]- as previously defined by Kerr and co-

[6] The use of programmed cell death (PCD) and apoptosis as synonymous terms by some researchers has brought considerable confusion in the field up to the point of apoptosis being considered a created concept, not a real entity (Sloviter, RS. Apoptosis: a guide for the perplexed. Trends Pharmacol Sci 2002; 23:19-24). "What exactly is apoptosis and PCD?". Several answers to this question have been proposed over the years. As recently noted by P. Nicotera "...apoptosis is a morphological phenotype of cell death, whereas PCD refers specifically to a cell autonomous genetic developmental death programme..." [*Moreover,*] "...this distinction is now redundant...cells doomed to die inevitably will have to process their intracellular organelles, the nucleus and cytoskeleton and will engage whichever execution system to that end. In development, the programme is genetically regulated and encoded. In adult cells the programme refers more to a defined series of biological processes that dismantle dying cells. In that sense, all forms of cell death is indeed programmed....We should perhaps consider that cells can engage multiple death routines, some being more efficient than others depending on the cell type. In fact, the execution of cell death is different depending on the differentiation programme, and while the death machinery exists in virtually every cell, it might be used differently..." (Melino, G; Knight, RA; Nicotera, P. How many ways to die?. How many different models of cell death?. Cell Death Differ 2005; 12: 1457-1462). Accordingly, cells may be eliminated by a number of different execution routines or mechanisms

workers (1972, 1995) and several others (Kroemer et al., 1997; Strasser et al., 2000; Reed, 2000; Zimmermann et al., 2001; Adams, 2003; Thorburn, 2004; Okada and Mak, 2004; Lorenzo and Susin, 2004; Zamzami and Kroemer, 2005; Kroemer and Martin, 2005; Bras et al., 2005), respectively. Apoptosis is originally a morphological phenomenon characterised by chromatin condensation and nuclear fragmentation, plasma membrane blebbing, cell shrinkage, and preservation of organelles such as mitochondria. These characteristics can be recognised in lymphocytes under the fluorescent microscopy (see Fig 1B and 1C) or electron microscopy (Sakahira et al., 1999; Marini et al., 2001). Noticeably, what causes these morphological changes that we recognize as apoptosis occurs through multiple independent pathways that are initiated either from triggering events within the cell ("the intrinsic pathway" which involves the release of mitochondrial proteins such as cytochrome C, second mitochondrial-derived activator of caspase/direct IAP-associated binding protein with low PI (Smac/DIABLO), apoptosis inducing factor (AIF), and Endonuclease G (Endo G)) or from outside the cell ("the extrinsic pathway" which involves Fas/FasL pathway, caspase-8 activation, bid degradation and releasing cytochrome C) converging on a common machinery of cell dismantling by a family of executer cysteine proteases known as Caspases, particularly caspase-3, that cleavage proteins at aspartate residues (Earnshaw et al., 1999; Zhivotovsky, 2003). In fact, caspase-3 degrades the inhibitor of caspase-activated DNase (ICAD/ DNA fragmentation Factor-45, DFF-45) protein releasing the caspase-activated DNase (CAD/ DFF-40) that result in DNA degradation ("DNA ladder pattern") from mouse T-cell lymphoma (Enari et al., 1998; Sakahira et al. 1998), Jurkat T cells (Liu et al, 1997) and HeLa cells (Halenbeck et al., 1998) under pro-apoptotic treatments. It is worth to mention that almost 18-years passed before an explanation could be drawn for one of the earliest well-recognized biochemical characteristics of apoptosis i.e. "DNA ladder pattern", from the time when Wyllie reported glucocorticoid-induced thymocytes apoptosis associated with endonuclease activation (Wyllie,1980). Unquestionably, both morphological and biochemical (and mechanistically) data (Chang et al., 2002; Polster and Fiskum, 2004; Dlamini et al., 2004) have helped considerably to enlighten, yet unsettled, the mode of neural cell death in AD (Roth, 2001; LeBlanc, 2005; Takuma et al., 2005), PD (Jenner and Olanow, 1998;

including autopaghy (Kroemer, G; Jaatela, M. Lysosomes and autophagy in cell death control. Nat Rev Cancer 2005; 5: 886-897), mitotic catastrophe (Castedo, M; Perfettini, JL; Roumier, T; Andreau, K; Medema, R; Kroemer, G. Cell death by mitotic catastrophe: a molecular definition. Oncogene 2004; 23: 2825-2837), paraptosis (Sperandio, S; deBelle, I; Bredesen, DE. An alternative, nonapoptotic from of programmed cell death. Proc Natl Acad Sci 2000; 97: 14376-14381), cornification (Candi, E; Schmidt, R; Melino, G. The cornified envelope: a model of cell death in the skin. Nat Rev Mol Cell Biol 2005; 6: 328-340), chondroptosis (Roach, HI; Aigner, T; Kouri, JB. Chondroptosis: a variant of apoptotic cell death in chondrocytes? Apoptosis 2004; 9: 265-277), autoschizis (Jamison, JM; Gilloteaux, J; Taper, HS; Calderon, PB; Summers, JL. Autoschizis: a novel cell death. Biochem Pharmacol 2002; 63: 1773-1783), apoptosis-like PCD, necrosis-like PCD, accidental necrosis/cell lysis (Leist and Jaattela, 2001), among others (Melino et al., 2005). Obviously, these types of cell death differ not only morphologically, but biochemically as well. Therefore, apoptosis might be considered as one of many ways a cell is committed to die. Furthermore, according to the Nomenclature Committee on Cell Death (NCCD) equating PCD= apoptosis= caspase-dependent cell death or accidental cell death= necrosis= caspase-independent cell death should be avoided and be replaced by a precise description of the parameters that are measured (Kroemer, G; El-Deiry, WS; Golstein, P; Peter, ME; Vaux, D; Vandenabeele, P; Zhivotovsky, B; Blagosklonny, MV; Malorni, W; Knight, RA; Piacentini, M, Nagata, S; Melino, G. Nomenclature Committee on Cell Death. Classification of cell death: recommendations of the Nomenclature Committee on Cell Death. Cell Death Differ 2005; 12 Suppl 2: 1463-1467).

Przedborki, 2005; Moore et al., 2005), and other neurodegenerative disorders (Leist and Jaattela, 2001; Graeber and Moran, 2002).

Third, the lymphocytes in human blood and neurons are post-mitotic cells, i.e. they become locked in a G_0 phase of the cell cycle. This is a remarkable biological feature to be cautiously considered when interpreting experimental data since evidence has accumulated that a cell division forced on a mature neuron leads to apoptosis rather than division (Herrup et al., 2004), but cell division is induced in lymphocytes. In other words, the use of cell lines instead of primary cultures could be confusing and /or misleading. For instance, NF-κB is a transcriptional factor (composed of a p50/p65 heterodimer protein that upon activation binds to specific DNA sequences in target genes, designated as κB-elements) involved in both cell cycle-regulation and cell death processes. In dividing cells, NF-κB transcribes cyclin D1 (Hinz et al., 1999; Guttridge et al., 1999), which in association with cyclin-dependent kinases, CDK4 and CDK6, promotes G1/S phase transition through CDK-dependent phosphorilation of retinoblastoma protein (pRb), thereby releasing the transcription factor E2F, required from the activation of S phase-specific genes. Indeed, constitutive activation of NF-κB is involved in the activation of primary adult T-cell leukemia cells (Arima et al., 1999). On the other hand, the regulatory roles of NF-κB on apoptosis suggest that NF-κB is acting on the upstream pathways of apoptosis, either negatively or positively (Shishodia and Aggarwal, 2004). However, it is possible that regulation of apoptotic molecules such as as p53 on NF-κB by cross-competition for a limiting pool of the co-activator p300/CREB-binding protein (CBP) for maximal activity (Wadgaonkar et al., 1999; Webster and Perkins, 1999; Culmsee et al., 2003), and/or stimulation of NF-κB through induction of its target gene, $p21^{Waf}$, which inhibits cyclin E/Cdk2 activity and blocks its ability to compete with NF-κB for co-factors, such as p300 and CBP (Perkins et al., 1997). Appreciably, in non-dividing cells, these confounding matters connected with the role of NF-κB in apoptosis and cell-cycle control might not be an important issue given that NF-κB function can eventually be studied independently from the cell cycle function. Thus, G_0 represents not simply the absence of signals for mitosis but an active repression of the genes needed for mitosis. Of interest, Bullido et al. (1996) have found that APP was weakly expressed on the cell surface of resting human lymphocytes and monocytes, but it could be induced to the surface of those cells upon stimulation such as mitogenic lectins, calcium ionophores, phosphatase inhibitors, and anti mu-chain or anti-CD3 antibodies in B and T cells, respectively.

Fourth, lymphocytes express APP751 and APP770 mRNA isoforms (Ebstein et al., 1996), presenilin- 1 & 2 proteins (Mirinics et al., 2002; Takahashi et al., 2003), express ubiquitin proteosomal system (UPS), including E3 ligase parkin protein (Sunada et al., 1998), and express various neural-related proteins such as μ-;δ-;κ-opiates receptors (Bidlack, 2000); calcium, sodium and potassium channels (Gomes et al., 2004; Panyi et al., 2004; Diss et al., 2004; Krasznai, 2005). Fifth, although lymphocytes and neurons develop from different embryologic origin (i.e. lymphocytes derive from the mesoderm and neurons derive from the ectoderm), Hao and colleagues (2003) have demonstrated that human haematopoietic stem cells can differentiate into neural stem cells and astrocytes. Furthermore, several reports have shown both in vitro and in vivo that neural cell can be obtained by trans-differentiation of either mesenchymal stem cells (MSCs) from bone marrow (Muñoz-Elias et al., 2004; Deng et al., 2005; Long et al., 2005; Crigler et al., 2006) or from umbilical cord blood cells (UCBCs;

Sanchez-Ramos, 2002; Chen et al., 2005). These amazing observations suggest the possibility that these cells may serve as a source for stem cell-based regenerative medicine to treat neurological disorders.

It is worth to note that more than 200 scientific papers have been published during the past five years (according to a search in www.scopus.com 2001-2006) concerning lymphocytes, AD and PD. Taken together this information shows peripheral blood lymphocytes as valid and important experimental tool for cellular and molecular AD and PD research.

Oxidative Stress in Alzheimer's and Parkinson's Diseases: On the Crossroad

Oxidative stress (OS) refers to a state in which free radicals (defined as any atom or molecule that has one or more unpaired electrons in its outer shell such as anion superoxide radical ($^{\cdot}O_2$), hydroxyl radical ($^{\cdot}OH$) and their products (H_2O_2)) are in excess of antioxidant defence mechanism (e. g. superoxide dismutase (SOD), glutathion peroxidase (GPx), catalase, vitamin C and E). As a result of this imbalance, these free radical and others (e. g. nitric oxide, peroxynitrite) are capable of reacting with lipids, proteins, nucleic acids, and other molecules and altering their structure and function. Accordingly, OS can lead to alterations in cells with an accumulation of oxidized products such as aldehydes and isoprostanes from lipid peroxidation, protein carbonyls from protein oxidation, and base adducts from DNA oxidation. Because the human brain is a high oxygen consumer organ, it is reasonable to assume that, under pathological conditions, it might be a target of permanent OS attack.

Over the last decade, oxidative stress has been proposed to play a critical role in the pathogenesis of AD (Markesbery, 1997) and PD (Fanh and Cohen, 1992; Jenner and Olanow, 1996), respectively. In fact, several markers of OS have been identified in post-mortem brain tissues including increased levels of DNA and RNA oxidation (e.g. 8-hydroxyl-2-deoxyguanosine (8OHdG) and 8-hydroxyl deoxyguanosine (8OHG); protein carbonyl levels, glycation and glycoxidation, lipid peroxidation, and high iron concentration (for a review, see Butterfield et al., 2001; Kaur and Andersen, 2004; Maynard et al., 2005). Moreover, given that $A\beta_{[1-42]}$, iron and DA generate reactive oxygen species (ROS), they have been implicated in the OS observed in AD (for a review, see Lahiri and Greig, 2004) and PD (Ben-Shachar et al., 2004; Asanuma et al., 2004; Jackson and Smeyre, 2005)[7]. Not surprisingly, lymphocytes have been used to test for oxidative stress in both disorders. For example, Migliore et al., (2005) have demonstrated a high level of primary DNA damage in leukocytes of AD and also of mild cognitive impairment (MCI is defined as a clinical condition between normal aging and AD, characterized by a memory deficit without loss of general cognitive and functional abilities) patients compared to control individuals. Based on these results, they have proposed OS as an earlier event in the pathogenesis of AD. In a previous report, the same research

[7] For an alternative view of the DA oxidative stress hypothesis in Parkinson's disease, see Ahlskog, JE. Challenging conventional wisdom: the etiologic role of dopamine oxidative stress in Parkinson's disease. Mov Disord 2005; 3: 271-282.

group has demonstrated an increase in the incidence of spontaneous micronuclei; single strand breaks; and oxidized purine bases in PD patients without treatment. These results clearly showed a primary DNA damage and oxidative DNA damage demonstrable in lymphocytes (Migliore et al., 2002). Taken together these data suggest that analysis of DNA in lymphocytes might be used as an early marker of the OS status (i.e. DNA oxidation) in AD and PD patients. Despite these unquestionable evidences, there are still major unresolved issues in the understanding of the molecular and cellular biology of AD and PD. For instance, a complete picture of the precise molecular cascade leading to cell death in a single cellular model in those disorders is still lacking. Therefore, we have been interested in investigating the oxidative stress phenomenon and apoptosis signalling in lymphocytes induced by DA, Fe^{2+}, Aβ.

Lymphocytes as Cellular Model to Study Oxidative Stress and Apoptosis in AD and PD

Previous work has shown that dopamine and Fe^{2+}, (the two major players in PD) induced-apoptosis in PC-12 cell line by an oxidative mechanism wherein Fe^{2+} increases the oxidation of monoamines such as serotonin (5-HT), DA and related toxins into oxidation products (i.e. O-quinones) that can undergo covalent binding to free sulfhydryl groups of proteins to form protein-conjugates (e.g. DA-O-quinone-conjugates), and high molecular protein aggregates (Velez-Pardo et al., 1997). These results supported the notion that iron/DA represent an in vitro model to understanding the mechanism by which dopaminergic neurons are eroded in PD (Velez-Pardo et al., 1998a and references within; Asanuma et al, 2004).
Based on these findings and the growing understanding of the importance of OS as an early event in the pathophysiology of AD and PD, we were initially interested to investigate the effect of DA and Fe^{2+} in peripheral blood lymphocytes (PBL), which resembles neurons (see justification above). We found that (1mM) DA/ (50μM) Fe^{2+} induces apoptosis (according to nuclei morphological changes evidenced by acridine orange/ ethidium bromide technique, Leite et al., 1999, Figure 1) in PBL via (H_2O_2)-mediated oxidative mechanism requiring RNA and de novo protein synthesis (Jimenez Del Rio and Velez-Pardo, 2001). These data further supported and validated our previous observations that DA/ Fe^{2+}/ H_2O_2 could be a mediator of OS in both neural and non-neural cells. These findings prompted us to investigate the mechanism of cell death induced by the neurotoxins 6-hydroxydopamine (6-ODHA, a dopaminergic related toxin, a model toxin for Parkinson (Bove et al., 2005)), 5,6-dihydroxytryptamine and 5,7-dihydroxytryptamine (5,6-& 5,7-DHT, serotonergic related toxins) and whether these toxins shared a common oxidative stress mechanism through activation of NF-κB, p53, c-Jun transcription factors which are involved in cell death process, and caspase-3, an executer protease. Effectively, we demonstrated that (50, 150, 250μM) monoamine related neurotoxins-induced apoptosis by a common oxidative mechanism involving the oxidation of toxins into quinones and production of H_2O_2, concurrent activation of NF-κB, p53, c-Jun transcription factors (Figure 2) and caspase-3 activation. These data clearly suggested H_2O_2 as an important molecule triggering a cascade of molecular signalling events leading to cell death (Figure 3; Jimenez Del Rio and Velez-Pardo, 2002).

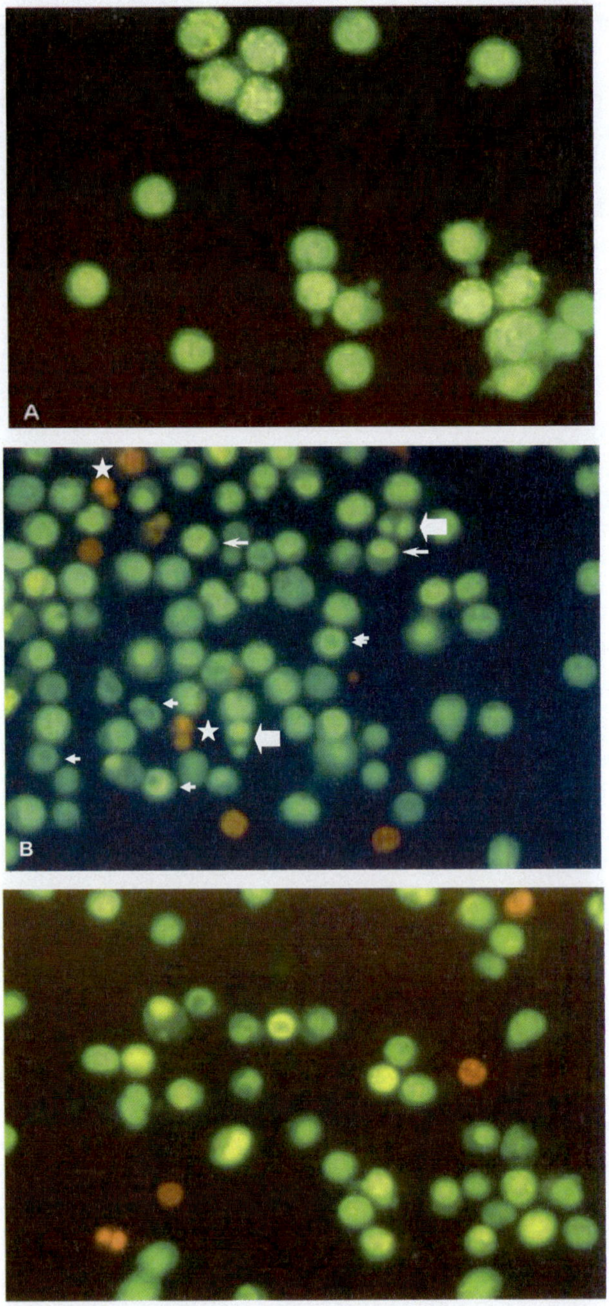

Figure 1. 6-OHDA-induced apoptosis in peripheral blood lymphocytes (PBL). PBL was incubated in absence (A), or in presence of 150μM 6-OHDA (B) for 24 h. The evaluation of apoptosis was performed according to Leite et al., 1999. Cells were stained with 1μl acridine orange / ethidium bromide (AO/EB 100 μg/ml) after incubation time. Photomicrograph (A) illustrates intact nuclei with bright green homogenous chromatin, (B) illustrates typical nuclear apoptotic morphology such as highly condensed chromatin (*arrows*), nuclear fragmentation (ring-like, horseshoe, half-moon forms, *arrowheads*), and apoptotic bodies (*double arrows*). Pictures represent one out of three independent observations. Similar morphological results were observed when PBL are incubated with 10μM $A\beta_{[25-35]}$. Magnifications: A, B X400. (Jimenez Del Rio and Velez-Pardo, 2002; Velez-Pardo et al., 2002). Reproduced with permission from Elsevier.

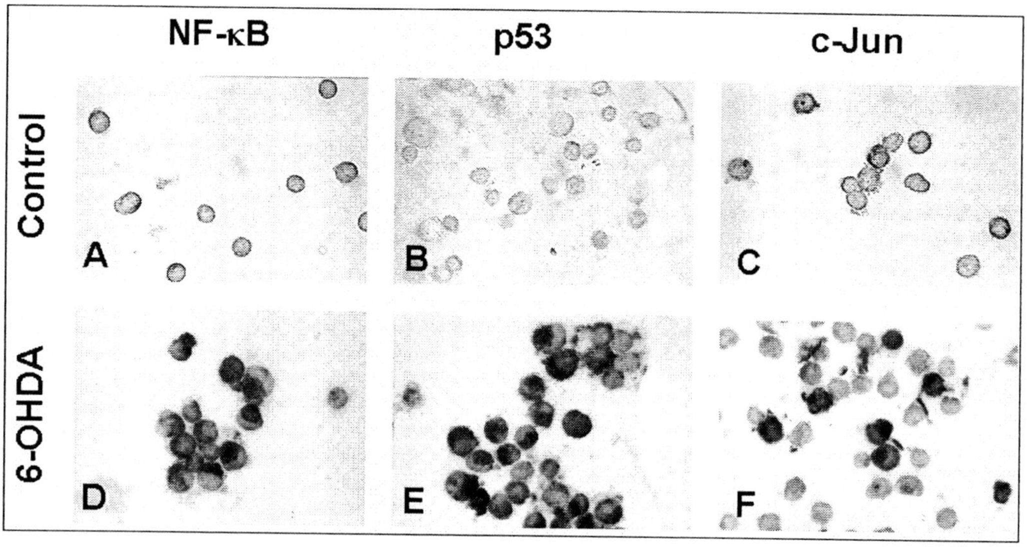

Figure 2. Dopaminergic and serotonergic-related toxins induced activation of NF-kappa B, p53 and c-Jun transcription factors. PBL cells were left untreated (A-C) or exposed to 6-OHDA (150 μM) alone (D-F) for 24 hr. After this time of incubation, cells were stained with anti-NF-kappa B-p65 (A, D), anti-p53 (B, E) and anti-c-Jun (C,F) antibodies according to Santa Cruz Biotechnology protocol. Notice that NF-kappa B, p53 and c-Jun positive-nuclei (dark brown colour) reflect their nuclear translocation and appear to correlate with the apoptotic nuclear morphology i.e. condensed/fragmented nuclei when compared with untreated cells (A-C). Similar observations were obtained with PBL cells treated with 250μM 5,6-DHT and 5,7-DHT toxins (photomicrographs not shown). Magnification of the photomicrographs 400x (A-F). (Jimenez Del Rio and Velez-Pardo, 2002). Reproduced with permission from Elsevier.

Given that Aβ neurotoxicity effect has previously been proposed to be partially mediate by an increasing production of ROS and H_2O_2 (Behl et al., 1994; Hensley et al., 1994), we wonder whether Aβ lethality could be explained by a similar oxidative mechanism triggered by the monoamine related neurotoxins. We demonstrated that 10μM $A\beta_{[25-35]}$ (i.e. the cytotoxic functional sequence of the amyloid peptide) induce apoptosis either alone or in presence of (25μM) iron in PBL cells in a concentration-dependent fashion by an OS mechanism resembling 6-OHDA, 5,6-& 5,7-DHT cytotoxic effect, i.e. $A\beta_{[25-35]}$ was able to generate H_2O_2, which in turn activates NF-κB (Figure 4), p53 and c-Jun transcription factors and caspase-3. Additionally, it was also demonstrated that the cytotoxic effect was dependent on de novo mRNA synthesis (Velez-Pardo et al., 2002). Taken together these results are consistent with the view that H2O2 generation precede the apoptotic process and that once H2O2 is generated by either monoamine related toxins or $A\beta_{[25-35]}$, it is able to trigger a specific cell death signalization (Figure 3). Indeed, several lines of evidence have supported our observations. McLlelan and co-workers (2003) have demonstrated a direct association between amyloid-β deposits and free radical production (e.g. O_2, H_2O_2) in vivo in live, transgenic mouse models of Alzheimer's disease and in analogous ex vivo experiments in human Alzheimer tissue. Recently, Tabner and colleagues (2005a) have shown H_2O_2 formation during the early stages of protein aggregation of the amyloid-β involved in Familial British dementia (FBD). Furthermore, Garcia-Ospina et al. (2003) have found a

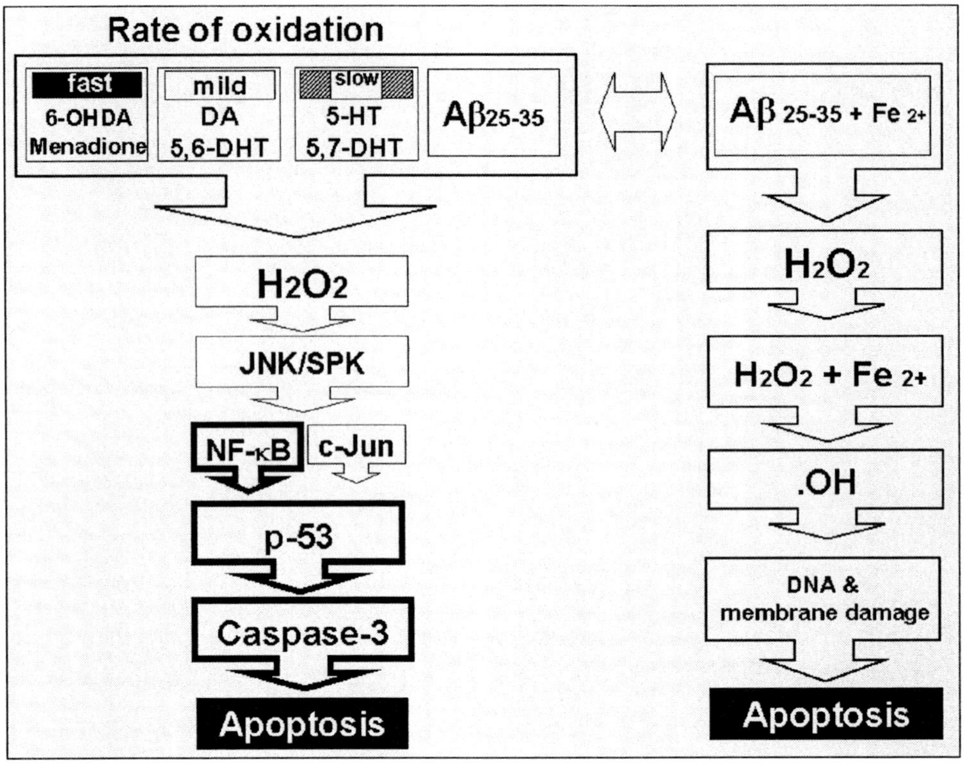

Figure 3. Schematic model of dopaminergic and serotonergic related toxins; $A\beta_{[25\text{-}35]}$ and Fe^{2+}-induced apoptosis by an oxidative stress mechanism in PBL. 6-OHDA; 5,6-& 5,7-DHT or both $A\beta_{[25\text{-}35]}$ alone or in combination with Fe^{2+} generate H_2O_2. This last compound might activates JNK/SAPK kinases pathway (Lander et al., 1995), which in turn activate in parallel both NF-κB and c-Jun transcription factors. NF-κB is able to activate the transcriptional factor p53 and subsequently it may activate the pro-apoptotic Bax protein (Miyashita and Reed, 1995), which induces cytochrome c release from mitochondria (Jurgensmeier et al., 1998) to activate the apoptosome complex (Li et al., 1997). This complex activates caspase-3 leading PBL cells to cell death by apoptosis. Alternatively, H_2O_2 reacts with Fe^{2+} to produce highly reactive oxygen species (probably hydroxyl radicals by Fenton reaction), which directly damage nuclear DNA and provoke plasma membrane rupture leading to apoptosis. (Jimenez Del Rio and Velez-Pardo, 2002; Velez-Pardo et al., 2002). Reproduced with permission from Elsevier.

positive correlation between severity of DNA damage (evaluated according to terminal transferase-mediated dUTP-biotin nick labelling (TUNEL)-technique) and nuclear activation of the c-Jun, Par-4, p53 and NF-κB transcription factors from sporadic AD brains. Additionally, Aleyasin et al., (2004) have uniquely defined a signalling paradigm by which NF-κB modulates the p53-dependent pro-apoptotic response in neurons exposed to DNA damage. These findings support the notion that beta-plaques and/or Aβ oligomers may induce cell death by a molecular signalisation process in AD brain similar to non-neuronal model signalisation initiated by $A\beta_{[25\text{-}35]}$-exposure. Taken together these observations suggest that H_2O_2 might play a pivotal role in PD and AD, as proposed by our research group (Jimenez Del Rio and Velez-Pardo, 2004b) and others (Milton, 2004; Tabner et al, 2005a and references within). Of notice, although the initiator cytotoxic molecules are clearly different in nature, they evidently converged on a common activator compound, H_2O_2. It is, therefore, predictable that any chemical or biochemical compound capable to generate H_2O_2 could trigger an analogous molecular cascade observed with the monoamine related neurotoxins and Aβ. However, a more detailed study of DA and metals cytotoxic effect on PBL brought complementary and exciting data.

In a previous work reported by Pineda-Trujillo et al. (2001), a recessive Cys212Tyr (C212Y) mutation in parkin (exon 6) has been found in 2 juvenile Parkinsonism families (PFJ-1, PFJ3) with typical AR-JP features from an isolated population in North West Colombia. Because parkin deals with unfolded protein stress, and apoptosis is relevant to PD, it is reasonable to assume that if the E3 ubiquitin-mediated protein degradation is impaired in AR-JP patients, then exposure of cells to stressful conditions could shed new light on the molecular association between genetic factors, oxidative stress, and cell demise in PD. To address this, we exposed PBL cells, which express parkin isoform 3 (Sunada et al., 1998), from 3 homozygote patients bearing the C212Y mutation in parkin, 1 heterozygote (i.e., C212Y/C), and 4 control subjects, to (25, 50μM) H_2O_2/ (1mM) DA/ (50μM) Fe^{2+} stressors. It has been shown that homozygote Cys212Tyr parkin mutation in AR-JP patients renders lymphocytes sensitive to dopamine, iron and hydrogen peroxide stimuli. Furthermore, dopamine-caused apoptosis by four alternative mechanisms converging on caspase-3 activation and apoptotic morphology: (1) NF-κB-dependent pathway activated by dopamine oxidation by-product H_2O_2; mitochondrial dysfunction either by (2) direct H_2O_2 or (3) hydroxyl radicals (if Fe^{2+} is present) exposure and (4) increase of unfolded–protein stress (Jimenez Del Rio et al., 2004). These data together with our early observation that oxidation of dopamine could form protein DA-o-quinone-conjugates (Jimenez del Rio et al., 1993), provided evidence that DA might cause cytotoxicity by complex pathways, and that the homozygotic parkin C212Y mutation from AR-JP increases susceptibility to DA and iron-mediated apoptosis probably due to its failure to dispose of unfolded proteins provoked by OS (Imai et al., 2000). Interestingly, Jiang et al., (2004) have shown that overexpression of parkin protects human dopamine neuroblastoma cell line (SH-SY5Y) against apoptosis induced by DA or 6-OHDA. It has also been shown that PD-linked mutations of parkin significantly abrogated the protective effect of wild-type parkin, as well as its ability to suppress ROS and protein carbonylation. Furthermore, LaVoie and colleagues (2005) have recently reported that DA covalently modifies and functionally inactivates parkin protein in agreement with the previous notion that DA is able to form protein o-quinone conjugates with important target protein such as actin, "serotonin binding proteins" and α-synuclein, thereby modifying their structure and functionality (Velez-Pardo et al., 1998a and references within; Conway et al., 2001). These findings further support the hypothesis that OS could be a general mediator of cell death in PD. Moreover, these non-neural cells might be used to monitor antioxidant drug efficiency, specifically in C212Y parkin mutation.

Redox transition metals (especially Fe^{2+} and Cu^{2+}) have been implicated as crucial players in the pathogenesis of AD (Maynard et al., 2005) and PD (Kaur and Andersen, 2004). However, the intracellular signalling mechanism(s) responsible for OS and death in single-cell model exposed to metals has not yet been fully elucidated. Consequently, we were interested to determine the mechanism by which metals induced apoptosis in PBL. To achieve this goal, PBL cells were exposed to increasing concentration (50, 100, 250, 500, and 1000μM) of Fe^{2+}, Cu^{2+}, Mn^{2+}, and Zn^{2+}-(SO_4). Morphologic analysis showed that 500μM provoked maximal percentage of apoptosis (22–30% according to AO/EB technique) and minimal necrosis (3–7%), whereas low concentrations were innocuous but 1,000 μM induced mainly necrosis (40%). Noticeably, metals generated both H_2O_2 and (·OH) by Fenton reaction and all metals induced mitochondrial depolarisation (17–62% according to $D_iOc_6(3)$ non-fluorescent cells) and activated caspase-3 concomitantly with apoptotic morphology (25–32%). In contrast to DA, 6-OHDA and $A\beta_{[25\text{-}35]}$ observations, neither NF-κB nor p53 transcription factors showed activation. Taken together these results provide evidence that redox-active metals, without pre-existing H_2O_2,- can react with molecular dioxygen (O_2) to generate ($^{\cdot}O_2$), (H_2O_2) and ($^{\cdot}OH$) through Fenton reaction (Velez-Pardo et al., 1995; Qian and Buettner, 1999; Urbanski and Beresewicz, 2000) resulting in mitochondria depolarization, caspase-3 activation, and apoptotic morphology independent of NF-κB and p53 transcription factors activation (Jimenez Del Rio and Velez-Pardo, 2004a).

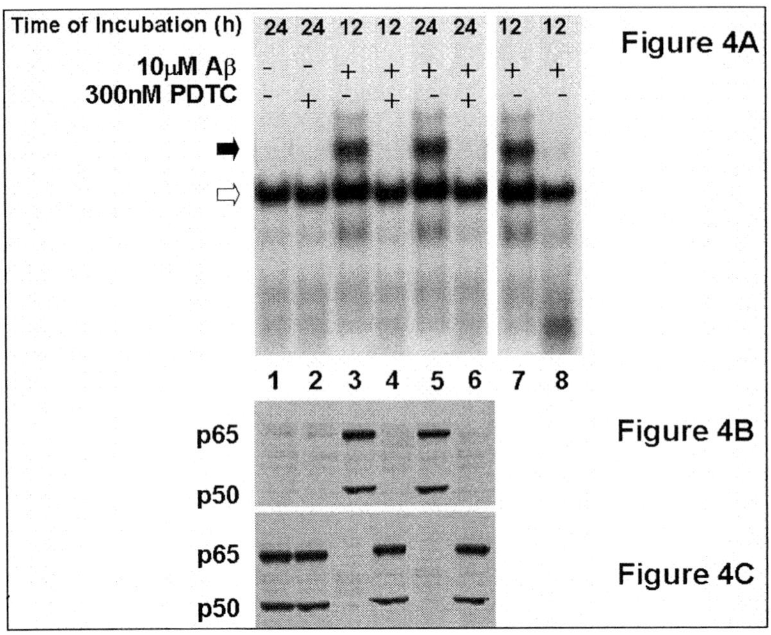

Figure 4. Induction of NF-κB DNA-binding activity by $A\beta_{[25\text{-}35]}$ in PBL cells. (A) Equal amounts of (10 μg) of nuclear extracts from untreated or treated PBL with 10 μM $A\beta_{[25\text{-}35]}$ in the presence or absence of PDTC were subjected to EMSAs with $^{32}\alpha P$-labeled oligonucleotide encompassing the NF-κB site. In lanes 1-2, a nuclear extract of untreated PBL or treated with PDTC was analysed, respectively. In lanes 3-6, extracts of PBL treated with $A\beta_{[25\text{-}35]}$ in the presence or absence of PDTC for 12-24 h were analysed. The NF-κB specific complex (filled arrow) and the free probe (open arrow) are indicated. In lanes 7-8, competition analysis was analysed. Reaction mixtures containing extracts of cells were either untreated (lane 7) or incubated with a 50-fold excess of unlabeled oligonucleotide (lane 8). (B-C) PBL cells were untreated or treated with 10 μM $A\beta_{[25\text{-}35]}$ in the presence or absence of PDTC for 12-24 h. NF-κB complexes in the PBL cells nuclear (B) and cytoplasmic extracts (C) were detected by western blot analysis. In lanes 1-2, nuclear and cytoplasmic extracts of untreated PBL or treated with PDTC was analysed, respectively. In lanes 3-6, nuclear and cytoplasmic extracts of PBL treated with $A\beta_{[25\text{-}35]}$ in the presence or absence of PDTC for 12-24 h were analysed. The p65 and p50 protein components of the NF-κB specific complex are indicated. (Velez-Pardo et al., 2002). Reproduced with permission from Elsevier.

In summary, we have clarified the major signalling events by which DA, monoamine related toxins (6-OHDA; 5,6-DHT; 5,7-DHT), redox metals such as Fe^{2+}, Cu^{2+}, Mn^{2+}, Zn^{2+}; and $A\beta_{[25-35]}$ might induced cell death in normal and/or mutated lymphocytes (e. g. C212Y in parkin) in AD and PD. Our data also highlight the potential use of lymphocytes as a model to screen antioxidant strategies designed to remove $(Fe^{2+})/(^{\cdot}O_2)/(H_2O_2)/(^{\cdot}OH)$ associated with oxidant stimuli and signalling.

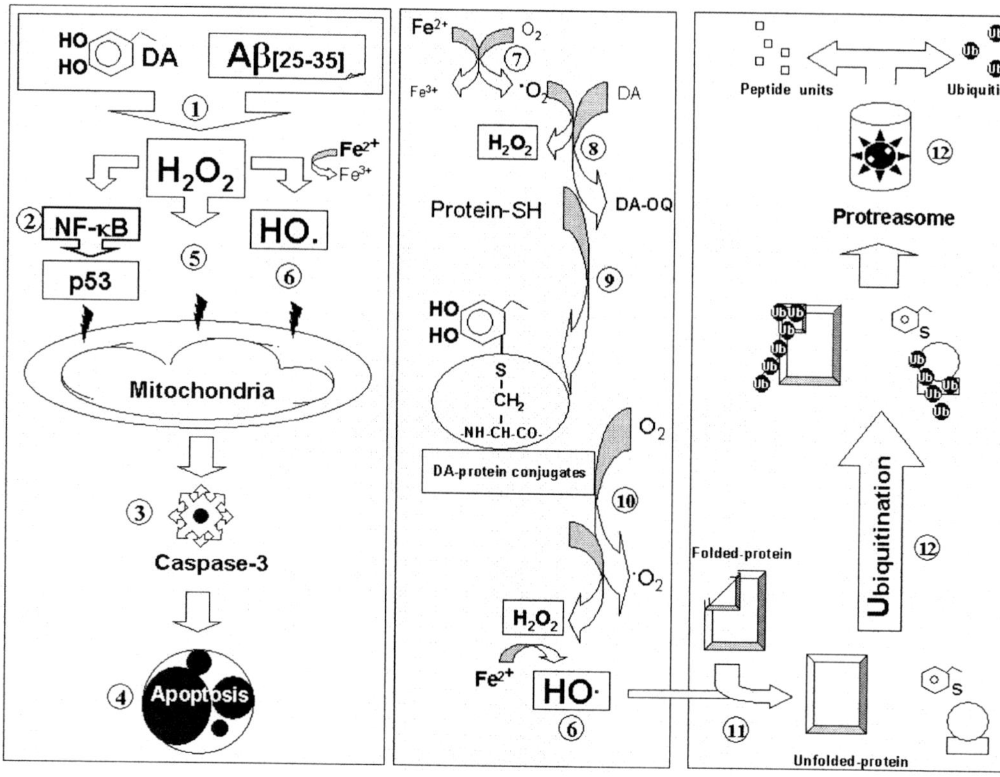

Figure 5. Proposed scheme for DA and $A\beta_{[25-35]}$-induced cytotoxicity. For detailed description, please see text. (Jimenez Del Rio and Velez-Pardo, 2004b) Reproduced with permission From Bentham Science Publishers Ltd.

Unified Molecular Mechanism of Cell Death in Alzheimer and Parkinson

Deciphering the Alzheimer's and Parkinson's disease cascade(s) may offer not only the possibility to scrutinize a basic cellular machinery of response to different deleterious stimuli, but also the possibility to predict novel therapies. One of the greatest advances in Biomedical research is the recognition that (putative) causative molecules of AD and PD such as DA, α-synuclein and $A\beta_{[25-35]}$ are able to generate ROS and H_2O_2 (for a review, see Tabner et al., 2002; Tabner et al., 2005b). Therefore, it is reasonable to think that these neurological entities share a common pathophysiologic process. Accordingly, we have over the years proposed a common molecular oxidative mechanism involved in AD (Velez-Pardo et al., 1998b) and PD (Jimenez Del Rio and Velez-Pardo, 2000). Based on our own and others in vitro, in situ and in vivo work, we have recently postulated an unified molecular cascade model wherein H_2O_2 is definitely a paramount molecule involved in intracellular signalisation that induces neuronal loss in both disorders (Jimenez Del Rio and Velez-Pardo, 2004b).

As shown in Figure 5, a hypothetical apoptotic cell death routine is proposed. Once H_2O_2 compound is produced (step 1) either by auto-oxidation of (1mM) dopamine, (10μM) $A\beta_{[25-35]}$ (or α-synuclein) , it can take part in both chemical reactions and in complex cellular signalling that lead to cell demise. Indeed, accumulated data have demonstrated that H_2O_2 can indirectly activates NF-κB (step 2) through activation of the multisubunit IκB Kinase (IKK) by p21-ras protein (Lander et al., 1995), and/or MEKK1/ MAPKK kinase (Lee et al., 1997), respectively. Once IκB is phosphorylated at Ser^{32}/Ser^{36}, it is degraded by the ubiquitin/proteosomal pathway-$E3RS^{I\kappa B}$ (Yaron et al., 1998). Alternatively, H_2O_2 also can induce dissociation of IκBα from p65 without degrading IκBα via Syk protein-tyrosine kinase by phosphorylation of p65 at Ser^{529} (Takada et al., 2003). Hence, the release of active NF-B dimmer (p50, p65) translocates into the nucleus and transcribes p53 transcription factor (Wu and Lozano, 1994). This last protein can activate, in turn, pro-apoptotic proteins such as PUMA, Noxa, Bim, Bid, Bax (for a review, see Zamzami and Kroemer, 2005; Yu and Zhang, 2005), which are able to permeabilize mitochondria, thus promoting the release of the apoptogenic cytochrome c (Jurgensmeier et al., 1998). As a result, cyt C together with Apaf 1, dATP, and procaspase-9 (i.e. apoptosome complex (Li et al., 1997) elicits caspase-3 protease activation (step 3; Zou et al., 1999), which degrades the inhibitor of caspase-activated DNase (ICAD/DFF45) to release CAD/DFF40 (Liu et al., 1997; Sakahira et al., 1998). Finally, CAD/DFF40 causes nuclear chromatin fragmentation, typical of apoptotic morphology (step 4). Interestingly, MEKK1/MAPKK kinase which also phosphorylates MKK4 /MAPK kinase (Yan et al., 1994) represents a cross talk between the JNK pathway and NF-κB pathway by phosphorylation of IκBα with subsequent activation of NF-κB. Noticeably, MKK4 phosphorylates JNK/SAPK (Yang et al., 1997), which in turn phosphorylates c-Jun transcription factor (Gupta et al., 1996) and p53 (Fuchs et al, 1998). Importantly, JNK/SAPK also can phosphorylate tau (τ) protein in vitro (Goedert et a., 1997; Reynolds et al., 1997). This last observation suggests that τ -pathology is a downstream process related to Aβ signalization. In other words, Aβ might induce hiperfosforylation of τ, which aggregated into paired helical filaments, and conform the neurofibrillary tangles, contributing to cell death. In summary, NF-κB, p53, c-Jun and caspase-3 are crucial protein molecules in mediating apoptosis-induced by Aβ.

Alternatively, if NF-κB pathway is blocked (e.g. Ammonium pyrrolidinedithiocarbamate, PDTC), $A\beta_{[25-35]}$, DA and 6-OHDA may still cause cell death by (i) direct H_2O_2–induced mitochondria permeability transition and cytochrome release (step 5; Takeyama et al., 2002) and subsequent caspase-3 activation (step 3; DiPietrantonio et al., 1999), or (ii) by $^{\cdot}OH$ radicals generated by Fenton reaction when iron is present (step 6). Additionally to this last reaction, Fe^{2+} is also able to generate ($^{\cdot}O_2$) out of molecular oxygen (step 7) and increases DA oxidation (step 8). These oxidized products could bind covalently to proteins (e.g. α-synuclein, parkin, actin), and form cystenyl-S-adducts. Due to their redox cycling capability, these adducts are highly H_2O_2 producers (steps 9-10, Velez-Pardo et al., 1998a and references within). Strikingly, the hydroxyl radicals (step 6) may interact with intracellular (folded) proteins to generate unfolding proteins (step 11), which then are degraded by the ubiquitin-proteasome system (UPS, step 12). Yet, an excess of unwanted proteins or unfolded proteins overcome this physiological UPS of protein degradation system (Betarbet et al., 2005), this metabolic disruption might accelerate and amplify the toxic dopamine effects on nigral neurons in PD (step 4; Jimenez Del Rio et al., 2004).

Alternative Therapies for Alzheimer's and Parkinson's Diseases: A Mechanistic IGF-1, and Cannabinoids Proposal

Based on recent progress in delineating the disease cascade and cell death process (Blum et al., 2001; Wirths et al., 2004; Green and Kroemer, 2005), discrete types of potentially disease modifying treatment could be administered for both diseases. Indeed, during the last five years or so, several therapeutic molecules have been tested either in clinical research or in animal modelling involving β- (Burton, 2005; Thompson et al., 2005) and γ-secretase inhibitors (Tomita and Iwatsubo, 2006), prevention of Aβ aggregation (De Felice et al., 2004), Aβ vaccination (Levites et al., 2006), anti-inflammatory administration (Prasad et al., 2002; Hernan et al., 2006), among others (Wu and Fruch, 2005; Kapkova et al., 2006). Although providing substantial benefit, adverse effects and long-term complications burden these agents. Therefore, other therapeutic advances are urgently needed. Lastly, much attention has been paid to antioxidants and/or restorative approaches as promising therapy for AD and PD (Kedar, 2003; Viña et al., 2004; Hajieva and Behl, 2006). In this regard, our interest has been to screen molecules with possible anti-oxidant properties. As depicted in Fig 3 and 5, these mechanistic pathways may be of potential use for screening pharmacologically chemical libraries containing hundreds to thousands of compounds each that could modulate or control sensible molecules critical in cell fate (e. g. NF-κB, Pande and Ramos, 2005; p53, Komarova and Gudkov, 2000; caspases, Lavrik et al., 2005). Accordingly, it has been demonstrated that 17β-estradiol and testosterone protect lymphocytes against dopamine and iron-induced apoptosis by quenching H_2O_2 and other by-products of Fenton reaction (Jimenez Del Rio and Velez-Pardo, 2001; Jimenez Del Rio et al., 2004). However, estrogen has been declared carcinogenic (see Report on Carcinogens, 2002; and for a review, see Yager and Davidson 2006).

Recently, neurotrophic factors have come into focus as potential therapy in AD and PD (Siegel and Chauhan, 2000; Fumagalli et al., 2006). Specifically, insulin-like growth factor-I (IGF-1) has been demonstrated to protect against $A\beta_{[25\text{-}35]}$ insult (Dore et al., 1997), though the exact molecular mechanism of action involved in neuroprotective properties of IGF-1 remains to be clarified. One clue of its neuroprotective capability comes from the fact that IGF-1 is able to activate NF-κB against H_2O_2 oxidative stress (Heck et al., 1999). However, it has also been shown that NF-κB activation is involved in H_2O_2-induced apoptosis (Kutuk and Basaga, 2003). Therefore, the molecular mechanism(s) that may explain the dual role of NF-κB as an attenuator or promoter of apoptosis and the IGF-1´s molecular mechanism of neuroprotection still remain to be established.

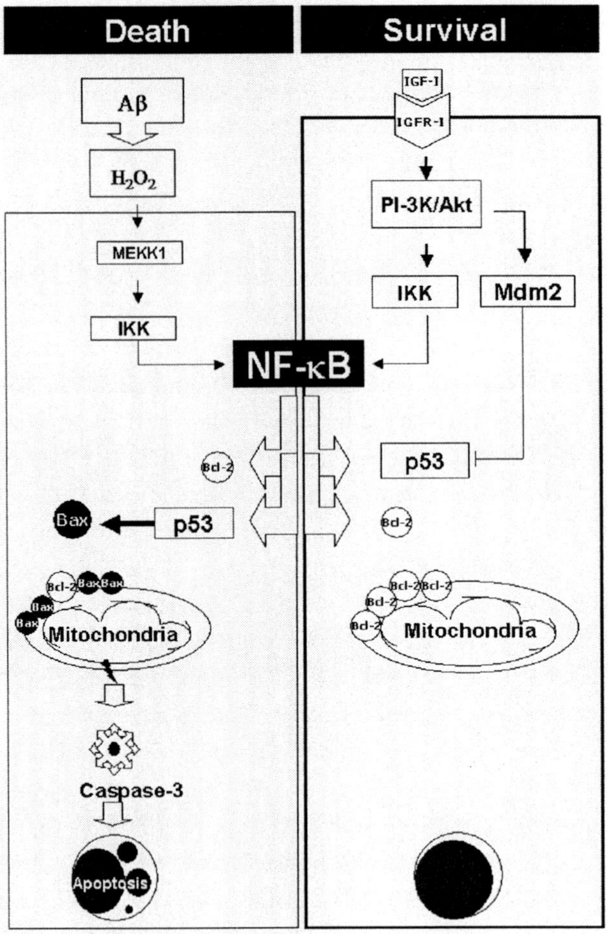

Figure 6. Schematic model of $A\beta_{[25-35]}$-induced apoptosis and IGF-1 survival signaling in lymphocytes. $A\beta_{[25-35]}$ (10μM) generates H_2O_2, which in turn indirectly activates NF-κB through activation of the multisubunit IκB Kinase (IKK) by p21-ras protein and MEKK1/ MAPKK kinase, respectively. Hence, the released of active NF-B dimmer (p50, p63) translocates into the nucleus and transcribes p53 transcription protein. Consequently, this protein activates the pro-apoptotic Bax protein. By out-numbering Bcl-2 proteins (e.g. Bcl-xL), Bax, which is able to permeabilize mitochondria, promotes the release of the apoptogenic cytochrome c. As a result, cyt C together with Apaf 1, dATP, and procaspase-9 (i.e. apoptosome complex) elicits caspase-3 protease activation leading to typical nuclei apoptotic morphology (Jimenez Del Rio and Velez-Pardo, 2004b and references within). IGF-1 can also activate NF-κB through IKK activation via PI-3K/Akt pathway (Kane et al., 1999; Sizemore et al., 2002). Then, NF-κB translocates into the nucleus and transcribes both anti-apoptotic Bcl-2 proteins and p53. Interestingly, p53 proteins could be indirectly regulated by PI-3K/Akt which phosphorylated mdm2 at S^{186}, thus enhancing its binding to p53 (Ogawara et al., 2003) and promoting p53 degradation. In fact, Mdm2 is an ubiquitin ligase that binds to the N-terminus and transfers ubiquitin moieties to several sites of p53. Ubiquitinated p53 is then rapidly exported from the nucleus and degraded by the ubiquitin proteosome system (UPS, Pickart, 2001). As a result, p53 dependent pro-apoptotic proteins are not transcribed (e. g. Bax), and the ratio of anti-apoptotic versus pro-apoptotic is balanced in favor of the former proteins, providing mitochondrial membrane stability. Hence, normal nuclei morphology and cellular functions are preserved. (Jimenez Del Rio and Velez-Pardo, 2006). Reproduced with permission From Taylor and Francis Group.

Taken advantage of the fact that human PBL express IGF-1 receptors (Tapson et al., 1988; Kooijman et al., 1992) and IGF-1 appears to be of potential therapeutic use against AD (Carro and Torres-Aleman, 2004), we were interested in the understanding of the molecular events that are thought to be downstream of IGF-1, in relation to the role played by NF-κB in survival and death-signalisation against $A\beta_{[25-35]}$ and H_2O_2 in lymphocytes, as a single cell model. We have found that IGF-1 protects lymphocytes from (10, 30μM) $A\beta_{[25-35]}$ and (25, 50, 100μM) H_2O_2-induced apoptosis through NF-κB activation and p53 down regulation involving the phosphoinositide 3-kinase (PI-3K)–dependent pathway. Interestingly, IGF-1, PDTC (a NF-κB inhibitor) and pifithrin- α (PFT, a p53 inhibitor) were able to protect and rescue lymphocytes pre-exposed to 10μM $A\beta_{[25-35]}$, even when the three compounds were added up-to 12 h post- $A\beta_{[25-35]}$, exposure. Overall these observations suggest that survival/rescue of lymphocytes from $A\beta_{[25-35]}$ toxicity is determined by p53 inactivation via IGF-1/ PI-3K pathway (Jimenez Del Rio and Velez-Pardo, 2006).

Which molecular mechanism(s) explain the dual role of NF-κB as an attenuator or promoter of apoptosis?. NF-κB has been reported to activate both pro-apoptotic genes such as p53 transcription factor gene (Wu and Lozano, 1994; Hellin et al., 1998; Jimenez Del Rio and Velez-Pardo, 2002; Velez-Pardo et al., 2002; Aleyasin et al., 2004), which in turn activates the expression of several genes that directly control or regulate the process of apoptosis such as Bax (a pro-apoptotic Bcl-2 protein family, Miyashita and Reed, 1995; Deng et al., 2000), and anti-apoptotic genes such as Bcl-2, Bcl-X_L, X-linked inhibitor of apoptosis (for a review, see Burstein and Duckett, 2003). Therefore, one prevailing model proposes that when the molecular ratio of pro-survival (e.g. Bcl-2, Bcl-xL, Bcl-w) to pro-death Bcl-2 family members (e.g. Bax, Bad, Bak, Bid) is biased towards pro-death Bcl-2 family members (either through changes in expression level, localization or activity), the outer mitochondrial membrane becomes permeable to apoptogenic proteins resulting in the activation of a cascade of effector caspases, such as caspase-3, that kill the cells by irreversible proteolysis of critical nuclear and cytoplasmic constituents (for a review, see Cory et al., 2003). In this vein, our data suggest that IGF-1 might promote gene transcription of survival genes via NF-κB activation and suppresses gene transcription of pro-apoptotic proteins through p53 inactivation. How then p53 turn-off could be related with IGF-1 citoprotection? One piece of information comes from the work by Ogawara and colleagues (2002) showing that Akt enhances the ubiquitinization-promoting function of Mdm2 (murine double minute) by phosphorylation of S^{186}, which results in reduction of p53 protein. Furthermore, Feng and colleagues (2004) have recently shown that PKB/Akt induces phosphorylation of Mdm2 at Ser^{166} and Ser^{188} resulting in Mdm2 protein stabilization. Based on our data and this information, it is reasonable to assume that p53 is modulated by IGF-1 through PI3K-Akt pathway. In fact, our findings reveal that p53 but not NF-κB is the critical transcription factor that may possibly balances the expression of pro-death proteins towards intracellular death decision under oxidative noxious stimuli (Lu, 2005; Culmsee and Mattson, 2005). Therefore, an ideal pharmacological compound (i.e. natural or synthetic) would be one that efficiently function as an antioxidant (e.g. 17β-estradiol, vitamin E, etc.), and simultaneously, as a survival-signalling molecule (e.g. IGF-1). To our surprise, those molecules exhibiting both features might come from the glandular hairs of Cannabis sativa or marijuana, and they are actually known as cannabinoids.

Cannabinoids are a group of C_{21} terpenophenolic compounds (Elsohly and Slade, 2005) which exert their effects by binding to specific plasma membrane G-protein-couple receptors, termed CB1 (Matsuda et al., 1990) and CB2 (Munro et al., 1993) receptors. Activation of these receptors has been shown to trigger several $G_i/_o$-protein-mediated signalling pathways (for a review, see Demuth and Molleman, 2006). Although, it is currently accepted that CB1 receptors are specially abundant in basal ganglia, hippocampus, cerebellum, and cortical structures; and CB2 receptors are restricted to cell types related to the immune function such as spleen macrophages, tonsils, B cells and natural killer cells, monocytes, neutrophils, and T cells (for a review, see Pazos et al., 2005), it has also been demonstrated the existence of CB2 receptors in purkinje cerebellar neurons (Skaper et al., 1996), microglia (Klegeris et al., 2003), oligodendrocytes (Molina-Holgado et al., 2002) and brainstem neurons (Van Sickle et al., 2005). Moreover, both receptors elicit similar signalling pathways such as inhibition of adenylate cyclase, stimulation of extracellular-signal-regulated kinase (Demuth and Molleman, 2006) and activation of phosphoinositide 3-kinase/PKB (Gomez Del Pulgar et al., 2000; 2002; Molina-Holgado et al., 2002; Sanchez et al., 2003). The physiological significance of these common characteristics is still unknown.

Cannabinoids have been proposed as potential therapeutic agents against AD (Ramírez et al., 2005) and PD (Venderova et al., 2004) thanks to their involvement in control of cell death/ survival decision and in neuroprotection (van der Stelt and Di Marzo, 2005). However, the mechanism of both actions by cannabinoids is far from clear. In fact, there are numerous, sometimes contradictory, reports of cannabinoids effects on cell fate (Guzman, 2003). Moreover, cannabinoids have been shown to function as antioxidant compounds via receptor-independent (Hampson et al., 1998; Chen et al., 2000; Marsicano et al., 2002) or receptor-dependent mechanisms (Nagayama et al., 1999; Kim et al., 2005). Although CB antagonists (e.g. SR141716A) have been used to elucidate the neuroprotective mechanism of cannabinoids, they have not been conclusive (see Marsicano et al., 2002 versus Nagayama et al., 1999; Kim et al., 2005). Therefore, the molecular mechanism(s) of cannabinoids effect on cells is a complex and still controversial issue (Sarne and Mechoulam, 2005).

Despite intense investigation, the detailed intracellular mechanism(s) involved in cannabinoids survival effect remains to be elucidated. Because CB2 cannabinoid receptor is linked to activation of PI3K (Sanchez et al., 2003), and the non-classical cannabinoid (-)-CP55,940 (a CB1 and CB2 agonist) and JWH-015 (a CB2 agonist) are commercially available, we wanted to elucidate the molecular signalling downstream of CB2 receptor linked to the role played by NF-κB and p53 in survival and death-signalisation against $A\beta_{[25-35]}$ and H_2O_2 stimuli. We have found that both synthetic agonists protect and rescue PBL from (10μM) $A\beta_{[25-35]}$ and (50μM) H_2O_2-induced apoptosis by two alternative mechanisms: (1) receptor-independent pathway, as demonstrated by no-dihydrorhodamine oxidation into fluorescent rhodamine 123 (R-123) as a result of cannabinoid inhibition of Aβ-generated H_2O_2; (2) receptor-dependent pathway through NF-κB activation and p53 down regulation involving phosphoinositide 3-kinase (PI-3K), as demonstrated by using either (25 μM) LY294002 (a PI-3K inhibitor), (50nM) pifithrin-α (PFT) or by using immunocytochemistry detection of NF-κB and p53 transcription factors activation. Importantly, cannabinoid agonists and PFT were able to protect and rescue lymphocytes pre-exposed to toxicants-, even when the three compounds were added up-to 12 h post-$A\beta_{[25-35]}$ /(H_2O_2) exposure

(Velez-Pardo and Jimenez Del Rio, 2006). In agreement with our previous observations with IGF-1, these results suggest that CP55,940 /(JWH-015) protection/rescue of PBL from noxious stimuli is determined by p53 inactivation. Unexpectedly, IGF-1 and CP55,940 have shown a weak protection against DA. Moreover, CP55,940 has shown to increase cell death when co-incubated with DA (Jimenez Del Rio and Velez-Pardo, unpublished observations). These findings could be explained by the fact that DA promotes apoptosis in PBL by at least five alternative mechanisms (Jimenez del Rio et al., 1993; Jimenez Del Rio et al., 2004), and/or CP55,940 might act as a pro-oxidant rather than as an anti-oxidant when co-incubated with DA. Therefore, further investigation is needed to classify cannabinoids molecules (Padgett, 2005; Thakur et al., 2005) with effective anti-oxidant from those with pro-oxidant actions.

Conclusion

As noted by S. Brenner (2002) "...choosing the right organism for one's research is as important as finding the right problems to work on...". In this regard, peripheral blood lymphocytes (PBL) as model system are well validated and permit totally controlled experiments, are relatively low cost and ease to use; but most importantly, they resemble neuronal cells. As any other model (e.g. animal or human tissue and cell lines), their limitation is your removal from the reality of the whole, integrated physiologic system. Despite this drawback, it turns out that their use in complex biologic investigations such as the one presented in this chapter, introduce lymphocytes as a unique opportunity to integrate oxidative stress, cell death, and cell survival signalling in a single-cell model (Figure 6). Moreover, lymphocytes could be used to monitor pharmacological therapies. Taken collectively our findings suggest that molecules with receptor-dependent and antioxidant properties such as cannabinoids may offer potential first-line pharmacologic intervention to retard the progression and/or deterioration of neurons in early onset FAD, PD and AR-JP disorders.

Acknowledgements

This work was supported by grants from "Comité para el Desarrollo y la Investigación CODI-UdeA" (# 9805; #9835; #9961); "Instituto para el Desarrollo de la Ciencia y la Tecnología en Colombia-Colciencias" (#1115-04-10231), "Proyecto Investigación de las enfermedades neurodegenerativas" #8780.

References

Adams, JM. Ways of dying: multiple pathways to apoptosis. *Genes & Develop* 2003; 17: 2481-2495.

Aleyasin, H; Cregan, SP; Lyirhiaro, G; O'Hare, MJ, Callaghan, SM; Slack, RS; Park DS. Nuclear factor-(kappa)B modulates the p53 response in neurons exposed to DNA damage. *J Neurosci* 2004; 24: 2963-2973.

Amenta, F; Bronzetti, E; Felici, L; Ricci, A; Tayebati, SK. Dopamine D2-like receptors on human peripheral blood lymphocytes: a radioligand binding assay and immunocytochemical study. *J Auton Pharmacol* 1999; 19:151-159.

Amenta, F; Bronzetti, E; Cantalamessa, F; El-Assouad, D; Felici, L; Ricci, A; Tayebati, SK. Identification of dopamine plasma membrane and vesicular transporters in human peripheral blood lymphocytes. *J Neuroimmunol* 2001; 117 (1-2): 133-142.

Arima, N; Matsushita, K; Obata, H; Ohtsubo, H; Fujiwara, H; Arimura, K; Kukita, T; Suruga, Y; Wakamatsu, S; Hidaka, S; Tei, C. NF-kappaB involvement in the activation of primary adult T-cell leukemia cells and its clinical implications. *Exp Hematol* 1999; 27:1168-1175.

Arcos-Burgos, OM; Muenke, M. Genetics of population isolates. *Clin Genet* 2002; 61: 233--247.

Ardila, A; Lopera, F; Rosselli, M; Moreno, S; Madrigal L, Arango-Lasprilla, JC; Arcos, M; Murcia, C; Arango-Viana, JC; Ossa, J; Goate, A; Kosik, KS. Neuropsychological profile of a large kindred with familial Alzheimer's disease caused by the E280A single presenilin-1 mutation. *Arch Clin Neuropsychol.* 2000; 15: 515-528.

Asanuma, M; Miyazaki, I; Diaz-Corrales, FJ; Ogawa, N. Quinone formation as dopaminergic neuron-specific oxidative stress in the pathogenesis of sporadic Parkinson's disease and neurotoxin-induced parkinsonism. *Acta Med Okayama* 2004; 58: 221-233.

Barbanti, P; Fabbrini, G; Ricci, A; Cerbo, R; Bronzetti, E; Caronti, B; Calderaro, C; Felici, L; Stocchi, F; Meco, G; Amenta, F; Lenzi, GL. Increased expression of dopamine receptors on lymphocytes in Parkinson's disease. *Mov Disord* 1999; 14: 764-771.

Barbanti, P; Fabbrini, G; Ricci, A; Bruno, G; Cerbo, R; Bronzetti, E; Amenta, F; Luigi Lenzi, G . Reduced density of dopamine D2-like receptors on peripheral blood lymphocytes in Alzheimer's disease. *Mech Ageing Dev* 2000; 120(1-3): 65-75.

Behl, C; Davis, JB; Lesley, R; Schubert, D. Hydrogen peroxide mediates amyloid β-protein toxicity. *Cell* 1994; 77: 817-827.

Ben-Shachar D, Zuk R, Gazawi H, Ljubuncic P. Dopamine toxicity involves mitochondrial complex I inhibition: implications to dopamine-related neuropsychiatric disorders. Biochem Pharmacol 2004; 67: 1965-1974.

Betarbet, R; Sherer, TB; Greenamyre, JT. Ubiquitin-proteasome system and Parkinson's disease. *Exp Neurol* 2005; 191 Suppl 1: S17-S27.

Bidlack, JM. Detection and function of opioid receptors on cells from the immune system. *Clin Diagn Lab Immunol* 2000; 7:719-723.

Bove, J; Prou, D; Perier, C; Przedborski, S. Toxin-induced Models of Parkinson' disease. *NeuroRx* 2005; 2:484-494.

Blum, D; Torch, S; Lambeng, N; Nissou, M; Benabid, AL; Sadoul, R; Verna, JM. Molecular pathways involved in the neurotoxicity of 6-OHDA, dopamine and MPTP: contribution to the apoptotic theory in Parkinson's disease. *Prog Neurobiol* 2001; 65: 135-172.

Bras, M; Queenan, B; Susin, SA. Programmed cell death via mitochondria: different modes of dying. *Biochemistry (Mosc).* 2005; 70: 231-239.

Brenner, S. Nature's gift to science. Nobel lecture, December 8, 2002. (www.nobelprize.com).

Bullido, MJ; Muñoz-Fernandez, MA; Recuero, M; Fresno, M; Valdivieso, F. Alzheimer's amyloid precursor protein is expressed on the surface of hematopoietic cells upon activation. *Biochim Biophys Acta* 1996; 21: 54-62.

Burstein, E; Duckett, CS. Dying for NF-kappaB? Control of cell death by transcriptional regulation of the apoptotic machinery. *Curr Opin Cell Biol* 2003; 15: 732-737.

Burton, A. RNA inhibition of beta-secretase reverts AD in mice. *Lancet Neurol.* 2005;4: 698.

Butterfield, DA; Drake, J; Pocernich, Ch; Castegna, A. Evidence of oxidative damage in Alzheimer's brain: central role for amyloid β-peptide. *Trends Mol Med* 2001; 7: 548-554.

Caronti, B; Tanda, G; Colosimo, C; Ruggieri, S; Calderaro, C; Palladini, G; Pontieri, FE; DiChiara, G. Reduced dopamine in peripheral blood lymphocytes in Parkinson's disease. *Neuroreport* 1999; 10: 2907-2910.

Carrillo-Vico, A; Calvo, JR; Abreu, P; Lardone, PJ; Garcia-Maurino, S; Reiter, RJ; Guerrero; JM. Evidence of melatonin synthesis by human lymphocytes and its physiological significance: possible role as intracrine, autocrine, and/or paracrine substance. *FASEB J* 2004; 18: 537-539.

Carro, E; Torres-Aleman, I. Insulin-like growth factor I and Alzheimer's disease: therapeutic prospects?. *Exp Rev Neurotherapeutics* 2004; 4: 79-86.

Chang, LK; Putcha, GV; Desmukh, M; Jonson, EM. Mitochondrial involvement in the point of no return in neuronal apoptosis. *Biochemie* 2002; 84:223-231.

Chen, Y; Buck, J. Cannabinoids protect cells from oxidative cell death: a receptor-independent mechanism. *J Pharmacol Exp Ther* 2000; 293: 807-812.

Chen, N; Hudson, JE; Walczak, P; Misiuta, I; Garbuzova-Davis, S; Jiang, L; Sanchez-Ramos, J; Sanberg, PR; Zigova, T; Willing, AE. Human Umbilical Cord Blood Progenitors: The Potential of These Hematopoietic Cells to Become Neural. *Stem Cells* 2005; 23: 1560-1570.

Clark, RF; Hutton, M; Fuldner, RA; Froelich, S; Karran, E; Talbot, C; Crook, R; Lendon, C; Prihar, G; He, C; Korenblat, K; Martinez, A; Wragg, M; Busfield, F; Behrens, MI; Myers, A; Norton, J; Morris, J; Mehta, N; Pearson; C; Lincoln, C; Baker, M; Duff, K; Zehr, C; Perez-Tur, J; Houlden, H; Ruiz A; Ossa, J; Lopera, F; Arcos, M; Madrigal, L; Collinge, J; Humphreys, C; Ashworth, A; Sarner, S; Fox, N; Harvey, R; Kennedy, A; Roques, P; Cline, RT; Phillips, CA; Venter; JC; Forsell, L; Axelman, K; Lilius, L; Johnston, J; Cowburn, R; Viitanen, M; Winblad, B; Kosik, K; Haltia, M; Poyhonen, M; Dickson, D; Mann, D; Neary, D; Snowdon, J; Lantos, P; Lannfelt, L; Rossor, M; Roberts, GW; Adams, MD; Hardy, J; Goate, A. The structure of the presenilin 1 (S182) gene and identification of six novel mutations in early onset AD families. *Nature Genetics* 1995; 11: 219-222.

Conway, KA; Rochet, JC; Bieganski, RM; Lansbury, PT Jr. Kinetic stabilization of the alpha-synuclein protofibril by a dopamine-alpha-synuclein adducts. *Science* 2001; 294: 1346-1349.

Cookson, MR. The biochemistry of Parkinson's disease. *Annu Rev Biochem* 2005; 74: 29-52.

Cory, S; Huang, DC; Adams, JM. The Bcl-2 family: roles in cell survival and oncogenesis. *Oncogene* 2003; 22: 8590-607.

Culmsee, C; Siewe, J; Junker, V; Retiounskaia, M; Schwarz, S; Camandola, S; El-Metainy, S; Behnke, H; Mattson, MP; Krieglstein, J. Reciprocal inhibition of p53 and nuclear factor-kappaB transcriptional activities determines cell survival or death in neurons. *J Neurosci* 2003; 23: 8586-8595.

Culmsee, C; Mattson, MP. p53 in neuronal apoptosis. *Biochem Biophys Res Commun* 2005; 331: 761-777.

Crigler, L; Robey, RC; Asawachaicharn, A; Gaupp, D; Phinney, DG. Human mesenchymal stem cell subpopulations express a variety of neuro-regulatory molecules and promote neuronal cell survival and neuritogenesis. *Exp Neurol* 2006; 198: 54-64.

De Felice, FG; Vieira, MN; Saraiva, LM; Figueroa-Villar, JD; Garcia-Abreu, J; Liu, R; Chang, L; Klein, WL; Ferreira, ST. Targeting the neurotoxic species in Alzheimer's disease: inhibitors of Abeta oligomerization. *FASEB J* 2004; 18: 1366-1372.

Deng, Y; Wu, X. Peg3/Pw1 promotes p53-mediated apoptosis by inducing Bax translocation from cytosol to mitochondria. *Proc Natl Acad. Sci USA* 2000; 97: 12050-12055.

Deng, J; Petersen, BE; Steindler, DA; Jorgensen, ML; Laywell, ED. Mesenchymal stem cells spontaneously express neural proteins in culture, and are neurogenic after transplantation. *Stem Cells* 2006; 24: 1054-1064.

Demuth, DG; Molleman, A. Cannabinoid signalling. *Life Sci* 2006; 78: 549-563.

Dexter, DT; Wells, FR; Lees, AJ; Agid, F; Agid, Y; Jenner, P; Marsden, CD. Increased nigral iron content and alterations in other metal ions occurring in brain in Parkinson's disease. *J Neurochem* 1989; 52:1830-1836.

DiPietrantonio, AM; Hsieh, T; Wu, JM. Activation of caspase-3 in HL-60 cells exposed to hydrogen peroxide. *Biochem Biophys Res Commun* 1999; 255: 477–482.

Diss, JKL; Fraser, SP; Djamgoz, MBA. Voltage-gated Na^+ channels: multiplicity of expression, plasticity, functional implications and pathophysiological aspects. *Eur Biophys J* 2004; 33: 180-193.

Dore, S; Kar, S; Quirion, R. Insulin-like growth factor I protects and rescues hippocampal neurons against beta-amyloid- and human amylin-induced toxicity. *Proc Natl Acad Sci USA 94*; 1997: 4772–4777.

Dlamini, Z; Mbita, Z; Zungu, M. Genealogy, expression, and molecular mechanisms in apoptosis. *Pharmacol & Therap* 2004; 101: 1-15.

Ebstein, RP; Nemanov, L; Lubarski, G; Dano, M; Trevis, T; Korczyn, AD. Changes in expression of lymphocyte amyloid precursor protein mRNA isoforms in normal aging and Alzheimer's disease. *Brain Res Mol Brain Res* 1996; 35(1-2): 260-268.

Elsohly, MA; Slade, D. Chemical constituents of marijuana: the complex mixture of natural cannabinoids. *Life Sci* 2005; 78: 539-548.

Enari, M; Sakahira, H; Yokoyama, H; Okawa, K; Iwamatsu, A; Nagata, S. A caspase-activated DNase that degrades DNA during apoptosis, and its inhibitor ICAD. *Nature* 1998; 391: 43-50. Erratum in: *Nature* 1998; 393:396.

Earnshaw, WC; Martins, LM; Kaufmann, SH. Mammalian caspases: structure, activation, substrates, and functions during apoptosis. *Annu Rev Biochem* 1999; 68: 383-424.

Esiri, MM. The neuropathology of Alzheimer's disease. In: Dawbarn, D; Allen, SJ., editors. *Neurobiology of Alzheimer's disease*. Oxford: Oxford University Press; 2001; pp. 33-53.

Fahn; S; Cohen, G. The oxidant stress hypothesis in Parkinson´s disease: evidence supporting it. *Ann Neurol* 1992; 32: 804-812.

Faraj, BA; Olkowski, ZL; Jackson, RT. Binding of [3H]-dopamine to human lymphocytes: possible relationship to neurotransmitter uptake sites. *Pharmacology* 1991; 42: 135-141.

Feng, J; Tamaskovic, R; Yang, Z; Brazil, DP; Merlo, A; Hes, D; Hemmings, BA. Stabilization of Mdm2 via decreased ubiquitination is mediated by protein Kinase/ Akt-dependent phosphorylation. *J Biol Chem* 2004; 279: 35510-35517.

Ferri, CP; Prince, M; Brayne, C; Brodaty, H; Fratiglioni, L; Ganguli, M; Hall, K; Hasegawa, K; Hendrie, H; Huang, Y; Jorm, A; Mathers, C; Menezes, PR; Rimmer, E; Scazufca, M. Alzheimer's disease International. Global prevalence of dementia: a Delphi consensus study. *Lancet* 2005; 366: 2112-2117.

Forno, LS. Neuropathology of Parkinson's disease. *J Neuropathol Exp Neurol* 1996; 55:259-272.

Fuchs, SY; Adler, V; Pincus, MR; Ronai, Z. MEKK1/JNK signaling stabilizes and activates p53. *Proc Natl Acad Sci USA* 1998; 95: 10541-10546.

Fumagalli, F; Racagni, G; Riva, MA. The expanding role of BDNF: a therapeutic target for Alzheimer's disease?. *Pharmacogenomics J* 2006; 6: 8-15.

Garcia-Ospina, G; Jimenez Del Rio, M; Lopera, F; Velez-Pardo C. [Neuronal DNA damage correlates with a positive detection of c-Jun, nuclear factor kB, p53 and par-4 transcription factors in Alzheimer´s disease]. (Spanish) *Rev Neurol* 2003; 36:1004-1010.

Goedert, M; Hasegawa, M; Jakes, R; Lawler, S; Cuenda, A; Cohen, P. Phosphorylation of microtubule-associated protein tau by stress-activated protein kinases. *FEBS Lett* 1997; 409: 57-62.

Gomes, B; Savignac M; Moreau, M; Leclerc, C; Lory, P; Guery, JC, Pelletier L. Lymphocyte calcium signaling involves dihydropyridine-sensitive L-type calcium channels: facts and controversies. *Crit Rev Immunol* 2004; 24:425-447.

Gomez Del Pulgar, T; Velasco, G; Guzman, M. The CB 1 cannabinoid receptor is coupled to the activation of protein kinase B/Akt. *Biochem J* 2000; 347: 369-373.

Gomez Del Pulgar, T; de Ceballos, ML; Guzman, M; Velasco, G. Cannabinoids Protect astrocytes from ceramide-induced apoptosis through the phosphatidylinositol 3-kinase/protein kinase B pathway. *J Biol Chem* 2002; 277: 36527-36533.

Gomez-Isla, T; Growdon, WB; McNamara, MJ; Nochlin, D; Bird, TD; Arango, JC; Lopera, F; Kosik, KS; Lantos, PL; Cairns, NJ; Hyman, BT. The impact of different presenilin 1 andpresenilin 2 mutations on amyloid deposition, neurofibrillary changes and neuronal loss in the familial Alzheimer's disease brain: evidence for other phenotype-modifying factors. *Brain* 1999; 122 (Pt 9): 1709-1719.

Gupta, S; Barrett, T; Whitmarsh, AJ; Cavanagh, J; Sluss, HK; Derijard, B; Davis, RJ. Selective interaction of JNK protein kinase isoformas with transcription factors. *EMBO J* 1996; 15:2760–2770.

Guttridge, DC; Albanese, C; Reuther, JY; Pestell, RG; Baldwin, AS Jr. NF-kappaB controls cell growth and differentiation through transcriptional regulation of cyclin D1. *Mol Cell Biol* 1999; 19: 5785-5799.

Guzman, M. Neurons on cannabinoids: dead or alive?. *British J Pharmacol* 2003; 140: 439-440.

Graeber, MB; Moran, LB. Mechanism of cell death in neurodegenerative diseases: fashion, fiction, and facts. *Brain Pathol* 2002; 12:385-390.

Green, DR; Kroemer, G. Pharmacological manipulation of cell death: clinical applications in sight?. *J Clin Inv* 2005; 115: 2610-2617.

Griffiths, PD; Dobson, BR; Jones, GR; Clarke, DT. Iron in the basal ganglia in Parkinson's disease. An in vitro study using extended X-ray absorption fine structure and cryo-electron microscopy. *Brain* 1999; 122 (Pt 4): 667-673.

Haass, C. Take five-BACE and the γ-secretase quartet conduct Alzheimer's amyloid β-peptide generation. *EMBO J* 2004; 23: 483-488.

Hajieva, P; Behl, C. Antioxidants as a potential therapy against age-related neurodegenerative diseases: amyloid beta toxicity and Alzheimer's disease. *Curr Pharm Des* 2006; 12: 699-704.

Halenbeck, R; MacDonald, H; Roulston, A; Chen, TT; Conroy, L; Williams, LT. CPAN, a human nuclease regulated by the caspase-sensitive inhibitor DFF45. *Curr Biol* 1998; 8: 537-540.

Hampson, AJ; Grimaldi, M; Axelrod, J; Wink, D. Cannabidiol and (-) Delta9-tetrahydrocannabinol are neuroprotective antioxidants. *Proc Natl Acad Sci USA* 1998; 95: 8268-8273.

Hao, HN; Zhao, J; Thomas, RL; Parker, GC; Lyman, WD. Fetal human hematopoietic stem cells can differentiate sequentially into neural stem cells and then astrocytes in vitro. *J Hematother Stem Cell Res* 2003; 12: 23-32.

Hardy, JA; Higgins, GA. Alzheimer's disease: the amyloid cascade hypothesis. *Science* 1992; 256: 184-185.

Heck, S; Lezoualc'h, F; Engert, S; Behl, C. Insulin-like growth factor-1-mediated neuroprotection against oxidative stress is associated with activation of nuclear factor κB. *J Biol Chem* 1999; 274:9828–9835.

Hellin, AC; Calmant, P; Gielen, J; Bours, V; Merville, MP. Nuclear factor-kappaB-dependent regulation of p53 gene expression induced by daunomycin genotoxic drug. *Oncogene* 1998; 16: 1187-1195.

Hernan, MA; Logroscino, G; Garcia-Rodriguez, LA. Nonsteroidal anti-nflammatory drugs and the incidence of Parkinson disease. *Neurology* 2006; 66: 1097-1099.

Herrup, K; Neve, R; Ackerman, SL; Copani, A. Divide and die: cell cycle events as triggers of nerve cell death. *J Neurosci* 2004; 24: 9232-9239.

Hinz, M; Krappmann, D; Eichten, A; Heder, A; Scheidereit, C; Strauss, M. NF-kappaB function in growth control: regulation of cyclin D1 expression and G_0/G_1-to-S-phase transition. *Mol Cell Biol.* 1999; 19: 2690-2698.

Howlett, DR; Allsop, D; Karran, EH. The biology and molecular neuropathology of β-amyloid protein. In: Dawbarn, D; Allen, SJ., editors. *Neurobiology of Alzheimer's disease*. Oxford: Oxford University Press; 2001; pp. 75-102.

Imai, Y; Soda, M; Takahashi, R. Parkin suppresses unfolded protein stress-induced cell death through its E3 ubiquitin-protein ligase activity. *J Biol Chem* 2000; 275: 35661-35664.

Imai, Y; Takahashi, R. How do parkin mutations result in neurodegeneration. *Curr Opin Neurobiol* 2004; 14: 384-389.

Jackson-Lewis, V; Smeyne, RJ. MPTP and SNpc DA neuronal vulnerability: role of dopamine, superoxide and nitric oxide in neurotoxicity. *Neurotox Res* 2005; 7: 193-202.

Janca, A. Parkinson's disease from WHO perspective and a public health point of view. *Parkinsonism Relat Disord* 2002; 9: 3-6.

Jenner, P; Olanow, W. Oxidative stress and the pathogenesis of Parkinson´s disease. *Neurology* 1996; 47 (Suppl. 3): S161-S170.

Jenner, P; Olanow, CW. Understanding cell death in Parkinson's disease. *Ann Neurol* 1998; 44 (3 Suppl 1): S72-S84.

Jiang, H; Ren, Y; Zhao, J; Feng, J. Parkin protects human dopaminergic neuroblastoma cells against dopamine-induced apoptosis. *Human Mol Genet* 2004; 13: 1745–1754.

Jimenez Del Rio, M; Velez-Pardo; Pinxteren, J; DePotter, WP; Ebinger, G; Vauquelin, G. Binding of serotonin and dopamine to "serotonin binding proteins" in bovine frontal cortex: evidence for iron-induced oxidative mechanism. *Eur J Pharmacol* 1993; 247:11-21.

Jimenez Del Rio, M; Velez-Pardo, C. Molecular mechanism of monoamine toxicity in Parkinson's disease: A hypothetical cell death model. *Medic Hypotheses* 2000; 54: 269-274.

Jimenez Del Rio, M; Velez-Pardo, C. 17β-Estradiol protects lymphocytes against dopamine and iron-induced apoptosis by a genomic-independent mechanism Implication in Parkinson's disease. *Gen Pharmacol* 2001; 35: 1– 9.

Jimenez Del Rio, M; Velez-Pardo, C. Monoamine neurotoxin-induced apoptosis in lymphocytes by a common mechanism: involvement of hydrogen peroxide (H_2O_2), caspase-3, and nuclear factor kappa-B (NF-κB), p53, c-Jun transcription factor. *Biochem Pharmacol* 2002; 63: 677-688.

Jimenez Del Rio M; Moreno, S; Garcia-Ospina, G; Buritica, O; Uribe, CS; Lopera, F; Velez-Pardo, C. Autosomal recessive juvenile parkinsonism Cys212Tyr mutation in parkin renders lymphocytes susceptible to dopamine and iron-mediated apoptosis. *Mov Disord* 2004; 19: 324-330.

Jimenez Del Rio, M; Velez-Pardo, C. Transition metals-induced apoptosis in lymphocytes via hydroxyl radical generation, mitochondria dysfunction and caspase-3 activation: an *in vitro* model for neurodegeneration. *Arch Medic Res* 2004a; 35:185-193.

Jimenez Del Rio, M; Velez-Pardo, C. The hydrogen peroxide and its importance in the Alzheimer's and Parkinson's disease. *Current Medical Chemistry- Central Nervous System Agents* 2004b; 4: 279-285.

Jimenez Del Rio, M; Velez-Pardo, C. Insulin-like growth factor-1 prevents $A\beta_{[25\text{-}35]}$ / (H_2O_2)-induced apoptosis in lymphocytes by reciprocal NF-κB activation and p53 inhibition via PI3K-dependent pathway. *Growth Factors* 2006; 24: 67-78.

Johnson, KA; Lopera, F; Jones, K; Becker, A; Sperling, R; Hilson, J; Londono, J; Siegert, I; Arcos, M; Moreno, S; Madrigal, L; Ossa, J; Pineda, N; Ardila, A; Roselli, M; Albert, MS; Kosik, KS; Rios, A. Presenilin-1-associated abnormalities in regional cerebral perfusion. *Neurology* 2001; 56 (11): 1545-1551.

Jurgensmeier, JM; Xie, Z; Deveraux, Q; Ellerby, L; Bredesen, D; Reed, JC. Bax directly induces release of cytochrome c from isolated mitochondria. *Proc Natl Acad Sci USA* 1998; 95:4997–5002.

Kane, LP; Shapiro, VS; Stokoe, D; Weiss, A. Induction of NF-kappaB by the Akt/PKB kinase. *Curr Biol* 1999; 9: 601-604.

Kaur, D; Andersen, J. Does cellular iron dysregulation play a causative role in Parkinson's disease? *Ageing Res Rev* 2004; 3: 237-343.

Kapkova, P; Alptuzun, V; Frey, P; Erciyas, E; Holzgrabe, U. Search for dual function inhibitors for Alzheimer's disease: synthesis and biological activity of acetylcholinesterase inhibitors of pyridinium-type and their Abeta fibril formation inhibition capacity. *Bioorg Med Chem* 2006; 14: 472-478.

Kawashima, K; Fujii, T. Expression of non-neuronal acetylcholine in lymphocytes and its contribution to the regulation to the regulation of immune function. *Frontiers Biosci* 2004; 9:2063-2085.

Kedar, NP. Can we prevent Parkinson's and Alzheimer's disease?. *J Postgrad Med* 2003; 49: 236-245.

Kerr, JFR; Wyllie, AH; Currie, AR. Apoptosis: a basic biological phenomenon with wide ranging implications in tissue kinetics. *Br J Cancer* 1972; 26: 239-257.

Kerr, JFR; Gobe, GC; Winterford, CM; Harmon, BV. Anatomical methods in cell death. In: Schwartz, LM; Osborne, BA; editors. *Methods in cell biology: cell death*. New York: Academic Press; 1995; pp. 1–27.

Kim, SH; Won, SJ; Mao, XO; Jin, K; Greenberg, DA. Involvement of protein kinase A in cannabinoid receptor-mediated protection from oxidative neuronal injury. *J Pharmacol Exp Ther* 2005; 313: 88-94.

Kimberly, WT; LaVoie, MJ; Ostaszewski, BL; Ye, W; Wolfe, MS; Selkoe, DJ. Gamma-secretase is a membrane protein complex comprised of presenilin, nicastrin, Aph-1, and Pen-2. *Proc Natl Acad Sci USA* 2003; 100: 6382-6387.

Kitada, T; Asakawa, S; Hattori, N; Matsumine, H; Yamamura, Y; Minoshima, S; Yokochi, M; Mizuno, Y; Shimizu, N. Mutations in the parkin gene cause autosomal recessive juvenile parkinsonism. *Nature* 1998; 392:605-608.

Kooijman, R; Willems, M; DeCarla, HJC; Rijkers, GT; Schuurmans, ALG; Van Buul-Offers, SC; Heijnen, CJ; Zegers, BJM. Expression of type I insulin-like growth factor receptors on human peripheral blood mononuclear cells. *Endocrinol* 1992; 131: 2244-2250.

Komarova, EA; Gudkov, AV. Suppression of p53: a new approach to overcome side effects of antitumor therapy. *Biochemistry (Mosc)* 2000; 65: 41-48.

Kutuk, O; Basaga, H. Aspirin prevents apoptosis and NFkappaB activation induced by H_2O_2 in HeLa cells. *Free Radic Res* 2003; 37:1267–1276.

Klegeris, A; Bissonnette, CJ; McGeer, PL. Reduction of human monocytic cell neurotoxicity and cytokine secretion by ligands of the cannabinoid-type CB2 receptor. *Br J Pharmacol* 2003; 139: 775-786.

Krasznai, Z. Ion channels in T cells: from molecular pharmacology to therapy. Arch Immunol Ther Exp 2005; 52:127-135.

Kriesberg N. Animals as Models. http://ori.dhhs.gov/education/products/Herkert%20(NC%20State)/models.htm.

Kroemer, G; Zamzami, N; Susin, SA. Mitochondrial control of apoptosis. *Immunology Today* 1997; 18: 44-51.

Kroemer, G; Martin, SJ. Caspase-independent cell death. *Nat Med* 2005; 11: 725-730.

Lander, HM; Jason, S; Ogiste, JS; Kenneth, K; Teng, KK; Novogrodsky, A. $p21^{ras}$ as a Common Signaling Target of Reactive Free Radicals and Cellular Redox Stress. *J Biol Chem* 1995; 270: 21195–21198.

Lahiri, DK; Greig, NH. Lethal weapon: amyloid-β-peptide, role in the oxidative stress and neurodegeneration of Alzheimer's disease. *Neurobiol Aging* 2004; 25: 581-587.

LaVoie, MJ; Ostaszewski, BL; Weihofen, A; Schlossmacher, MG; Selkoe, DJ. Dopamine covalently modifies and functionally inactivates parkin. *Nat Med* 2005; 11: 1214-1221.

Lavrik, IN; Golks, A; Krammer, PH. Caspases: pharmacological manipulation of cell death. *J Clin Invest* 2005; 115: 2665-2672.

LeBlanc, AC. The role of apoptotic pathways in Alzheimer's disease neurodegeneration and cell death. *Curr Alzh Res* 2005; 2: 389-402.

Lee, FS; Hagler, J; Chen, ZJ; Maniatis, T. Activation of the IB kinase complex by MEKK1, a kinase of the JNK pathway. *Cell* 1997; 88: 213–22.

Lemere, CA; Lopera, F; Kosik, KS; Lendon, CL; Ossa, J; Saido, TC; Yamaguchi, H; Ruiz, A; Martinez, A; Madrigal, L; Hincapie, L; Arango, JC; Anthony, DC; Koo, EH; Goate, AM; Selkoe, DJ; Arango, JC. The E280A presenilin 1 Alzheimer mutation produces increased A beta 42 deposition and severe cerebellar pathology. *Nat Med* 1996; 2: 1146-1150.

Leist, M; Jaattela, M. Four deaths and a funeral: from caspases to alternative mechanisms. *Nat Rev Mol Cell Biol.* 2001; 2: 589-598.

Leite, M; Quinta-Costa, M; Leite OS; Guimarães, JE. Critical evaluation of techniques to detect and measure cell death- study in a model of UV radiation of the leukaemic cell line HL60. *Anal Cell Pathol* 1999; 19 (3-4): 139–151.

Levites, Y; Das, P; Price, RW; Rochette, MJ; Kostura, LA; McGowan, EM; Murphy, MP; Golde, TE. Anti-Abeta42- and anti-Abeta40-specific mAbs attenuate amyloid deposition in an Alzheimer disease mouse model. *J Clin Invest* 2006; 116: 193-201.

Li, PD; Nijhawan, D; Budihardjo, I; Srinivasula, SM; Ahmad, M; Alnemri, ES; Wang, X. Cytochrome c and dATP-dependent formation of Apaf-1/caspase-9 complex initiates an apoptotic protease cascade. *Cell* 1997; 91: 479–489.

Liu, X; Zou, H; Slaughter, C; Wang, X. DFF, a heterodimeric protein that functions downstream of caspase-3 to trigger DNA fragmentation during apoptosis. *Cell* 1997; 89:175-184.

Lombardi, G; Dianzani, C; Miglio, G; Canonico, PL; Fantozzi, R. Characterization of ionotropic glutamate receptors in human lymphocytes. *Br J Pharmacol* 2001; 133: 936-944.

Lombardi, G; Miglio, G; Dianzani, C; Mesturini, R; Varsaldi, F; Chiocchetti, A; Dianzani, U; Fantozzi, R. Glutamate modulation of human lymphocyte growth: in vitro studies. *Biochem Biophys Res Commun* 2004; 318: 496-502.

Long, X; Olszewsk, M; Huang, W; Kletzel, M. Neural cell differentiation in vitor from adult human bone marrow mesenchymal stem cells. *Stem Cells Dev* 2005; 14: 65-69.

Lopera, F; Ardilla, A; Martinez, A; Madrigal, L; Arango-Viana, JC; Lemere, CA; Arango-Lasprilla, JC; Hincapie, L; Arcos-Burgos, M; Ossa, JE; Behrens, IM; Norton, J; Lendon,

C; Goate, AM; Ruiz-Linares, A; Rosselli, M; Kosik, KS. Clinical features of early-onset Alzheimer disease in a large kindred with an E280A presenilin-1 mutation. *JAMA* 1997; 277: 793-799.

Lorenzo, HK; Susin, SA. Mitochondrial effectors in caspase-independent cell death. *FEBS Lett* 2004; 557(1-3): 14-20.

Lu, Y. p53: a heavily dictated dictator of life and death. Curr Opin Genet Dev 2005; 15: 27-33.

Markesbery, WR. Oxidative stress hypothesis in Alzheimer's disease. *Free Rad Biol Med* 1997; 23: 134-147.

Marini, M; Frabetti, F; Canaider, S; Dini, L; Falcieri, E; Poirier, GG. Modulation of caspase-3 activity by zinc ions and by the cell redox state. *Exp Cell Res* 2001; 266:323-332.

Marino, F; Cosentino, M; Bombelli, R; Ferrari, M; Lecchini, S; Frigo, G. Endogenous catecholamine synthesis, metabolism, storage, and uptake in human peripheral blood mononuclear cells. *Exp Hematol* 1999; 27:489-495.

Marsicano, G; Moosmann, B; Hermann, H; Lutz, B; Behl, C. Neuroprotective properties of cannabinoids against oxidative stress: role of the cannabinoid receptor CB1. *J Neurochem.* 2002; 80: 448-456.

Matsuda, LA; Lolait, SJ; Brownstein, M; Young, A; Bonner, TI. Structure of a cannabinoid receptor and functional expression of the cloned cDNA. *Nature (London)* 1990; 346 : 561-564.

Maynard, CJ ; Bush, AI ; Master, CL ; Cappai, R ; Li, Q-X. Metals and amyloid-β in Alzheimer`s disease. *Int J Exp Pathol* 2005 ; 86 : 147-159.

Miglio, G; Varsaldi, F; Dianzani, C; Fantozzi, R; Lombardi, G. Stimulation of group I metabotropic glutamate receptors evokes calcium signals and c-jun and c-fos gene expression in human T cells. *Biochem Pharmacol* 2005; 70:189-199.

Migliore, L; Petrozzi, L; Lucetti, C; Gambaccini, G; Bernardini, S; Scarpato, R; Trippi, F; Barale, R; Frenzilli, G; Rodilla, V; Bonuccelli, U. Oxidative damage and cytogenetic analysis in leukocytes of Parkinson's disease patients. *Neurology* 2002; 58: 1809-1815.

Migliore, L; Fontana, I; Trippi, F; Colognato, R; Coppede, F; Tognoni, G; Nucciarone, B; Siciliano, G. Oxidative DNA damage in peripheral leukocytes of mild cognitive impairment and AD patients. *Neurobiol Aging* 2005; 26: 567-573.

Milton, NG. Role of hydrogen peroxide in the aetiology of Alzheimer's disease: implications for treatment. *Drugs Aging* 2004; 21: 81-100.

Mirinics, ZK; Calafat, J; Udby, L; Lovelock, J; Kjeldsen, L; Rothermund, K; Sisodia, SS; Borregaard, N; Corey, SJ. Identification of the presenilins in hematopoietic cells with localization of presenilin 1 to neutrophil and platelet granules. *Blood Cells Mol Dis* 2002; 28:28-38.

Miyashita, T; Reed, JC. Tumor suppressor p53 is a direct transcriptional activator of the human bax gene. *Cell* 1995; 80: 293–299.

Molina-Holgado, E; Vela, J.M; Arevalo-Martin, A; Almazan, G; Molina-Holgado, F; Borrell, J; Guaza, C. Cannabionoids promote oligodendrocyte progenitor survival: involvement of cannabinoid receptors and phospahatidylinositol-3 kinase/Akt signaling. *J Neurosc* 2002; 22: 9742-9753.

Moore, DJ; West, AB; Dawson, VL; Dawson, TM. Molecular pathophysiology of Parkinson's disease. *Annu Rev Neurosci* 2005; 28: 57-87.

Munro, S; Thomas, KL; Abu-Shaar, M. Molecular characterization of a peripheral receptor for cannabinoids. *Nature (London)* 1993; 365: 61-65.

McLellan, ME; Kajdasz, ST; Hyman, BT; Bacskai, BJ. *In vivo* imaging of reactive oxygen species specifically associated with thioflavine S-positive amyloid plaques by multiphoton microscopy. *J Neurosci* 2003; 23: 2212–2217.

McKenna, F; McLaughlin, PJ; Lewis, BJ; Sibbring, GC; Cummerson, JA; Bowen-Jones, D; Moots, RJ. Dopamine receptor expression on human T- and B-lymphocytes, monocytes, neutrophils, eosinophils and NK cells: a flow cytometric study. *J Neuroimmunol* 2002; 132 (1-2): 34-40.

Muñoz-Elias, G; Marcus, AJ; Coyne, TM; Woodbury, D; Black, IB. Adult bone marrow stromal cells in the embryonic brain: engrafment, migration, differentiation, and long-term survival. *J Neurosci* 2004; 24: 4585-4595.

Nagayama, T; Sinor, AD; Simon, RP; Chen, J; Graham, SH; Jin, K; Greenberg, DA. Cannabinoids and neuroprotection in global and focal cerebral ischemia and in neuronal cultures. *J Neurosci* 1999; 19: 2987-2995.

Ogawara, Y; Kishishita, S; Obata, T; Isazawa, Y; Suzuki, T; Tanaka, K; Masuyama, N; Gotoh, Y. Akt enhances Mdm2-mediated Ubiquitination and degradation of p53. *J Biol Chem* 2002; 277: 21843-21850.

Okada, H; Mak, WM. Pathways of apoptotic and non-apoptotic death in tumour cells. *Nat Rev Cancer* 2004; 4: 592-603.

Padgett, LW. Recent developments cannabinoid ligands. *Life Sci* 2005; 77: 1767-1798.

Pande, V; Ramos, MJ. NF-κB in human disease: current inhibitors and prospects for *De Novo* structure based deign of inhibitors. *Curr Med Chem* 2005; 12: 357-374.

Panyi, G; Varga, Z; Gaspar, R. Ion channels and lymphocyte activation. *Immunol Lett* 2004; 92(1-2): 55-66.

Pastor, P; Roe, CM; Villegas, A; Bedoya, G; Chakraverty, S; Garcia, G; Tirado, V; Norton, J; Rios, S; Martinez, M; Kosik, KS; Lopera, F; Goate, AM. Apolipoprotein Eepsilon4 modifies Alzheimer's disease onset in an E280A PS1 kindred. *Ann Neurol* 2003; 54: 163-169.

Pazos, MR; Nunez, E; Benito, C; Tolon, RM; Romero, J. Functional neuroanatomy of the endocannabinoid system. *Pharmacol Biochem Behav* 2005; 81: 239-247.

Perkin, DG. *An atlas of Parkinson's disease and related disorders*. New York, London: The Parthenon Publishing Group; 1998.

Perkins, ND; Felzien, LK; Betts, JC; Leung, K; Beach, DH; Nabel, GJ. Regulation of NF-kappaB by cyclin-dependent kinases associated with the p300 coactivator. *Science* 1997; 275: 523-527.

Pickart, CM. Mechanism underlying ubiquitination. *Ann Rev Biochem* 2001; 70: 503–533.

Pineda-Trujillo, N; Carvajal-Carmona, LG; Buritica, O; Moreno, S; Uribe, C; Pineda, D; Toro, M; Garcia, F; Arias, W; Bedoya, G; Lopera, F; Ruiz-Linares, A. A novel Cys212Tyr founder mutation in parkin and allelic heterogeneity of juvenile Parkinsonism in a population from North West Colombia. *Neurosci Lett* 2001; 298: 87-90.

Polster, BM; Fiskum, G. Mitochondrial mechanisms of neural cell apoptosis. *J Neurochem* 2004; 90: 1281-1289.

Polymeropoulos, MH; Lavedan, C; Leroy; E, Ide, SE; Dehejia, A; Dutra, A; Pike, B; Root, H; Rubenstein, J; Boyer, R; Stenroos, ES; Chandrasekharappa, S; Athanassiadou, A; Papapetropoulos, T; Johnson, WG; Lazzarini, AM; Duvoisin, RC, Di Iorio G, Golbe LI, Nussbaum RL. Mutation in the alpha-synuclein gene identified in families with Parkinson's disease. *Science* 1997; 276: 2045-2047.

Prasad, KN; Cole, WC; Prasad, KC. Risk factors for Alzheimer's disease: role of multiple antioxidants, non-steroidal anti-inflammatory and cholinergic agents alone or in combination in prevention and treatment. *J Am Coll Nutr* 2002; 21: 506-522.

Przedborski, S. Pathogenesis of nigral cell death in Parkinson's disease. *Parkinsonism Relat Disord.* 2005; 11 (Suppl 1): S3-S7.

Qian, YS; Buettner, GR. Iron and dioxygen chemistry is an important route to initiation of biological free radical oxidations: An electron paramagnetic resonance spin trapping study. *Free Rad Biol Med* 1999; 26: 1447-1456.

Ramírez, BG; Blázquez, C; Gomez Del Pulgar, T; Guzmán, M; de Ceballos, ML. Prevention of Alzheimer's disease pathology by cannabinoids: neuroprotection mediated by blockade of microglial activation. *J Neurosci* 2005; 25: 1904-1913.

Reed, JC. Mechanisms of Apoptosis. *Am J Pathol* 2000; 157: 1415-1430.

Reynolds, CH; Utton, MA; Gibb, GM; Yates, A; Anderton BH. Stress-activated protein kinase/ c-jun N-terminal kinase phosphorylates tau protein. *J Neurochem* 1997; 68: 1736-1744.

Report on carcinogens, Tenth Edition. Washington, DC: U.S. Department of Health and Human Service, National Toxicology Program, December 2002. On line at http://eph.niehs.nih.roc/toc10html.

Ricci, A; Amenta, F. Dopamine D5 receptors in human peripheral blood lymphocytes: a radioligand binding study. *J. Neuroimmunol* 1994; 53: 1-7.

Ricci, A; Veglio, F; Amenta, F. Radioligand binding characterization of putative dopamine D3 receptor in human peripheral blood lymphocytes with {3H} 7OH-DPAT. *J Neuroimmunol* 1995; 58: 139-144.

Ricci, A; Bronzetti, E; Felici, L; Tayebati, SK; Amenta, F. Dopamine D4 receptor in human peripheral blood lymphocytes: a radioligand binding assay study. *Neurosci Lett* 1997; 229: 130-134.

Riederer, P; Dirr, A; Goetz, M; Sofic, E; Jellinger, K; Youdim, MB. Distribution of iron in different brain regions and subcellular compartments in Parkinson's disease. *Ann Neurol* 1992; 32 Suppl: S101-S104.

Rosselli, MC; Ardila, AC; Moreno, SC; Standish, VC; Arango-Lasprilla, JC; Tirado, VM; Ossa, JM; Goate, AM; Kosik, KS; Lopera, F. Cognitive decline in patients with familial Alzheimer's disease associated with E280A presenilin-1 mutation: A longitudinal study. *J Clin Exp Neuropsychol* 2000; 22: 483-495.

Roth, KA. Caspases, apoptosis, and Alzheimer disease: causation, correlation, and confusion. *J Neuropathol Exp Neurol* 2001; 60: 829-38.

Sanchez, MG; Ruiz-Llorente, L; Sanchez, AM; Diaz-Leviada, I. Activation of phosphoinositide 3-kinase/PKB pathway by CB(1) and CB(2) cannabioid receptors expressed in prostate PC-3 cells. Involvement in Raf-1 stimulation and NGF induction. *Cell Signal* 2003; 15: 851-859.

Sanchez, JL; Buritica-Henao, O; Pineda Salazar, DA; Ribe, CS; Palacio-Baena, LG. Prevalence of Parkinson´s disease and Parkisnonism in a Colombian popoualation using the capture recapture methos. *Int J Neurosci* 2004; 113:175-182.

Sanchez-Ramos, JR. Neural cells derived from adult bone marrow and umbilical cord blood. *J Neurosci Res* 2002; 69: 880-893.

Sanders, VM. The role of norepinephrine and beta-2 adrenergic receptor stimulation in the modulation of Th1, Th2, and B lymphocyte function. *Adv Exp Med Biol* 1998; 437:269-278.

Sakahira, H; Enari, M; Nagata, S. Cleavage of CAD inhibitor in CAD activation and DNA degradation during apoptosis. *Nature* 1998; 391: 96-99.

Sakahira, H; Enari, M; Ohsawa, Y; Uchiyama, Y; Nagata, S. Apoptotic nuclear morphological change without DNA fragmentation. *Curr Biol* 1999; 9: 543-546.

Sarne, Y; Mechoulam, R. Cannabinoids: between neuroprotection and neurotoxicity. *Curr Drug Targets CNS Neurol Disord* 2005; 4: 677-684.

Serafeim, A; Gordon, J. The immune system gets nervous. *Curr Opin Pharmacol* 2001; 1: 398-403.

Siegel, GJ; Chauhan, NB. Neurotrophic factors in Alzheimer's and Parkinson's disease brain. *Brain Res Brain Res Rev* 2000; 33 (2-3): 199-227.

Sizemore, N; Lerner, N; Dombrowski, N; Sakurai, H; Stark, GR. Distinct roles of the Ikappa B kinase alpha and beta subunits in liberating nuclear factor kappa B (NF-kappa B) from Ikappa B and in phosphorylating the p65 subunit of NF-kappa B. *J Biol Chem* 2002; 277: 3863-3869.

Sofic, E; Paulus, W; Jellinger, K; Riederer, P; Youdim, MB. Selective increase of iron in substantia nigra zona compacta of parkinsonian brains. *J Neurochem* 1991; 56: 978-982.

Sunada, Y; Saito, F; Matsumura, K; Shimizu, T. Differential expression of the parkin gene in the human brain and peripheral leukocytes. *Neurosci Lett* 1998; 254: 180-182.

Shishodia, S; Aggarwal, BB. Nuclear factor kB: a friend or a foe in cancer?. *Biochem Pharmacol* 2004; 68: 1071-1080.

Shimura, H; Hattori, N; Kubo, S; Mizuno, Y; Asakawa, S; Minoshima, S; Shimizu, N; Iwai, K; Chiba, T; Tanaka, K; Suzuki, T. Familial Parkinson disease gene product, parkin is a ubiquitin-protein ligase. *Nat Genet* 2000; 25:302-305.

Skaper, SD; Buriani, A; Dal Toso, R; Petrelli, L; Romanello, S; Facci, L; Leon, A. The Aliamide palmitoylethanolamine and cannabinoids, but not anandamide, are protective in a delayed postglutamate paradigm of exocitotoxic death in cerebellar granule neurons. *Proc Natl Acad Sci USA* 1996; 93: 3984-3989.

Stefulj, J; Jernej, B; Cicin-Sain, L; Rinner, I; Schauenstein, K. mRNA expression of serotonin receptors in cells of the immune tissues of the rat. *Brain Behav Immun* 2000; 14:219-224.

Strasser, A; O'Connor, L; Dixit, VM. Apoptosis signalling. *Annu Rev Biochem* 2000; 69: 217-245.

Tabner, BJ; Turnbull, S; El-Agnaf, OM; Allsop, D. Formation of hydrogen peroxide and hydroxyl radicals from Aβ and α-synuclein as a possible mechanism of cell death in Alzheimer's disease and Parkinson's disease. *Free Radic Biol Med* 2002; 32: 1076-1083.

Tabner, BJ; El-Agnaf, OM; Turnbull, S; German, MJ; Paleologou, KE; Hayashi, Y; Cooper, LJ; Fullwood, NJ; Allsop, D. Hydrogen peroxide is generated during the very early stages of aggregation of the amyloid peptides implicated in Alzheimer disease and familial British dementia. *J Biol Chem* 2005a; 280: 35789-35792.

Tabner, BJ; Turnbull, S; Fullwood, NJ; German, M; Allsop, D. The production of hydrogen peroxide during early-stage protein aggregation: a common pathological mechanism in different neurodegenerative diseases? *Biochem Soc Trans.* 2005b; 33 (Pt 4): 548-550.

Takada, Y; Mukhopadhyay, A; Kundu, GC; Mahabeleshwar, GH; Singh, S; Aggarwal, BB. Hydrogen peroxide activates NF-kappa B through tyrosine hosphorylation of I kappa B alpha and serine phosphorylation of p65: evidence for the involvement of I kappa B alpha kinase and Syk protein-tyrosine kinase. *J Biol Chem.* 2003; 278: 24233-24241.

Takanashi, M; Mochizuchi, H; Yokomizo, K; Hattori, N; Mori, H; Yamamura, Y; Mizuno, Y. Iron accumulation in the substantia nigra of autosomal recessive juvenile parkinsonism (ARJP). *Parkinsonism Relat Disord* 2001; 7: 311–314.

Takahashi, N; Kariya, S; Hirano, M; Ueno, S. Two novel spliced presenilin 2 transcripts in human lymphocyte with oxidant stress and brain. *Mol Cell Biochem* 2003; 252(1-2): 279-283.

Takeyama, N; Miki, S; Hirakawa, A, Tanaka, T. Role of the mitochondrial permeability transition and cytochrome C release in hydrogen peroxide-induced apoptosis. *Exp Cell Res* 2002; 274: 16-24.

Takuma, K; Yan, SS; Stern, DM; Yamada, K. Mitochondrial dysfunction, endoplasmic reticulum stress, and apoptosis in Alzheimer's disease. *J Pharmacol Sci* 2005; 97: 312-316.

Tanaka, K; Suzuki, T; Chiba, T; Shimura, H; Hattori, N; Mizuno, Y. Parkin is linked to the ubiquitin pathway. *J Mol Med.* 2001; 79: 482-494.

Tapson, VF; Schenetzler, B; Pilch, PF; Center, DM; Berman, JS. Structural and functional characterization of the human T lymphocyte receptor for insulin-like growth factor I in vitro. *J. Clin Invest* 1988; 82: 950-957.

Tillakaratne, NJ; Medina-Kauwe, L; Gibson, KM. Gamma Aminobutyric acid (GABA) metabolism in mammalian neural and non-neural tissues. *Comp Biochem Physiol A Physiol* 1995; 112(2): 247-263.

Tomita, T; Iwatsubo, T. gamma-secretase as a therapeutic target for treatment of Alzheimer's disease. *Curr Pharm Des* 2006; 12: 661-670.

Thompson, LA; Bronson, JJ; Zusi, FC. Progress in the discovery of BACE inhibitors. *Curr Pharm Des* 2005; 11: 3383-3404.

Thakur, GA; Duclos, RI Jr; Makriyannis, A. Natural cannabinoids: Templates for drug discovery. *Life Sci* 2005; 78: 454-466.

Thorburn, A. Death receptor-induced cell killing. *Cellular Signalling* 2004; 16: 139-144.

Urbanski, NK; Beresewicz, A. Generation of *OH initiated by interaction of Fe^{2+} and Cu^{2+} with dioxygen: comparison with the Fenton chemistry. *Acta Biochem Pol* 2000; 47: 951-962.

Van der Stelt, M; Di Marzo, V. Cannabinoid receptors and their role in neuroprotection. *Neuromolecular Med* 2005; 7(1-2): 37-50.

Van Sickle, MD; Duncan, M; Kingsley, PJ; Mouihate, A; Urbani, P; Mackie, K; Stella, N; Makriyannis, A; Piomelli, D; Davison, JS; Marnett, LJ; Di Marzo, V; Pittman, QJ; Patel, KD; Sharkey, KA. Identification and functional characterization of brainstem cannabinoid CB2 receptors. *Science* 2005; 310:329-332.

Velez-Pardo, C; Jimenez Del Rio, M; Ebinger, G; Vauquelin, G. Manganese and copper promote the binding of dopamine to "serotonin binding proteins" in bovine frontal cortex. *Neurochem Int* 1995; 26: 615-622.

Velez-Pardo, C; Jimenez Del Rio, M; Verschueren, H; Ebinger, G; Vauquelin, G. Dopamine and iron induced apoptosis in PC12 cells. *Pharmacol Toxicol* 1997; 80: 76-84.

Velez-Pardo, C; Jimenez Del Rio, M; Ebinger, G; Vauquelin, G. Monoamine and iron-related toxicity: from "serotonin binding proteins" to lipid peroxidation and apoptosis in PC12 cells. *Gen Pharmacol* 1998a; 31: 19-24.

Velez-Pardo, C, Jimenez Del Rio, M; Lopera, F. Familial Alzheimer's disease: Oxidative stress, β-amyloid, presenilins & cell death. *Gen Pharmacol* 1998b; 31: 675-681.

Velez-Pardo, C; Garcia Ospina, G; Jimenez Del Rio, M. $A\beta_{[25-35]}$ peptide and iron promote apoptosis in lymphocytes by a common oxidative mechanism: involvement of hydrogen peroxide (H_2O_2), caspase-3, NF-kappa B, p53 and c-Jun. *NeuroToxicology* 2002; 23: 351-365.

Velez-Pardo, C; Jimenez Del Rio, M. Avoidance of $A\beta_{[25-35]}$/ (H_2O_2) -induced apoptosis in lymphocytes by the cannabinoid agonists CP55,940 and JWH-015 via receptor-independent and PI3K-dependent mechanisms: Role of NF-κB and p53. *Medicinal Chem* 2006; 2: 471-479.

Venderova, K; Ruzicka, E; Vorisek, V; Visnovsky P. Survey on cannabis use in Parkinson's disease: subjective improvement of motor symptoms. *Mov Disord* 2004; 19: 1102-1106.

Vetrivel, KS; Thinakaran, G. Amyloidogenic processing of beta-amyloid precursor protein in intracellular compartments. *Neurology* 2006; 66 (2 Suppl 1): S69-S73.

Viña, J; Lloret, A; Orti, R; Alonso, D. Molecular bases of the treatment of Alzheimer's disease with antioxidants: prevention of oxidative stress. *Mol Aspects Med* 2004; 25: 117-123.

Wadgaonkar, R; Phelps, KM; Haque, Z; Williams, AJ; Silverman, ES; Collins, T. CREB-binding protein is a nuclear integrator of nuclear factor-kappaB and p53 signaling. *J Biol Chem* 1999; 274:1879-1882.

Webster, GA; Perkins, ND. Transcriptional cross talk between NF-kappaB and p53. *Mol Cell Biol* 1999; 19: 3485-3495.

Wirths, O; Multhaup, G; Bayer, T. A modified β-amyloid hypothesis: intraneural accumulation of the β-amyloid peptide- the first step of a fatal cascade. *J Neurochem* 2004; 91:513-520.

Wischik, CM; Theuring, F; Harrington, ChR. The molecular basis of tau protein pathology in Alzheimer's disease and related neurodegenerative dementias. In: Dawbarn, D; Allen, SJ., editors. *Neurobiology of Alzheimer's disease*. Oxford: Oxford University Press; 2001; pp. 103-206.

Wu, H; Lozano, G. NF-κB activation of p53. *J Biol Chem* 1994; 269: 20067–20074.

Wu, SS; Frucht, SJ. Treatment of Parkinson's disease: what's on the horizon? *CNS Drugs* 2005; 19: 723-743.

Wyllie, AH. Glucocorticoid-induced thymocytes apoptosis associated with endogenous endonuclease activation. *Nature* 1980; 284: 555-556.

Yager, JD; Davidson, NE. Estrogen carcinogenesis in breast cancer. *N Engl J Med* 2006; 354:270-282.

Yan, M; Dai, T; Deak, JC; Kyriakis, JM; Zon, LI; Woodgett; JR; Templeton, DJ. Activation of stress-activated protein kinase by MEKK1 phosphorylation of its activator SEK1. *Nature* 1994; 372: 798–800.

Yang, D; Tournier, C; Wysk, M; Lu, H-T; Xu, J; Davis, RJ; Richard, A. Targeted disruption of the MKK4 gene causes embryonic death, inhibition of c-Jun NH2-terminal kinase activation, and defects in AP-1 transcriptional activity. *Proc Natl Acad Sci USA* 1997; 94: 3004–3009.

Yaron, A; Hatzubai, A; Davis, M; Lavon, I; Amit, S; Manning, AM; Andersen, JS; Mann, M; Mercurio, F; Ben-Neriah, Y. Identification of the receptor component of the kappaBalpha-ubiquitin ligase. *Nature* 1998; 396: 590-594.

Yu, J; Zhang, L. The transcriptional targets of p53 in apoptosis control. *Biochem Biophys Res Commun* 2005; 331: 851-858.

Zamzami, N; Kroemer, G. p53 in apoptosis: An introduction. *Biochem Biophys Res Commun* 2005; 331: 685-687.

Zimmermann, KC; Bonzon, C; Green, DR. The machinery of programmed cell death. *Pharmacol Ther* 2001; 92: 57-70.

Zou, H; Li, Y; Liu, X; Wang, X. An APAF-1 cytochrome c multimeric complex is a functional apoptosome that activates procaspase-9. *J Biol Chem.* 1999; 274: 11549-11556.

Zhang, Y; Dawson, VL; Dawson, TM. Parkin: clinical aspects and neurobiology. *Clin Neurosci Res* 2001; 1: 467– 482.

Zhivotovsky, B. Caspases: the enzymes of death. *Essays Biochem* 2003; 39: 25-40.

In: Focus on Cell Apoptosis Research
Editor: Daniel W. Cho, pp. 119-135

ISBN: 1-60021-433-9

Chapter IV

ROLE OF ADVANCED GLYCATION END PRODUCTS (AGES) IN CELL GROWTH AND DEATH

Sho-ichi Yamagishi[1*], Kazuo Nakamura[1], Takanori Matsui[1], Hiroyoshi Inoue[2], Teruhiko Makino[3], Tadamichi Shimizu[3] and Masayoshi Takeuchi[4]

[1]Department of Internal Medicine, [2]Radioisotope Institute for Basic and Clinical Medicine, Kurume University School of Medicine, Kurume 830-0011;
[3]Department of Dermatology, Faculty of Medicine, University of Toyama, 930-0194;
[4]Department of Pathophysiological Science, Hokuriku University, Kanazawa 920-1181, Japan

ABSTRACT

Non-enzymatic modification of proteins by reducing sugars, a process that is also known as Maillard reaction, leads to the formation of advanced glycation end products (AGEs) *in vivo*. There is a growing body of evidence to show that formation and accumulation of AGEs progress during normal aging, and at an extremely accelerated rate under diabetes, thus being involved in the pathogenesis of various diseases such as diabetic vascular complications, neurodegenerative diseases and cancers. Recently, engagement of their receptor, RAGE, is shown to activate its down-stream signaling and subsequently evoke proliferative or proapoptotic responses in various types of cells. Therefore, in this study, we summarize the role for AGEs-RAGE system in the pathogenesis of various disorders, especially focusing on cell growth and death. We

* Correspondence concerning this article should be addressed to Sho-ichi Yamagishi MD, PhD, Department of Internal Medicine III, Kurume University School of Medicine, Kurume 830-0011, Japan. Tel; +81-942-31-7580; Fax; +81-942-31-7707; E-mail; shoichi@med.kurume-u.ac.jp.

further review here the potential therapeutic strategy that may specially target the AGEs-RAGE system in these devastating disorders.

Keywords: advanced glycation end products, apoptosis, cancer, diabetic vascular complication, oxidative stress, neurodegenerative disease

Introduction

The modification, aggregation, and deposition of proteins are a prominent part of many pathological processes and can play a direct role in tissue damage. The pathological role of the non-enzymatic modification of proteins by reducing sugars, a process that is known as glycation (also called the "Maillard reaction"), has become increasingly evident in various types of diseases [1,2]. It is now well established that early glycation products undergo progressive modification over time *in vivo* to the formation of irreversible cross-links, after which these molecules are termed "advanced glycation end-products (AGEs)". AGEs have been implicated in the development of many of the pathological sequelae of diabetes and aging, such as atherosclerosis and diabetic microangiopathy [1,2]. Furthermore, there is a growing body of evidence that RAGE is a signal-transducing receptor for AGEs and that engagement of RAGE by AGEs evokes proliferative or proapoptotic responses in vascular wall cells as well [3-7]. Recently, it has become clear that the AGE-RAGE system also has a role in neurodegenerative diseases and cancers including Alzheimer's disease (AD), Parkinson's disease, Creutzfeldt-Jakob disease, amyotrophic lateral sclerosis, and melamoma [8-12]. In this study, we summarize the role for AGEs-RAGE system in the pathogenesis of various disorders, especially focusing on cell growth and death. We further review here the potential therapeutic strategy that may specially target the AGEs-RAGE system in these devastating disorders.

AGEs

Reactive derivatives from non-enzymatic glucose-protein condensation reactions, as well as lipids and nucleic acids exposed to reducing sugars, form a heterogeneous group of irreversible adducts called "AGEs". AGEs were originally characterized by a yellow-brown fluorescent color and an ability to form cross-links with and between amino groups [13-15], but the term is now used for a broad range of advanced products of the glycation process (also called the "Maillard reaction"), including N-carboxymethyllysine (CML) and pyrraline, which show neither color nor fluorescence and do not cross-link proteins. CML can be formed from the precursors glyoxal and glycolaldehyde by an intra-molecular Cannizzaro reaction, a process that is largely independent of glucose autoxidation [16]. The concept that CML is a marker of oxidation rather than glycation has recently attracted support [17].

The formation of AGEs *in vitro* and *in vivo* is dependent on the turnover rate of the chemically modified target, the time available, and the sugar concentration. The structures of the various cross-linked AGEs that are generated *in vivo* have not yet been completely determined. Because of their heterogeneity and the complexity of the chemical reactions involved, only some AGEs have been structurally characterized *in vivo*. The structural identity of AGEs with cytotoxic properties remains unknown.

DIABETIC RETINOPATHY

(1) Pathophysiology

Diabetic retinopathy is a common and potentially devastating microvascular complication in diabetes and is a leading cause of acquired blindness among the people of occupational age. The lesions of diabetic retinopathy can be grouped into those associated with background, preproliferative and proliferative retinopathy [18,19].

The earliest histopathological hallmark of diabetic retinopathy is loss of pericytes [20]. In parallel with loss of pericytes, several characteristic changes including thickening of the basement membrane, hyperpermeability, and formation of microaneurysm are observed [21]. These structural and functional abnormalities are followed by microvascular occlusion in the retinas, which ultimately progresses to proliferative changes associated with neovascularization [22]. Since retinal pericytes play an important role in the maintenance of microvascular homeostasis [23-25], it has been postulated that many of these pathological changes are the consequent of loss of pericytes.

(2) Role of AGEs

Retinal pericytes accumulate AGEs during diabetes, which would be expected to have a detrimental influence on pericyte survival and function [26,27]. AGEs are toxic to retinal pericytes *in vitro*. We have found that AGEs not only induce growth retardation and apoptotic cell death of, but also exert an immediate toxicity to, cultured retinal pericytes [28,29]. Antisense DNA complementary mRNA coding for a receptor for AGEs (RAGE) reversed the AGE-induced decrease in viable cell number of pericytes, while overexpression of RAGE potentiated the deleterious effects of AGEs. These results suggest that the AGE effects on pericytes could be mediated through the interactions with RAGE. Further, we found that ROS generation elicited by the AGE-RAGE interaction mediated these deleterious effects of AGEs [30,31]. Indeed, *N*-acetylcysteine (NAC) or pigment epithelium-derived factor (PEDF), one of the superfamily of serine protease inhibitors with potent neuronal differentiating activity in human retinoblastoma cells, was found to protect cultured retinal pericytes against the AGE-induced apoptosis via its anti-oxidative properties [30].

AGEs up-regulated RAGE mRNA levels in pericytes through the intracellular ROS generation as well [31]. These positive feedback loops transduce the AGE signals further, thereby exacerbating the cytopathic effects of AGEs on pericytes. With regard to a source of ROS generation elicited by AGEs, we have recently found that beraprost sodium, a prostacyclin analogue, or forskolin, an activator of adenylate cyclase, protects against the AGE-induced pericyte apoptosis by suppressing ROS generation and subsequent RAGE overexpression [31]. Since cyclic AMP elevating agents were known to block ROS generation in neutrophils by inhibiting NADPH oxidase activity [32], NADPH oxidase might be a source of ROS production elicited by AGEs and also be a target of beraprost sodium.

Pericyte dysfunction has also been considered one of the characteristic changes of early phase of diabetic retinopathy. AGEs act on pericytes to stimulate vascular endothelial growth factor (VEGF) expression both *in vitro* and *in vivo* [29,33]. VEGF is a specific mitogen to ECs, also known as vascular permeability factor, and is generally thought to be involved in the pathogenesis of proliferative diabetic retinopathy. Indeed, some clinical observations have demonstrated that VEGF level in ocular fluid is positively correlated with the activity of neovascularization in diabetic retinopathy [34,35]. Recently, VEGF level was also found to be associated with the breakdown of the blood-retinal barrier, thus being involved in microvascular hyperpermeability in background retinopathy [36-38]. These observations suggest that AGE-RAGE-elicited ROS generation might be involved in the development of early diabetic retinopathy by inducing VEGF expression in pericytes as well.

The local RAS is activated under diabetes [39]. We have recently found that angiotensin II (AII) stimulated intracellular ROS generation in retinal pericytes through an interaction with type 1 receptor. Further, AII decreased DNA synthesis and simultaneously up-regulated VEGF mRNA levels in pericytes, both of which were blocked by treatment with telmisartan, a newly developed AII type 1 receptor antagonist, or an anti-oxidant NAC [40,41]. These results suggest that AII-type 1 receptor interaction could induce pericyte loss and dysfunction through intracellular ROS generation, thus being involved in diabetic retinopathy. Furthermore, we have very recently found that AII potentiates the deleterious effects of AGEs on pericytes by inducing RAGE protein expression [42-44]. The observations suggest the crosstalk between the AGE-RAGE system and the RAS in the pathogenesis of pericyte loss and dysfunction in diabetic retinopathy. Since AII induces the VEGF receptor KDR (kinase insert domain-containing receptor) expression in retinal microvascular ECs, the retinal RAS might augment the permeability- and angiogenesis-inducing activity of VEGF, thus implicated in the progression of diabetic retinopathy as well [45].

AGEs directly stimulate growth and tube formation of microvascular ECs, the key steps of angiogenesis [46-49]. The angiogenic activity of AGEs was significantly enhanced in RAGE-overexpressed ECs, while it was completely inhibited by treatment of antisense DNA against RAGE mRNA. The observations suggest that AGEs could elicit angiogenesis through the interaction with RAGE. We found that it was autocrine VEGF production in ECs that mainly mediated the angiogenic activity of AGEs. Although the molecular mechanisms of VEGF overexpression elicited by AGEs are not fully understood, our recent investigation has shown that the AGE-RAGE interaction could increase VEGF gene transcription in microvascular ECs by NADPH oxidase-mediated ROS generation and the subsequent nuclear factor-κB (NF-κB) activation via Ras-mitogen activated protein kinase (MAPK) pathway [47-51]. Recently, we have found that PEDF inhibits the AGE-induced angiogenesis and EC activation by suppressing NADPH oxidase activity [52,53]. Since oxidative stress down-regulates PEDF expression in vascular cells [54-57] and that the levels of vitreal PEDF are decreased in angiogenic eye diseases such as proliferative diabetic retinopathy [58], substitution of PEDF may disrupt inappropriate retinal cell responses to AGEs, thus being a promising strategy for treatment of patients with diabetic retinopathy.

Microthrombosis formation contributes to capillary obliteration and retinal ischemia, thus being involved in the progression of diabetic retinopathy [59]. AGEs inhibit prostacyclin production and induce plasminogen activator inhibitor-1 (PAI-1) in microvascular ECs through an interaction with RAGE [60,61]. These observations suggest that AGE have the ability to cause platelet aggregation and fibrin stabilization, resulting in a predisposition to thrombogenesis and thereby contributing to the promotion of diabetic retinopathy. Retinal ischemia due to microthrombus formation may trigger VEGF expression in retinal cells, thus further promoting diabetic retinopathy [62-64]. Since AGEs decrease the intracellular cyclic AMP concentrations in ECs and that cyclic AMP agonists such as beraprost sodium and forskolin reduce the AGE-induced PAI-1 production, cyclic AMP elevating agents may have a therapeutic potential in the treatment of diabetic retinopathy.

DIABETIC NEPHROPATHY

(1) Pathophysiology

Diabetic nephropathy is a leading cause of end-stage renal failure. Development of diabetic nephropathy is characterized by glomerular hyperfiltration and thickening of glomerular basement membranes, followed by an expansion of extracellular matrix in mesangial areas. Diabetic nephropathy ultimately progresses to glomerular sclerosis associated with renal dysfunction [65]. Proximal tubular cell atrophy and tubulointerstitial fibrosis are also important histopathological changes in terms of renal prognosis in diabetic nephropathy [66].

(2) Role of AGEs

We have recently found that AGEs induce apoptotic cell death and VEGF expression in human cultured mesangial cells, as the case in pericytes [67]. In this study, AGEs induced mesangial cell apoptotis through p53-dependent Bax overexpression. Since superoxide was reported to induce p53 and Bax accumulation during apoptosis and that NAC completely inhibited the pro-apoptotic signals of AGEs, the AGE-elicited ROS generation might also be involved in mesangial cell apoptosis. Mesangial cells occupy a central anatomical position in the glomerulus, playing crucial roles in maintaining structure and function of glomerular capillary tufts [67]. They actually provide structural support for capillary loops and modulate glomerular filtration by its smooth muscle activity [68-70]. Therefore, the AGE-induced mesangial apoptosis and dysfuction may contribute in part to glomerular hyperfiltration, an early renal dysfunction in diabetes. Recently, antibodies against VEGF have been found to improve hyperfiltration and albuminuria in experimental diabetes, supproting our speculation [71].

AGEs stimulated ROS generation and subsequently inhibited *de novo* protein synthesis in proximal tubular cells [72]. Since AGEs did not affect DNA synthesis and viable cell numbers in these cells, AGEs might be involved in the pathogenesis of tubular atrophy, one of the representative tubulointerstitial changes in diabetic nephropathy. Further, AGEs induced transforming growth factor-β (TGF-β) in proximal tubular cells, both at mRNA and protein levels. TGF-β is a major etiologic agent in the pathogenesis of tubulointerstitial fibrosis in diabetic nephropathy [73-77]. These observations suggest the involvement of AGEs in proximal tubular cell injury in diabetes.

RAGE-overexpressing diabetic mice have been recently found to show progressive glomerulosclerosis with renal dysfunction, compared with diabetic littermates lacking the RAGE transgene [78]. Further, AGEs including glycoxidation or lipoxidation products such as N^{ε}-(carboxymethyl)lysine, pentosidine, malondialdehyde-lysine accumulate in the expanded mesangial matrix and thickened glomerular basement membranes of early diabetic nephropathy, and in nodular lesions of advanced disease [79]. Taken together, these findings further support the concept that oxidative stress generation elicited by the AGE-RAGE system could be involved in diabetic nephropathy [80,81].

A number of studies have demonstrated that aminoguanidine decreased AGE accumulation and plasma protein trapping in the glomerular basement membrane [4,5]. In streptozocin-induced diabetic rats, aminoguanidine treatment for 32 weeks dramatically reduced the level of albumin excretion and prevented the development of mesangial expansion [82]. Furthermore, aminoguanidine treatment was found to prevent albuminuria in diabetic hypertensive rats without affecting blood pressure [83].

We have found that OPB-9195, a synthetic thiazolidine derivative and novel inhibitor of AGEs, prevented the progression of diabetic nephropathy by lowing serum concentrations of AGEs and their deposition of glomeruli in Otsuka-Long-Evans-Tokushima-Fatty rats, a type II diabetes mellitus model animal [84]. OPB-9195 was also found to retard the progression of diabetic nephropathy by blocking type IV collagen production and suppressing overproduction of two growth factors, TGF-β and VEGF.

Recently, Baynes *et al.* reported that pyridoxamine inhibited the progression of renal disease, and decreases hyperlipidemia and apparent redox imbalances in diabetic rats [85]. Pyridoxamine and aminoguanidine had similar effects on parameters measured, supporting a mechanism of action involving AGE inhibition [85]. Although the results of AGE inhibitors in animal models of diabetic nephropathy are promising, effectiveness of these AGE inhibitors must be confirmed by multicenter, randomized, double-blind clinical studies.

There is a growing body of evidence to suggest that that the RAS plays an important role in the regulation of glomerular hemodynamics and renal expression of cytokines, thus being involved in glomerular hyperfiltration and mesangial expansion in diabetic nephropathy [86,87]. AII stimulates TGF-β expression in mesangial cells as well, which may contribute to the development of glomerular sclerosis in diabetes [88-91]. Angiotensin converting enzyme inhibitors attenuate the development of glomerulosclerosis and tubulointerstitial fibrosis in several animal models and slow disease progression in type 1 diabetic patients [92-94]. Further, three major renal outcome studies have provided evidence to recommend AII type 1 receptor antagonists as the treatment of choice in type 2 diabetic patients with nephropathy [95-97].

There are several reports to suggest an intreaction between the RAS and AGEs in diabetic nephropathy. Angiotensin converting enzyme inhibition reduces the accumulation of renal and serum AGEs, probably via effects on oxidative pathways [98]. Long-term treatment with AII receptor 1 antagonist may exert salutary effects on AGEs levels in the rat remnant kidney model, probably due to improved renal function [99]. Ramipril administration has been recently shown to result in a mild decline of fluorescent non-carboxymethllysine-AGEs and malondialdehyde concentrations in non-diabetic nephropathy patients [100]. In addition, we have very recently found that AGE-RAGE-mediated ROS generation activates TGF-β-Smad signaling and subsequently induces mesangial cell hypertrophy and fibronectin synthesis by autocrine production of AII [101]. Taken together, these data may provide an important mechanistic link between metabolic and haemodynamic factors in promoting the development and progression of diabetic nephropathy.

ALZHEIMER'S DISEASE (AD)

(1) Role of AGEs

Recently discovered links between type 2 diabetes and AD are indications of AGE and increased AGE receptor expression in brains of patients with AD [102,103]. Ott *et al.* reported that diabetes mellitus almost doublet the risk of dementia and AD [104]. As AGEs are elevated and involved in vascular derangements in diabetes, diabetes might influence AD brain pathology as well.

The possibility of the involvement of glycation in AD was first suggested in several reports published successively during 1994-1995 [105-107]. Senile plaques and neurofibrillary tangles were positively stained with anti-pyrraline and anti-pentosidine antibodies. Further, Sasaki *et al.* reported that senile plaques and neurofibrillary tangles were positively stained by an antiserum against glucose-derived AGEs [108]. These findings support the possibility that glycation may be involved in the initial stage of plaque formation. In addition, we have recently found that AGE- and RAGE-positive granules were present in the perikaryon of hippocampal neurons (especially forms CA3 and CA4) in all subjects (AD and DM patients) [109]. In AD brains, most of astrocytes (approximately 70-80 %) contained both AGE- and RAGE-positive granules, thus suggesting the involvement of the system in AD.

In *vitro*-studies also support the active participation of AGEs in AD; AGEs derived from diabetic patients on hemodialysis cause cortical neuronal cell death [8]. In our study, antibodies directed against glyceraldehydes-derived AGEs neutralized the neurotoxic properties of AGEs [8]. These observations suggest a causal role of this type of AGEs in AD. We have recently found that this type of AGEs have the strongest binding affinity to RAGE *in vitro* [11].

The enzyme glyceraldehyde-3-phosphate dehydrogenase (GAPDH) is of particular interest because it has been reported to interact strongly with proteins implicated in the pathogenesis of AD [110]. Further, accumulating evidence suggests that GAPDH protein is multifunctional, being involved in apoptosis as well [111,112]. Since glyceraldehyde-derived AGEs decrease GAPDH activity in cultured cortical neuronal cells [8], AGEs may elicit neuronal cell death, at least in part, via suppression of GAPDH activity. Ishitani *et al.*

reported that GAPDH could play an important role in neuronal cell apoptosis by working as a "killing protein" [113], further supporting our speculation.

MELANOMA

(1) Role of AGEs

Melanoma has become a major public health problem in many countries. Since the mid 1960s, melanoma incidence has risen by 3–8% per year in most people of European origin, with the greatest increases in elderly men [114]. Despite this increase, and an overall rise in mortality due to melanoma, the survival rate has improved substantially. Roughly 60% of those diagnosed with melanoma in the 1960s died of the disease, compared with just 11% more recently, an improvement attributed mainly to early detection [115].

New blood vessel formation is an important and prominent feature of human cutaneous melanomas, indicating that these tumors have angiogenic activity [116,117]. A rise in the mean vascular density was correlated with melanoma progression in subsequent histochemical studies of cutaneous melanomas [118,119], and a number of retrospective histological studies have reported an inverse correlation between tumor microvessel density and disease-free and overall survival of melanoma patients [120,121]. A prospective study of cutaneous melanoma patients reported that tumor vascularity, as determined by analysis of routine histologic staining, was the most important determinant of overall patient survival [122].

Recently, we have demonstrated that RAGE was expressed in human malignant melanoma cells. Furthermore, glyceraldehyde-derived AGEs enhanced proliferation, migration and invasion of this malignant melanoma cell line *in vitro* [12]. In addition, treatment by intraperitoneal injection of neutralizing anti-RAGE antibody reduced tumor formation, prolonged survival rate and inhibited lung metastases in nude mice [12]. As described above, AGEs could stimulate various steps of angiogenesis both *in vitro* and *in vivo* [33, 46-49]. Further, AGE formation in the extracellular matrix of skin is accelerated by ultraviolet-induced oxidation [123], whereas AGEs themselves generate active oxygen species in the skin during ultraviolet irradiation [124]. These observations suggest the presence of a vicious cycle of AGEs formation in the skin during ultraviolet irradiation.

CONCLUSION

There is a growing body of evidence that the AGE-RAGE system is implicated in the pathogenesis of various diseases such as diabetic vascular complications, neurodegenerative disorders, and melanoma. Therefore, inhibition of AGE formation or blockade of the RAGE signaling may be a promising therapeutic target for the treatment of AGE-related disorders.

ACKNOWLEDGEMENT

This study was supported in part by Grants of Venture Research and Development Centers from the Ministry of Education, Cultures, Sports, Science and Technology, Japan (S. Y.).

REFERENCES

[1] Yamagishi S, Nakamura K, Imaizumi T. Advanced glycation end products (AGEs) and diabetic vascular complications. *Curr Diabetes Rev* 2005; 1: 93-106.

[2] Kikuchi S, Shinpo K, Takeuchi M, Yamagishi S, Makita Z, et al. Glycation- A sweet prompter for neuronal death. *Brain Res Rev* 2003; 41: 306-23.

[3] Yamagishi S, Takeuchi M, Inagaki Y, Nakamura K, Imaizumi T. Role of advanced glycation end products (AGEs) and their receptor (RAGE) in the pathogenesis of diabetic microangiopathy. *Int J Clin Pharm Res* 2003; 23: 129-34.

[4] Yamagishi S, Nakamura K, Takeuchi M, Imaizumi T. Molecular mechanism for accelerated atherosclerosis in diabetes and its potential therapeutic intervention. *Int J Clin Pharm Res* 2004; 24: 129-34.

[5] Takeuchi M, Yamagishi S, Iwaki M, Nakamura K, Imaizumi T. Advanced glycation end product (AGE) inhibitors and their therapeutic implications in diseases. *Int J Clin Pharm Res* 2004; 24: 95-101.

[6] Yamagishi S, Imaizumi T. Diabetic vascular complications: pathophysiology, biochemical basis and potential therapeutic strategy. *Curr Pharm Des* 2005; 11: 2279-99.

[7] Yamagishi S, Nakamura K, Jinnouchi Y, Takenaka K, Imaizumi T. Therapeutic implications of blockers of advanced glycation end products (AGEs)-their receptor (RAGE) system. *Int J Pharmacology* 2005; 1: 203-9.

[8] Takeuchi M, Watai T, Sasaki N, Choei H, Iwaki M, et al. Neurotoxicity of acetaldehyde-derived advanced glycation end-products for cultured cortical neurons. *J Neuropathol Exp Neurology*. 2003; 62: 486-96.

[9] Sasaki N, Takeuchi M, Chowei H, Kikuchi S, Hayashi Y, et al. Advanced glycation end products (AGE) and their receptor (RAGE) in the brain of patients with Creutzfeldt-Jakob disease with prion plaques. *Neurosci Lett* 200; 326: 117-20.

[10] Choei H, Sasaki N, Takeuchi M, Yoshida T, Ukai W, et al. Glyceraldehyde-derived advanced glycation end products in Alzheimer's disease. *Acta Neuropathol (Berl)* 2004; 108: 189-93.

[11] Takeuchi M, Kikuchi S, Sasaki N, Suzuki T, Wati T, et al. Involvement of advanced glycation end-products (AGEs) in Alzheimer's disease. *Cur Alzheimer Res* 2004; 1: 39-46.

[12] Abe R, Shimizu T, Sugawara H, Watanabe H, Nakamura H, et al. Regulation of human melanoma growth and metastasis by AGE-AGE receptor interactions. *J Invest Dermatol* 2004; 122: 461-7.

[13] Brownlee M, Cerami A, Vlassara H. Advanced glycosylation end products in tissue and the biochemical basis of diabetic complications. *N Engl J Med* 1988; 318: 1315-21.

[14] Dyer DG, Blackledge JA, Thorpe SR, Baynes JW. Formation of pentosidine during nonenzymatic browning of proteins by glucose. Identification of glucose and other carbohydrates as possible precursors of pentosidine in vivo. *J Biol Chem* 1991; 266: 11654-60.

[15] Grandhee SK, Monnier VM. Mechanism of formation of the Maillard protein cross-link pentosidine. Glucose, fructose, and ascorbate as pentosidine precursors. *J Biol Chem* 1991; 266: 11649-53.

[16] Glomb MA, Monnier VM. Mechanism of protein modification by glyoxal and glycolaldehyde, reactive intermediates of the Maillard reaction. *J Biol Chem* 1995; 70: 10017-26.

[17] Requena JR, Fu MX, Ahmed MU, Jenkins AJ, Lyons TJ, et al. Lipoxidation products as biomarkers of oxidative damage to proteins during lipid peroxidation reactions. *Nephrol Dial Transplant* 1996; Suppl 5: 48-53.

[18] L'Esperance FA, James WA, Judson PH. In: Lifkin H, Porte D Ed, *The eye and diabetes mellitus. Ellenberg and Rifkin's Diabetes Mellitus, Theory and Practice*. New York, Elsevier, 1990; 661-83.

[19] Klein R, Klein BE, Moss SE, Davis MD, DeMets DL. The Wisconsin epidemiologic study of diabetic retinopathy. II. Prevalence and risk of diabetic retinopathy when age at diagnosis is less than 30 years. *Arch Ophthalmol* 1984; 102: 520-6.

[20] Cogan DG, Toussaint D, Kuwabara T. Retinal vascular patterns. IV. Diabetic retinopathy. *Arch Ophthalmo* 1961; 66: 366-78.

[21] Kuwabara T, Cogan DG. Retinal vascular patterns. VI. Mural cells of the retinal capillaries. *Arch Ophthalmo* 1963; 69: 492-502.

[22] Mandarino LJ. Current hypotheses for the biochemical basis of diabetic retinopathy. *Diabetes Care* 1992; 15: 1892-901.

[23] Yamagishi S, Kobayashi K, Yamamoto H. Vascular pericytes not only regulate growth, but also preserve prostacyclin-producing ability and protect against lipid peroxide-induced injury of co-cultured endothelial cells. *Biochem Biophys Res Commun* 1993; 190: 418-25.

[24] Yamagishi S, Hsu CC, Kobayashi K, Yamamoto H. Endothelin 1 mediates endothelial cell-dependent proliferation of vascular pericytes. *Biochem Biophys Res Commun* 1993; 191: 840-6.

[25] Hammes HP, Lin J, Renner O, Shani M, Lundqvist A, et al. Pericytes and the pathogenesis of diabetic retinopathy. *Diabetes* 2002; 51: 3107-12.

[26] Stitt AW. The role of advanced glycation in the pathogenesis of diabetic retinopathy. *Experimental and Molecular Pathology* 2003; 75: 95-108.

[27] Stitt AW, Li YM, Gardiner TA, Bucala R, Archer DB, et al. Advanced glycation end products (AGEs) co-locarize with AGE receptors in the retinal vasculature of diabetic and AGE-infused rats. *Am J Pathol* 1997; 150: 523-31.

[28] Yamagishi S, Hsu CC, Taniguchi M, Harada S, Yamamoto Y, et al. Receptor-mediated toxicity to pericytes of advanced glycosylation end products: a possible mechanism of pericyte loss in diabetic microangiopathy. *Biochem Biophys Res Commun* 1995; 213: 681-7.

[29] Yamagishi S, Amano S, Inagaki Y, Okamoto T, Koga K, et al. Advanced glycation end products-induced apoptosis and overexpression of vascular endothelial growth factor in bovine retinal pericytes. *Biochem Biophys Res Commun* 2002; 290: 973-8.

[30] Yamagishi S, Inagaki Y, Amano S, Okamoto T, Takeuchi M, et al. Pigment epithelium-derived factor protects cultured retinal pericytes from advanced glycation end product-induced injury through its antioxidative properties. *Biochem Biophys Res Commun* 2002; 296: 877-82.

[31] Yamagishi S, Amano S, Inagaki Y, Okamoto T, Takeuchi M, et al. Beraprost sodium, a prostaglandin I_2 analogue, protects against advanced gycation end products-induced injury in cultured retinal pericytes. *Mol Med* 2002; 8: 546-50.

[32] Ottonello L, Morone MP, Dapino P, Dallegri F. Tumor necrosis factor-alpha-induced oxidative burst in neutrophils adherent to fibronectin: effects of cyclic AMP-elevating agents. *Br J Haematol* 1995; 91: 566-70.

[33] Lu M, Kuroki M, Amano S, Tolentino M, Keough K, et al. Advanced glycation end products increase retinal vascular endothelial growth factor expression. *J Clin Invest* 1998; 101: 1219-24.

[34] Adamis AP, Miller JW, Bernal MT, D'Amico DJ, Folkman J, et al. Increased vascular endothelial growth factor levels in the vitreous of eyes with proliferative diabetic retinopathy. *Am J Ophthalmol* 1994; 118: 445-50.

[35] Aiello LP, Avery RL, Arrigg PG, Keyt BA, Jampel HD, et al. Vascular endothelial growth factor in ocular fluid of patients with diabetic retinopathy and other retinal disorders. *N Engl J Med* 1994; 331: 1480-7.

[36] Murata T, Ishibashi T, Khalil A, Hata Y, Yoshikawa H, et al. Vascular endothelial growth factor plays a role in hyperpermeability of diabetic retinal vessels. *Ophthalmic Res* 1995; 27: 48-52.

[37] Stitt AW, Bhaduri T, McMullen CB, Gardiner TA, Archer DB. Advanced glycation end products induce blood-retinal barrier dysfunction in normoglycemic rats. *Mol Cell Biol Res Commun* 2000; 3: 380-8.

[38] Segawa Y, Shirao Y, Yamagishi S, Higashide T, Kobayashi M, et al. Upregulation of retinal vascular endothelial growth factor mRNAs in spontaneously diabetic rats without ophthalmoscopic retinopathy. A possible participation of advanced glycation end products in the development of thc early phase of diabetic retinopathy. *Ophthalmic Res* 1998; 30: 333-9.

[39] Anderson S. Role of local and systemic angiotensin in diabetic renal disease. *Kidney Int Suppl* 1997; 63: S107-10.

[40] Yamagishi S, Amano S, Inagaki Y, Okamoto T, Inoue H, et al. Angiotensin II-type 1 receptor interaction upregulates vascular endothelial growth factor messenger RNA levels in retinal pericytes through intracellular reactive oxygen species generation. *Drugs Exp Clin Res* 2003; 29: 75-80.

[41] Amano S, Yamagishi S, Inagaki Y, Okamoto T. Angiotensin II stimulates platelet-derived growth factor-B gene expression in cultured retinal pericytes through intracellular reactive oxygen species generation. *Int J Tissue React* 2003;25: 51-5.

[42] Yamagishi S, Takeuchi M, Matsui T, Nakamura K, Imaizumi T, et al. Angiotensin II augments advanced glycation end product-induced pericyte apoptosis through RAGE overexpression. *FEBS Lett* 2005; 579: 4265-70.

[43] Nakamura K, Yamagishi S, Nakamura Y, Takenaka K, Matsui T, et al. Telmisartan inhibits expression of a receptor for advanced glycation end products (RAGE) in angiotensin-II-exposed endothelial cells and decreases serum levels of soluble RAGE in patients with essential hypertension. *Microvasc Res* 2005; 70: 137-41.

[44] Yamagishi S, Nakamura K. Telmisartan, its potential therapeutic implications in cardiometabolic disorders. *Recent Patent Reviews on Cardiovascular Drug Discovery* 2006; 1: 79-83.

[45] Otani A, Takagi H, Suzuma K, Honda Y. Angiotensin II potentiates vascular endothelial growth factor-induced angiogenic activity in retinal microcapillary endothelial cells. *Circ Res* 1998; 82: 619-28.

[46] Yamagishi S, Yonekura H, Yamamoto Y, Katsuno K, Sato F, et al. Advanced glycation end products-driven angiogenesis in vitro. Induction of the growth and tube formation of human microvascular endothelial cells through autocrine vascular endothelial growth factor. *J Biol Chem* 1997; 272: 8723-30.

[47] Yamagishi S, Nakamura K, Matsui T, Sato T, Takeuchi M. Potential utility of statins, 3-hydroxy-3-methylglutaryl coenzyme A reductase inhibitors in diabetic retinopathy. *Med Hypotheses* 2006; 66: 1019-21.

[48] Okamoto T, Yamagishi S, Inagaki Y, Amano S, Koga K, et al. Angiogenesis induced by advanced glycation end products and its prevention by cerivastatin. *FASEB J* 2002; 16: 1928-30.

[49] Okamoto T, Yamagishi S, Inagaki Y, Amano S, Takeuchi M, et al. Incadronate disodium inhibits advanced glycation end products-induced angiogenesis in vitro. *Biochem Biophys Res Commun* 2002; 297: 419-24.

[50] Yamagishi S, Matsui T, Nakamura K, Takeuchi M. Minodronate, a nitrogen-containing bisphosphonate, inhibits advanced glycation end product-induced vascular cell adhesion molecule-1 expression in endothelial cells by suppressing reactive oxygen species generation. *Int J Tissue React* 2005; 27: 189-95.

[51] Yamagishi S, Nakamura K, Matsui T, Takeuchi M. Minodronate, a nitrogen-containing bisphosphonate, is a promising remedy for treating patients with diabetic retinopathy. *Med Hypotheses* 2006; 66: 273-5.

[52] Amano S, Yamagishi S, Inagaki Y, Nakamura K, Takeuchi M, Inoue H, Imaizumi T. Pigment epithelium-derived factor inhibits oxidative stress-induced apoptosis and dysfunction of cultured retinal pericytes. *Microvas Res* 2005; 69: 45-55.

[53] Inagaki Y, Yamagishi S, Okamoto T, Takeuchi M, Amano S. Pigment epithelium-derived factor prevents advanced glycation end products-induced monocyte chemoattractant protein-1 production in microvascular endothelial cells by suppressing intracellular reactive oxygen species generation. *Diabetologia* 2003; 46: 284-7.

[54] Yamagishi S, Matsui T, Inoue H. Inhibition by advanced glycation end products (AGEs) of pigment epithelium-derived factor (PEDF) gene expression in microvascular endothelial cells. *Drugs Exp Clin Res* 2005; 31: 227-32.

[55] Yamagishi S, Nakamura K, Ueda S, Kato S, Imaizumi T. Pigment epithelium-derived factor (PEDF) blocks angiotensin II signaling in endothelial cells via suppression of NADPH oxidase: a novel anti-oxidative mechanism of PEDF. *Cell Tissue Res* 2005; 320: 437-45.

[56] Yamagishi S, Inagaki Y, Nakamura K, Abe R, Shimizu T, et al. Pigment epithelium-derived factor (PEDF) inhibits TNF-α-induced interleukin-6 expression in endothelial cells by suppressing NADPH oxidase-mediated reactive oxygen species generation. *J Mol Cell Cardiol* 2004; 37: 497-506.

[57] Yamagishi S, Inagkai Y, Amano S, Okamoto T, Takeuchi M. Up-regulation of vascular endothelial growth factor and down-regulation of pigment epithelium-derived factor mRNA in leptin-exposed cultured retinal pericytes. *Int J Tissue React* 2002; 24: 137-142.

[58] Spranger J, Osterhoff M, Reimann M, Mohlig M, Ristow M, et al. Loss of the antiangiogenic pigment epithelium-derived factor in patients with angiogenic eye disease. *Diabetes* 2002; 50: 2641-5.

[59] Boeri D, Maiello M, Lorenzi M. Increased prevalence of microthromboses in retinal capillaries of diabetic individuals. *Diabetes* 2001; 50: 1432-9.

[60] Yamagishi S, Fujimori H, Yonekura H, Yamamoto Y, Yamamoto H. Advanced glycation endproducts inhibit prostacyclin production and induce plasminogen activator inhibitor-1 in human microvascular endothelial cells. *Diabetologia* 1998; 41: 1435-41.

[61] Yamagishi S, Yamamoto Y, Harada S, Hsu CC, Yamamoto H. Advanced glycosylation end products stimulate the growth but inhibit the prostacyclin-producing ability of endothelial cells through interactions with their receptors. *FEBS Lett.* 1996; 384: 103-6.

[62] Nomura M, Yamagishi S, Harada S, Hayashi Y, Yamashima T, et al. Possible participation of autocrine and paracrine vascular endothelial growth factors in hypoxia-induced proliferation of endothelial cells and pericytes. *J Biol Chem* 1995; 270: 28316-24.

[63] Yamagishi S, Kawakami T, Fujimori H, Yonekura H, Tanaka N, et al. Insulin stimulates the growth and tube formation of human microvascular endothelial cells through autocrine vascular endothelial growth factor. *Microvasc Res* 1999; 57: 329-39.

[64] Yamagishi S, Yonekura H, Yamamoto Y, Fujimori H, Sakurai S, et al. Vascular endothelial growth factor acts as a pericyte mitogen under hypoxic conditions. *Lab Invest.* 1999; 79: 501-9.

[65] Friedman EA. In: Lifkin H, Porte D Ed, Diabetic renal disease. Ellenberg and Rifkin's Diabetes Mellitus, Theory and Practice, New York, Elsevier, 1990; 684-709.

[66] Ziyadeh FN, Goldfarb S. The renal tubulointerstitium in diabetes mellitus. *Kidney Int* 1991; 39: 464-75.

[67] Yamagishi S, Inagaki Y, Okamoto T, Amano S, Koga K, et al. Advanced glycation end product-induced apoptosis and overexpression of vascular endothelial growth factor and monocyte chemoattractant protein-1 in human-cultured mesangial cells. *J Biol Chem* 2002; 277: 20309-15.

[68] Dworkin LD, Ichikawa I, Brenner BM. Hormonal modulation of glomerular function. *Am J Physiol* 1983; 244: F95-104.

[69] Kreisberg JI, Venkatachalam M, Troyer D. Contractile properties of cultured glomerular mesangial cells. *Am J Physiol* 1985; 249: F457-63.

[70] Schlondorff D. The glomerular mesangial cell: an expanding role for a specialized pericyte. *FASEB J* 1987; 1: 272-81.

[71] De Vriese AS, Tilton RG, Elger M, Stephan CC, Kriz W, et al. Antibodies against vascular endothelial growth factor improve early renal dysfunction in experimental diabetes. *J Am Soc Nephrol* 2001; 12: 993-1000.

[72] Yamagishi S, Inagaki Y, Okamoto T, Amano S, Koga K, et al. Advanced glycation end products inhibit de novo protein synthesis and induce TGF-beta overexpression in proximal tubular cells. *Kidney Int* 2003; 63: 464-73.

[73] Matsumura T, Yamagishi S, Brownlee M. In Leroith D, Taylor SI, Olefsky JM Ed, *Advanced glycation end products and the pathogenesis of diabetic complications. Diabetes Mellitus: A Fundamental and Clinical Text.* New York, Lippincott-Raven Publishers, 2000; 983-91.

[74] Yamagishi S, Takeuchi M, Makita Z. In: Tomino Y Ed, *Advanced glycation end products and the pathogenesis of diabetic nephropathy. Type-2 diabetic nephropathy in Japan, from bench to bedside.* Basel, Karger, 2001; 30-5.

[75] Yamagishi S, Koga K, Inagaki Y, Amano S, Okamoto T, et al. Dilazep hydrochloride, an antiplatelet drug, prevents progression of diabetic nephropathy in Otsuka Long-Evans Tokushima fatty rats. *Drugs Exp Clin Res* 2002; 28: 221-7.

[76] Raj DS, Choudhury D, Welbourne TC, Levi M. Advanced glycation end products: a Nephrologist's perspective. *Am J Kidney Dis* 2000; 35: 365-80.

[77] Ziyadeh FN, Hoffman BB, Han DC, Iglesias-De La Cruz MC, Hong SW, et al. Long-term prevention of renal insufficiency, excess matrix gene expression, and glomerular mesangial matrix expansion by treatment with monoclonal antitransforming growth factor-beta antibody in db/db diabetic mice. *Proc Natl Acad Sci USA* 2000; 97: 8015-20.

[78] Yamamoto Y, Kato I, Doi T, Tonekura H, Ohashi S, et al. Development and prevention of advanced diabetic nephropathy in RAGE-overexpressing mice. *J Clin Invest* 2001; 108: 261-8.

[79] Suzuki D, Miyata T, Saotome N, Horie K, Inagi R, et al. Immunohistochemical evidence for an increased oxidative stress and carbonyl modification of proteins in diabetic glomerular lesions. *J Am Soc Nephrol* 1999; 10: 822-32.

[80] Lal MA, Brismar H, Eklof AC, Aperia A. Role of oxidative stress in advanced glycation end product-induced mesangial cell activation. *Kidney Int* 2002; 61: 2006-14.

[81] Wendt TM, Tanji N, Guo J, Kislinger TR, Qu W, et al. RAGE drives the development of glomerulosclerosis and implicates podocyte activation in the pathogenesis of diabetic nephropathy. *Am J Pathol* 2003; 162: 1123-37.

[82] Soulis-Liparota T, Cooper M, Papazoglou D, Clarke B, Jerums G. Retardation by aminoguanidine of development of albuminuria, mesangial expansion, and tissue fluorescence in streptozocin-induced diabetic rat. *Diabetes* 1991; 40: 1328-34.

[83] Edelstein D, Brownlee M. Aminoguanidine ameliorates albuminuria in diabetic hypertensive rats. *Diabetologia* 1992; 35: 96-97.

[84] Tsuchida K, Makita Z, Yamagishi S, Atsumi T, Miyoshi H, et al. Suppression of transforming growth factor beta and vascular endothelial growth factor in diabetic nephropathy in rats by a novel advanced glycation end product inhibitor, OPB-9195. *Diabetologia.* 1999; 4: 579-88.

[85] Degenhardt TP, Alderson NL, Arrington DD, Beattie RJ, Basgen JM, et al. Pyridoxamine inhibits early renal disease and dyslipidemia in the streptozotocin-diabetic rat. *Kidney Int* 2002; 6: 939-50.

[86] Viberti GC. Early functional and morphological changes in diabetic nephropathy. *Clin Nephrol* 1979; 12: 47-53.

[87] Gilbert RE, Krum H, Wilkinson-Berka J, Kelly DJ. The renin-angiotensin system and the long-term complications of diabetes: pathophysiological and therapeutic considerations. *Diabet Med* 2003; 20: 607-21.

[88] Kagami S, Border WA, Miller DE, Noble NA. Angiotensin II stimulates extracellular matrix protein synthesis through induction of transforming growth factor-beta expression in rat glomerular mesangial cells. *J Clin Invest* 1994; 93: 2431-7.

[89] Singh R, Alavi N, Singh AK, Leehey DJ. Role of angiotensin II in glucose-induced inhibition of mesangial matrix degradation. *Diabetes* 1999; 48: 2066-73.

[90] Taniguchi M, Kim S, Zhan Y, Iwao H. Role of intrarenal angiotensin-converting enzyme in nephropathy of type II diabetic rats. *Hypertens Res* 2002; 25: 287-94.

[91] Koga K, Yamagishi S, Takeuchi M, Inagaki Y, Amano S, et al. CS-886, a New Angiotensin II Type 1 Receptor Antagonist, Ameliorates Glomerular Anionic Site Loss and Prevents Progression of Diabetic Nephropathy in Otsuka Long-Evans Tokushima Fatty Rats. *Mol Med* 2002; 8: 591-9.

[92] Lewis EJ, Hunsicker LG, Bain RP, Rohde RD. The effect of angiotensin-converting-enzyme inhibition on diabetic nephropathy. The Collaborative Study Group. *N Engl J Med* 1993; 329: 1456-62.

[93] Parving HH, Hommel E, Jensen BR, Hansen HP. Long-term beneficial effect of ACE inhibition on diabetic nephropathy in normotensive type 1 diabetic patients. *Kidney Int* 2001; 60: 228-34.

[94] Sharma K, Eltayeb BO, McGowan TA, Dunn SR, Alzahabi B, et al. Captopril-induced reduction of serum levels of transforming growth factor-beta1 correlates with long-term renoprotection in insulin-dependent diabetic patients. *Am J Kidney Dis* 1999; 34: 818-23.

[95] Lewis EJ, Hunsicker LG, Clarke WR, Berl T, Pohl MA, et al. Renoprotective effect of the angiotensin-receptor antagonist irbesartan in patients with nephropathy due to type 2 diabetes. *N Engl J Med* 2001; 345: 851-60.

[96] Brenner BM, Cooper ME, de Zeeuw D, Keane WF, Mitch WE, et al. Effects of losartan on renal and cardiovascular outcomes in patients with type 2 diabetes and nephropathy. *N Engl J Med* 2001; 345: 861-9.

[97] Parving HH, Lehnert H, Brochner-Mortensen J, Gomis R, Andersen S, et al. The effect of irbesartan on the development of diabetic nephropathy in patients with type 2 diabetes. *N Engl J Med* 2001; 345: 870-8.

[98] Forbes JM, Cooper ME, Thallas V, Burns WC, Thomas MC, et al. Reduction of the accumulation of advanced glycation end products by ACE inhibition in experimental diabetic nephropathy. *Diabetes* 2002; 51: 3274-82.

[99] Sebekova K, Schinzel R, Munch G, Krivosikova Z, Dzurik R, et al. Advanced glycation end-product levels in subtotally nephrectomized rats: beneficial effects of angiotensin II receptor 1 antagonist losartan. *Miner Electrolyte Metab* 1999; 25: 380-3.

[100] Sebekova K, Gazdikova K, Syrova D, Blazicek P, Schinzel R, et al. Effects of ramipril in nondiabetic nephropathy: improved parameters of oxidatives stress and potential modulation of advanced glycation end products. *J Hum Hypertens* 2003; 17: 265-70.

[101] Fukami K, Ueda S, Ymagishi S, Kato S, Inagaki Y, et al. AGEs activate mesangial TGF-beta-Smad signaling via an angiotensin II type 1 receptor interaction. *Kidney Int* 2004; 66: 2137-47.

[102] Munch G, Schinzel R, Loske C, Wong A, Durany N, et al. Alzheimer's disease—synergistic effects of glucose deficit, oxidative stress and advanced glycation endproducts. *J Neural Transm* 1998; 105: 439-61.

[103] Yan SD, Chen X, Fu J, Zhu H, Roher A, et al. RAGE and amyloid-beta peptide neurotoxicity in Alzheimer's disease. *Nature* 1996; 382: 685-91.

[104] Ott A, Stolk RP, Harskamp F, Pols HAP, Hofman A, et al. Diabetes mellitus and the risk of dementia—the Rotterdam study. *Neurology* 1999; 53: 1937-42.

[105] Smith MA, Taneda S, Richey PL, Miyata T, Yan SD, et al. Advanced Maillard reaction end products are associated with Alzheimer disease pathology. *Proc Natl Acad Sci USA* 1994; 91: 5710-4.

[106] Vitek MP, Bhattacharya K, Glendening JM, Stopa E, Vlassara H, et al. Advanced glycation end products contribute to amyloidosis in Alzheimer disease. *Proc Natl Acad Sci USA* 1994; 91: 4766-70.

[107] Yan SD, Chen X, Schmidt AM, Brett J, Godman G, et al. Glycated tau protein in Alzheimer disease: a mechanism for induction of oxidant stress. *Proc Natl Acad Sci USA* 1994; 91: 7787-91.

[108] Sasaki N, Fukatsu R, Tsuzuki K, Hayashi Y, Yoshida T, et al. Advanced glycation end products in Alzheimer's disease and other neurodegenerative diseases. *Am J Pathol* 1998; 153: 1149-55.

[109] Sasaki N, Toki S, Chowei H, Saito T, Nakano N, et al. Immunohistochemical distribution of the receptor for advanced glycation end products in neurons and astrocytes in Alzheimer's disease. *Brain Res* 2001; 888: 256-62.

[110] Shulze H, Schuler A, Stuber D, Dobeli H, Langen H, et al. Rat brain glyceraldehyde-3-phosphate dehydrogenase interacts with the recombinant cytoplasmic domain of Alzheimer's b-amyloid precursor protein. *J Neurochem* 1993; 60: 1915-22.

[111] Koshy B, Matilla T, Burright EN, Merry DE, Fischbeck KH, et al. Spinocerebellar ataxia type-1 and spinobulbar muscular dystrophy gene products interact with glyceraldehyde-3-phosphate dehydrogenase. *Hum Mol Genet* 1996; 5: 1311-8.

[112] Saunders PA, Chen RW, Chuang DM. Nuclear translocation of glyceraldehyde-3-phosphate dehydrogenase isoforms during neuronal apoptosis. *J Neurochem* 1999; 72: 925-32.

[113] Ishitani R, Sunaga K, Hirano A, Saunders P, Katsube N, et al. Evidence that glyceraldehyde-3-phosphate dehydrogenase is involved in age induced apoptosis in mature cerebellar neurons in culture. *J Neurochem* 1996; 66: 928-35.

[114] Thompson JF, Scolyer RA, Kefford RF. Cutaneous melanoma. *Lancet* 2005; 365: 687-701.

[115] Balch CM, Soong SJ, Gershenwald JE, Thompson JF, Reintgen DS, et al. Prognostic Factors Analysis of 17,600 Melanoma Patients: Validation of the American Joint Committee on Cancer Melanoma Staging System. *J Clin Oncol* 2001; 19: 3622-34.

[116] Streit, M, Detmar M. Angiogenesis, lymphangiogenesis, and melanoma metastasis. *Oncogene* 2003; 22: 3172-9.

[117] Mihm MC Jr, Clark WH Jr, Reed RJ. The clinical diagnosis of malignant melanoma. *Semin Oncol* 1975; 2: 1015-8.

[118] Erhard H, Rietveld FJ, van Altena MC, Brocker EB, Ruiter DJ, et al. Transition of horizontal to vertical growth phase melanoma is accompanied by induction of vascular endothelial growth factor expression and angiogenesis. *Melanoma Res* 1997; 7 Supple 2: S19-26.

[119] Marcoval J, Moreno A, Graells J, Vidal A, Escriba JM, et al. Angiogenesis and malignant melanoma. Angiogenesis is related to the development of vertical (tumorigenic) growth phase. *J Cutan Pathol* 1997; 24: 212-8.

[120] Srivastava A, Laidler P, Davies RP, Horgan K, Hughes LE. The prognostic significance of tumor vascularity in intermediate-thickness (0.76-4.0 mm thick) skin melanoma. A quantitative histologic study. *Am J Pathol* 1988; 133: 419-23.

[121] Vlaykova T, Laurila P, Muhonen T, Hahka-Kemppinen M, Jekunen A, et al. Prognostic value of tumour vascularity in metastatic melanoma and association of blood vessel density with vascular endothelial growth factor expression. *Melanoma Res* 1999; 9: 59-68.

[122] Kashani-Sabet M, Sagebiel RW, Ferreira CM, Nosrati M, Miller JR 3rd. Tumor vascularity in the prognostic assessment of primary cutaneous melanoma. *J Clin Oncol* 2002; 20: 1826-31.

[123] Mizutari K, Ono T, Ikeda K, Kayashima K, Horiuchi S. Photo-enhanced modification of human skin elastin in actinic elastosis by N(epsilon)-(carboxymethyl)lysine, one of the glycoxidation products of the Maillard reaction. *J Invest Dermatol* 1997; 108: 797-802.

[124] Masaki H, Okano Y, Sakurai H. Generation of active oxygen species from advanced glycation end-products (AGE) under ultraviolet light A (UVA) irradiation. *Biochem Biophys Res Commun* 1997; 235: 306-10.

In: Focus on Cell Apoptosis Research
Editor: Daniel W. Cho, pp. 137-147
ISBN: 1-60021-433-9

Chapter V

Antiapoptotic Treatment Targets in Patients with Congestive Heart Failure

Ali Khoynezhad*

Section of Cardiothoracic Surgery, University of Nebraska Medical Center, Omaha, NE

Introduction

National hospital discharge surveys estimate that 4.8 million Americans have heart failure [1]. Despite significant advances in medical and surgical treatment of heart failure this important challenge remains: During the last two decades, congestive heart failure has become an increasingly frequent reason for hospital admission and clearly represents a major health problem.

Within the past decade, there is accumulated evidence that apoptosis contributes significantly to the pathogenesis of heart failure. It is an important component of the remodeling process and the transition from adaptive myocardial condition to end-stage cardiac failure (figure 1). Apoptosis is a morphologically different mode of cardiocyte death, and differs substantially from necrosis (figure 2). It is an energy-requiring and programmed cell death. Apoptosis is executed by a family of ubiquitously expressed cysteine proteases termed caspases. Caspases are present in the cell as inactive pro-caspases that are activated in response to apoptotic stimuli (figure 1). Understanding of physiology of apoptosis and its clinical implications are important for the surgeon, because the therapeutic options may improve the outcome of patients with heart failure. In fact, the attenuation and/or prevention of apoptotic pathways are new frontiers of therapy for congestive heart failure, and this therapeutic target will enter clinical arena within next decade.

* Correspondence concerning this article should be addressed to Ali Khoynezhad, MD, PhD, Section of Cardiothoracic Surgery, University of Nebraska Medical Center, 982315 Nebraska Medical Center, Omaha NE 68198-2315. Telephone: 402-559-4424; Fax: 402-559-6913; E-mail: akhoynezhad@unmc.edu.

Cardiocyte apoptosis in heart failure has been the topic of research in many recent studies. However, a surgical group has contributed only to very few articles that are addressed directly to the surgeon, who is treating patients with congestive heart failure. The aim of this paper is to give a critical review of the studies on antiapoptotic therapy in conditions associated with congestive heart failure; to provide ramifications for surgeons treating patients with congestive heart failure; and to translate the current knowledge into clinical practice of modern heart failure surgery.

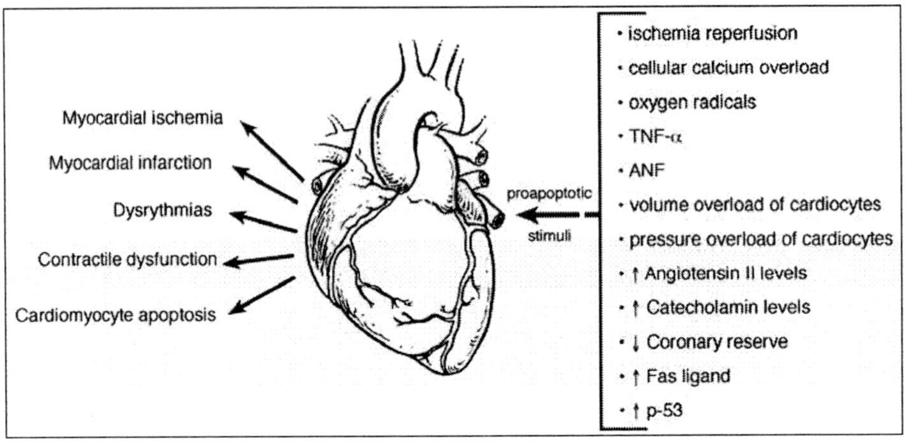

Figure 1. The most relevant myocardial conditions and agents that induce cardiomyocyte apoptosis are listed (right side). These can potentially lead to other clinical entities (left side).

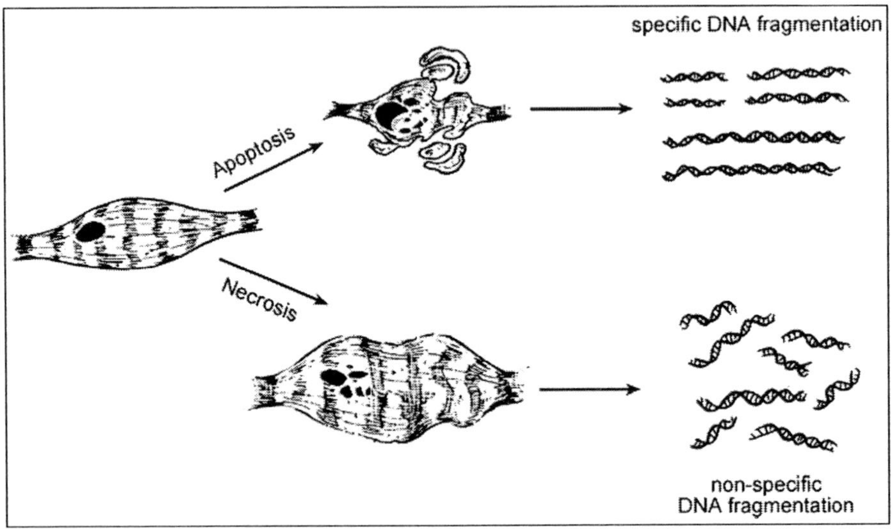

Figure 2. Cell shrinkage and aggregation of chromosomal DNA into small masses and preparation for exocytosis are microscopic hallmarks of early stages of apoptosis. These apoptotic bodies are membrane bounded and are subsequently phagocytosed by macrophages and neutrophils. Necrosis, on the other hand, is associated with loss of transmembrane ion gradient, and membrane disruption secondary to depletion of intracellular ATP causing an inflammatory reaction in surrounding tissue. The specific DNA fragmentation is a hallmark of apoptosis, and it is utilized as a detection method in the "DNA-laddering".

METHODS

A literature review using the "Ovid" search engine on "heart failure and apoptosis" and its subheadings resulted in 62008 publications. The search was subsequently limited to English language, and was focused on studies involving "human". The articles were assessed for their validity, importance and applicability to a surgeon's clinical practice. The detection methods were critically reviewed for methodological accuracy, since detection methods with poor sensitivity can lead to skewed results and incorrect conclusions (figure 3). Articles were reviewed and clinically important data was collected and incorporated.

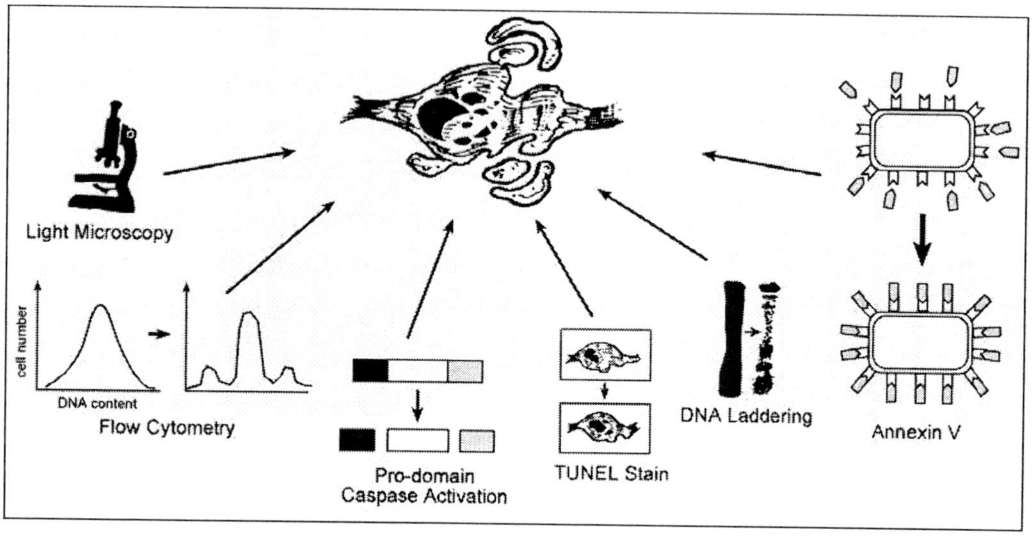

Figure 3. Most common and valid methods used for detection of apoptosis (TUNEL = terminal deoxynucleotidyl- transferase mediated dUTP nick end-labeling).

RESULTS

Cardiocyte apoptosis is part of the phenotype of the failing heart, and has been detected in animal and human studies [2-5]. Left ventricular (LV) hypertrophy and dilation, enhanced and sustained activity of the renin-angiotensin-aldosteron system (RAAS), intracellular calcium and increased sympathetic response are compensatory mechanisms and closely associated with failing myocardium. These mechanisms are also considered potent inducers of apoptosis and are responsible for remodeling and progression of heart failure [6-8]. Last but not least, myocardial infarction is one the prominent causes of heart failure. This can result from the acute loss of cardiocytes in the infracted area, but, more often, is precipitated by the delayed remodeling of the myocardium.

Valvular regurgitation is the leading cause of LV dilatation. The volume overload associated with aortic and mitral regurgitation increases the LV end-diastolic pressure and the isometric stretch of myocardium. In one prominent study utilizing a rodent model, the papillary muscle was subjected to 50 mN/mm isometric tension [9]. The authors found a 21-

fold higher apoptotic index, when compared to baseline tension (7-8 mN/mm) [9]. This evidence supports and explains the prevalence of cardiocyte apoptosis in volume-overloaded ventricle. Abbate and coworkers addressed the role of cardiocyte apoptosis in the genesis of LV dilatation and thinning of non-infarcted myocardium following large transmural myocardial infarction [10]. In this postmortem study, the apoptotic index in remote non-infarcted region correlated strongly with dilatation and thinning of the myocardium [10].

Ventricular hypertrophy represents another compensatory mechanism associated with heart failure; it can be the result of increased afterload, long-lasting hypertension and aortic or pulmonary stenosis. Cardiocyte apoptosis has been detected in hypertrophied heart regardless of etiology [11-12]. A rodent model of LV hypertrophy has convincingly demonstrated significantly higher apoptotic index in banded rats in comparison to the control group [12]. Using isolated adult human cardiocytes, Kajstura and coworkers showed that cardiocytes treated with angiotensin II (AT-II) revealed increased apoptotic index [7]. Apoptotic cell death was prevented in the same study with angiotensin converting enzyme (ACE) inhibitors. AT-II similarly activates proapoptotic p-53 transcription [13], and induces apoptogenic genes, such as cMyc, c-Fos, c-Jun, and up-regulation of genes encoding for proapoptotic atrial natriuretic factor [14-15]. Both ACE inhibitors and AT-I blockers can inhibit this effect, and reduce cardiocyte death due to apoptosis. Both agents counteract the effects of AT-II, including increase of afterload, ventricular remodeling, cardiocyte and endothelial necrosis and apoptosis, as well as diastolic heart failure due to fibroblast proliferation and collagen deposition in the LV wall [7,16]. There is level-one evidence supporting the use of ACE-inhibitors and AT-I blockers [17,18], especially after surgery on patients with poorly functioning or dilated ventricles.

The patients suffering from congestive heart failure have increased baseline intracellular calcium ion due to high levels of AT-II and catecholamines [7,14,19]. Increased cellular calcium ion is a proapoptotic stimulus [6]. Sodium-hydrogen exchanger inhibitors and calcium channel blockers reduce cellular calcium by decreasing cellular sodium, and, thereby could potentially decrease apoptotic index in the cardiocyte at risk. Oral administration of Sodium-hydrogen exchanger inhibitors has reduced mortality, arrhythmias, infarct size, apoptotic index and Bcl-2/Bax ratio in an ischemic rodent model [20]. Besides this promising study, a convincing experiment in a human model of heart failure is still missing in the literature. Calcium channel blockers have decreased direct toxic effects of norepinephrine by decreasing intracellular calcium [19], but no studies have been undertaken that might show decreased apoptosis in heart failure models. In randomized controlled studies on heart failure, there is no level-one evidence that calcium channel blockers are of benefit in people with heart failure [21]. This is explained by the fact that the calcium channel blockers do not influence the structural cardiac remodeling process, and, therefore fail to improve the outcome pf patients with congestive heart failure [22].

The level of catecholamine activity is increased in failing myocardium, and is associated with direct toxicity and apoptosis of cardiocytes [8,19]. The direct toxic and apoptogenic effects of norepinephrine were abolished with beta-adrenoreceptor antagonists, but not with an alpha-receptor antagonist in two independent studies [5,8]. The antiapoptotic effects of beta-adrenoreceptor antagonists on norepinephrine-induced apoptosis have been addressed and proven to be true for atenolol and carvedilol [23,24]. On the clinical side, the beta-adrenoreceptor antagonists reduce overall mortality and hospitalization in patients with congestive heart failure [25,26]. It is unknown if this improvement in outcome is secondary to antiapoptotic effects of beta-adrenoreceptor antagonists, or if the decreased apoptotic index is a microscopic marker for reverse cardiac remodeling.

Myocardial infarction is a prominent inducer of cardiocyte apoptosis and the leading cause of heart failure. Myocytes loss, due to apoptotic cell death during the acute stage of myocardial infarction, is well established in animal and human studies. Increased apoptotic index was demonstrated as early as three hours and up to one month after coronary occlusion in rat hearts [27]. Apoptotic human cardiocytes were predominantly found in the hypoperfused border zone between the central infarct area and non-ischemic cardiac tissue [28]. Olivetti et al. demonstrated 12% apoptotic cell death in infarct border zone and 1% in non-compromised myocytes distant from the infarct zone [28].

Broad-spectrum caspase inhibitors are agents that control the signal pathway of cardiocyte apoptosis. Such antiapoptotic agents have been used successfully in a rodent model of acute myocardial infarction [29,30]. Armstrong and coworkers found that IDN-6734 reduces the size of myocardial infarction in rats, if the poly-caspase-inhibitor is given prior (47% reduction) or one hour after (45% reduction) reperfusion of left anterior descending artery [30]. In another study, ZVAD.fmk has decreased infarction size with improved hemodynamics, and has attenuated apoptosis compared to control rats [29].

Discussion

Apoptosis is a precisely orchestrated process that is hardwired into all metazoan cells. Apoptosis contributes to the development of congestive heart failure at least in two distinct ways. First, it reduces the number of contractile cardiocytes through a programmed cell death. The frequency of cardiocytes apoptosis in remote myocardium following human myocardial infarction, as well as in dilated cardiomyopathy, is $80/10^5$ to $750/10^5$ [2,31,32]. This low, but abnormal, rate of cardiocytes apoptosis represents a significant loss for adult human myocardium, since adult human cardiocytes have lost the replicative potential of neonatal cardiocytes. Finally, apoptosis of cardiac non-myocytes may contribute significantly to the progressive nature of failing myocardium. Fujiwara and coworkers have supported the hypothesis that apoptosis of granulation cells is related to cardiac remodeling after myocardial infarction [33,34].

For the reasons mentioned above, apoptotic signaling pathway in cardiocytes and non-cardiocytes is an important factor for transition from compensated to decompensated maladaptive growth in heart failure. Antiapoptotic therapeutic interventions for congestive heart failure represent a very attractive platform for novel treatment options.

Numerous myocardial conditions and agents have been identified as inducers of cardiocyte apoptosis (figure 1). Table 1 represents a summary of most prominent antiapoptotic therapies, which are primarily investigational and without direct translation into the clinical praxis. The use of beta-adrenoreceptor antagonists, AT-II blockers and ACE inhibitors has been associated with decreased apoptotic activity. These medications are proven in randomized controlled studies to be of significant clinical benefit to patients with heart failure. It remains unclear whether the antiapoptotic characteristics of these medications are responsible for the associated survival benefit.

Table 1. Investigational treatment options in the failing myocardium

Medications	References
Caspase inhibitor zVAD.fmk	[29,30,45]
Sodium-hydrogen-exchanger inhibitors	[20]
Poly-ADP-ribose-synthetase inhibitors	[41,42]
Inhibitors of apoptosis protein	[46]
Upregulation of Insulin growth factor-1	[47]
Ischemic preconditioning	[48-53]
Antioxidants	[38-40]
TNF-α inhibitor	[54-56]
Cardiotrophin-1	[57]

Most of efforts in antiapoptotic treatment of heart failure are pharmaceutical in nature. The investigations are led mostly by the non-surgical colleagues and, therefore, rarely of translational use for the practice of surgeons dealing with heart failure population. Myocardial infarction, volume and pressure overload and clinical conditions associated with sustained or acute increase in levels of catecholamines are clinically relevant stimuli of apoptosis, which are pertinent to the surgeon. Therapeutic ramifications are, in some cases, already incorporated in clinical practice of modern heart failure surgery [table 2]. These include reducing myocardial ischemia time with prompt coronary artery revascularization, early valvular repair or replacement - prior to deterioration of ventricular function, and afterload reduction with intraaortic balloon pulsation, where cardiac function has been compromised following myocardial infarction or ischemic mitral regurgitation.

Table 2. "Translational" antiapoptotic strategies in treatment of failing myocardium

Strategies	References
Beta-adrenoreceptor antagonists	[23,24]
Angiotensin converting enzyme inhibitors	[7,13]
Angiotensin II antagonists	[7,13]
Increased use of phosphodiesterase-inhibitors	[8]
Valvular repair/replacement	[9,11,12]
Reducing use of catecholamines	[8,19,43]
Afterload reduction	[9,12]
Coronary revascularization	[27,28]
Cardiopulmonary bypass	[35]
Shortening myocardial ischemia	[27,28]

The use of cardiopulmonary bypass has been associated with increased apoptotic index. Blood contact with foreign surface results in increased serum level of Fas and Fas-ligand in humans undergoing cardiopulmonary bypass [35]. This increases apoptotic cell death in the myocardium and other organs through the death receptor pathway. The evidence of post-bypass syndrome, including cerebral dysfunction and renal insufficiency, may be secondary to apoptotic injury to other organs [36,37]. Although this apoptotic condition can be avoided while performing off-pump coronary artery bypass grafting, such a procedure is not feasible in the majority of currently performed coronary operations and all other open-heart procedures. However, if the proinflammatory and proapoptotic state of cardiopulmonary bypass can be lessened, an important therapeutic goal may be attained. Potential therapeutic implications are reduction of the foreign surface, shortening of the bypass time, and the addition of antiapoptotic medication to the priming solution in the cardiopulmonary bypass circuit. Myocardial protection can be also altered to fit antiapoptotic strategies. A myocardial protection strategy based heavily on hypothermia will result in longer bypass times and increased apoptotic index. Alternately, surgeons using normothermic techniques will have more prompt restoration of cardiac performance, decreased reperfusion time and diminished use of catecholamines. These strategies are beneficial in reducing the apoptotic rate in the failing myocardium.

The reperfusion injury after a cardioplegic-induced myocardial arrest is another apoptotic state with which the cardiac surgeon deals on a daily basis. This will be a treatment target in the pool of antiapoptotic strategies. Antioxidants, Poly-ADP-ribose-synthetase inhibitors and calcium channel blockers may be useful in decreasing reactive oxygen radicals, reducing neutrophil-mediated inflammatory cascade and attenuating cellular calcium overload [38-42]. However, all these agents are investigational. Once FDA approved, they should be studied as an adjunct of cardioplegic solution or cardiopulmonary bypass circuit to prove efficacy.

Sustained and long-term use of catecholamines induces apoptosis and has direct toxic effects on the heart [8,19,43]. In patients with low output syndrome after open-heart surgery, acute myocardial infarction or other conditions associated with increased endogenous or exogenous catecholamine, the inotropic support of the heart should be accomplished with alternative medications - such as milrinone. Experienced surgeons may consider being more liberate in insertion of intraaortic balloon pulsations or ventricular assist devises, in clinical low output states, as bridge to recovery or as destination therapy. The REMATCH trial has provided us level-one evidence for the survival benefits associated with ventricular assist devices [44]. However, the optimal time of initiation of circulatory assist has not been yet established. Late insertion of such devices - after a prolonged trial of external catecholamine - may produce a non-recoverable cardiac failure state, which is "burned out" due to progressive apoptotic cardiocyte death. Alternately, early insertion of ventricular assist devices may reduce the chronic apoptotic cell death associated with of increased levels of serum AT-II and catecholamines. Early ventricular assistance would, in all probability, increase the likelihood of myocardial recovery by limiting the apoptogenic stimuli and changing the strategy of destination therapy into bridge to recovery.

In order to give a critical overview of the current studies on apoptosis in failing myocardium, it is important to know that there are significant limitations of our current knowledge about cardiocyte apoptosis. It is unknown whether apoptosis is the primary event in pathological cardiac conditions or the secondary. Also further investigation is needed to determine whether inhibition of apoptosis will delay disease progression in vivo and in translational human studies. Antiapoptotic pharmaceutics have the highest potential to become clinically significant as a therapeutic option. These medications can be potentially administered preoperatively as an adjunct to the priming solution of cardiopulmonary circuit or added to cardioplegic solution. However, the safety and long-term consequences of these therapies are not adequately investigated; clinical integration of these strategies will require more studies. Such therapies potentially may possibly lead to alternative mode of cell death, such as necrosis or increase autoimmune and lymphoproliferative disorders. Also of concern is the inferiority of techniques used to detect cardiocyte apoptosis. The positive predictive value of terminal deoxynucleotidyl-transferase mediated dUTP nick end-labeling (TUNEL), which is the most common used method to detect apoptosis, is less than appropriate. The heterogeneity of examined tissue samples can deteriorate the accuracy of DNA-laddering, because it cannot specify the cell type undergoing apoptosis in a tissue sample with multiple cell contents. The transience and brevity of apoptosis is another culprit for the low sensitivity rates of the diagnostic methods. More accurate tools are needed for the future studies and are being currently developed.

Prevention or attenuation of cardiocytes apoptosis is a very attractive and efficient therapeutical goal in treatment of patients with congestive heart failure. Accumulating data will widen the spectrum of therapeutic strategies of management of these patients. Elucidation of anti- and proapoptotic interventions will translate into new treatment options in the field of heart failure surgery, and alter the management of patients with congestive heart failure. It is vital for the modern surgeon to understand these treatment options, and to incorporate the implications into the practice of modern heart failure surgery.

REFERENCES

[1] American Heart Association: Heart Disease and Stroke Statistics 2004 Update. Dallas, TX; American Heart Association 2003.

[2] Olivetti G, Abbi R, Quaini F, et al. Apoptosis in the failing human heart. *N Engl J Med* 1997;336(16):1131-41.

[3] Narula J, Haider N, Virmani R, et al. Apoptosis in myocytes in end-stage heart failure. *N Engl J Med* 1996;335(16):1182-9.

[4] Sharov VG, Sabbah HN, Shimoyama H, et al. Evidence of cardiocyte apoptosis in myocardium of dogs with chronic heart failure. *Am J Pathol* 1996;148:141-9.

[5] Yao M, Keogh A, Spratt P, et al. Elevated DNase I levels in human idiopathic dilated cardiomyopathy: an indicator of apoptosis? *J Mol Cell Cardiol* 1996;28:95-101.

[6] Orrenius S, McConkey DJ, Bellomo G, et al. Role of Ca2+ in toxic cell killing. *Trends Pharmacol Sci* 1989;10:281-5.

[7] Kajstura J, Cigola E, Malhotra A, et al. Angiotensin II induces apoptosis of adult ventricular myocytes in vitro. *J Mol Cell Cardiol* 1997;29(3):859-70.

[8] Communal C, Singh K, Pimentel DR, et al. Norepinephrine stimulates apoptosis in adult rat ventricular myocytes by activation of the beta-adrenergic pathway. *Circulation*. 1998;98:1329-34.

[9] Cheng W, Li B, Kajstura J, et al. Stretch-induced programmed myocyte cell death. *J Clin Invest* 1995;96(5):2247-59.

[10] Abbate A, Biondi-Zoccai GG, Bussani R, et al. Increased myocardial apoptosis in patients with unfavorable left ventricular remodeling and early symptomatic postinfarction heart failure. *J Am Coll Cardiol* 2003;41(5):753-60.

[11] Li Z, Bing OH, Long X, et al. Increased cardiomyocyte apoptosis during the transition to heart failure in the spontaneously hypertensive rat. *Am J Physiol* 1997; 272:H2313-9.

[12] Teiger E, Than VD, Richard L, et al. Apoptosis in pressure overload-induced heart hypertrophy in the rat. *J Clin Invest* 1996;97(12):2891-7.

[13] Leri A, Claudio PP, Li Q, et al. Stretch-mediated release of angiotensin II induces myocyte apoptosis by activating p53 that enhances the local renin-angiotensin system and decreases the Bcl-2-to-Bax protein ratio in the cell. *J Clin Invest* 1998;101(7):1326-42.

[14] Koglin J, Granville DJ, Glysing-Jensen T, et al. Attenuated acute cardiac rejection in NOS2 -/- recipients correlates with reduced apoptosis. *Circulation* 1999;99(6):836-42.

[15] Jiang L, Huang Y, Yuasa T, et al. Elevated DNase activity and caspase expression in association with apoptosis in failing ischemic sheep left ventricles. *Electrophoresis*. 1999;20:2046-52.

[16] Weber KT. Extracellular matrix remodeling in heart failure: a role for de novo angiotensin II generation. *Circulation*. 1997;96:4065-82.

[17] Garg R, Yusuf S. Overview of randomized trials of angiotensin-converting enzyme inhibitors on mortality and morbidity in patients with heart failure. *JAMA* 1995;273:1450-6.

[18] Sharma D, Buyse M, Pitt B, et al. Meta-analysis of observed mortality data from all-controlled, double-blind, multiple-dose studies of losartan in heart failure. Losartan Heart Failure Mortality Meta-analysis Study Group. *Am J Cardiol* 2000;85:187-92.

[19] Mann D L ,Kent RL, Parsons B, et al. Adrenergic effects on the biology of the adult mammalian cardiocyte. *Circulation* 1992;85:790-804.

[20] Humphreys RA, Haist JV, Chakrabarti S, et al. Orally administered NHE1 inhibitor cariporide reduces acute responses to coronary occlusion and reperfusion. *Am J Physiol* 1999;276:H749-57.

[21] Packer M, O'Connor CM, Ghali JK, et al. Effect of amlodipine on morbidity and mortality in severe chronic heart failure. Prospective Randomized Amlodipine Survival Evaluation Study Group. *N Engl J Med* 1996;335(15):1107-14.

[22] Cohn JN, Ferrari R, Sharpe N. Cardiac remodeling-concepts and clinical implications: a consensus paper from an international forum on cardiac remodeling. Behalf of an International Forum on Cardiac Remodeling. *J Am Coll Cardiol* 2000;35:569-82.

[23] Rossig L, Haendeler J, Mallat Z, et al. Congestive heart failure induces endothelial cell apoptosis: protective role of carvedilol. J Am Coll Cardiol 2000;36(7):2081-9.

[24] Romeo F, Li D, Shi M, et al. Carvedilol prevents epinephrine-induced apoptosis in human coronary artery endothelial cells: modulation of Fas/Fas ligand and caspase-3 pathway. *Cardiovasc Res* 2000;45:788-94.

[25] Packer M, Bristow MR, Cohn JN, et al. The effect of carvedilol on morbidity and mortality in patients with chronic heart failure. U.S. Carvedilol Heart Failure Study Group. *N Engl J Med* 1996;334(21):1349-55.

[26] Lechat P, Packer M, Chalon S, et al. Clinical effects of beta-adrenergic blockade in chronic heart failure: a meta-analysis of double-blind, placebo-controlled, randomized trials. *Circulation* 1998;98:1184-91.

[27] Cheng W, Kajstura J, Nitahara JA, et al. Programmed myocyte cell death affects the viable myocardium after infarction in rats. *Exp Cell Res* 1996;226(2):316-27.

[28] Olivetti G, Quaini F, Sala R, et al. Acute myocardial infarction in humans is associated with activation of programmed myocyte cell death in the surviving portion of the heart. *J Mol Cell Cardiol* 1996;28(9):2005-16.

[29] Yaoita H, Ogawa K, Maehara K, et al. Attenuation of ischemia/reperfusion injury in rats by a caspase inhibitor. *Circulation* 1998;97(3):276-81.

[30] Armstrong RC, Li F, Smiley R, et al. Caspase inhibitors reduce infarct size when dosed post-perfusion in a rodent cardiac ischemia/reperfusion model. *Circulation* 2001; 104:II-12.

[31] Saraste A, Pulkki K, Kallajoki M, et al. Cardiomyocyte apoptosis and progression of heart failure to transplantation. *Eur J Clin Invest* 1999;29(5):380-6.

[32] Guerra S, Leri A, Wang X, et al. Myocyte death in the failing human heart is gender dependent. *Circ Res* 1999;85(9):856-66.

[33] Takemura G, Ohno M, Hayakawa Y, et al. Role of apoptosis in the disappearance of infiltrated and proliferated interstitial cells after myocardial infarction. *Circ Res* 1998;82(11):1130-8.

[34] Hayakawa K, Takemura G, Kanoh M, et al. Inhibition of granulation tissue cell apoptosis during the subacute stage of myocardial infarction improves cardiac remodeling and dysfunction at the chronic stage. *Circulation.* 2003;108(1):104-9.

[35] Kawahito K, Misawa Y, Fuse K. Transient rise in serum soluble Fas (APO-1/CD95) in patients undergoing cardiac surgery. Artif Organs 2000; 24(8):628-31.

[36] Baumgartner WA, Walinsky PL, Salazar JD, et al. Assessing the impact of cerebral injury after cardiac surgery: will determining the mechanism reduce this injury? *Ann Thorac Surg* 1999;67(6):1871-3.

[37] Meldrum DR, Donnahoo KK. Role of TNF in mediating renal insufficiency following cardiac surgery: evidence of a postbypass cardiorenal syndrome. *J Surg Res* 1999; 85(2):185-99.

[38] Oskarsson HJ, Coppey L, Weiss RM, et al. Antioxidants attenuate myocyte apoptosis in the remote non-infarcted myocardium following large myocardial infarction. *Cardiovasc Res* 2000;45(3):679-87.

[39] Dhalla NS, Elmoselhi AB, Hata T, et al. Status of myocardial antioxidants in ischemia-reperfusion injury. *Cardiovasc Res* 2000;47(3):446-56.

[40] Zhao ZQ, Nakamura M, Wang NP, et al. Reperfusion induces myocardial apoptotic cell death. *Cardiovasc Res* 2000;45(3):651-60.

[41] Yaoita H, Ogawa K, Maehara K, et al. Apoptosis in relevant clinical situations: contribution of apoptosis in myocardial infarction. *Cardiovasc Res* 2000; 45(3):630-41.

[42] Thiemermann C, Bowes J, Myint FP, et al. Inhibition of the activity of poly(ADP ribose) synthetase reduces ischemia-reperfusion injury in the heart and skeletal muscle. *Proc Natl Acad Sci USA* 1997;94(2):679-83.

[43] Colucci WS, Sawyer DB, Singh K, et al. Adrenergic overload and apoptosis in heart failure: implications for therapy. *J Card Fail* 2000;6(2 Suppl 1):1-7.

[44] Rose EA, Gelijns AC, Moskowitz AJ, et al; Randomized Evaluation of Mechanical Assistance for the Treatment of Congestive Heart Failure (REMATCH) Study Group. Long-term mechanical left ventricular assistance for end-stage heart failure. *N Engl J Med* 2001;345(20):1435-43.

[45] Abbate A, Biondi-Zoccai G, Baldi F. Pathophysiologic Role of myocardial Apoptosis in Post-Infarction left Ventricular Remodeling. *J Cell Physiol* 2002;193:145-53.

[46] Deveraux QL, Reed JC. IAP family proteins--suppressors of apoptosis. *Genes Dev* 1999;13(3):239-52.

[47] Li Q, Li B, Wang X, et al. Overexpression of insulin-like growth factor-1 in mice protects from myocyte death after infarction, attenuating ventricular dilation, wall stress, and cardiac hypertrophy. *J Clin Invest* 1997;100(8):1991-9.

[48] Yellon DM, Alkhulaifi AM, Pugsley WB. Preconditioning the human myocardium. *Lancet* 1993;342(8866):276-7.

[49] Przyklenk K, Kloner RA. Ischemic preconditioning: exploring the paradox. *Prog CardioVasc Dis* 1998;40(6):517-47.

[50] Brocheriou V, Hagege AA, Oubenaissa A, et al. Cardiac functional improvement by a human Bcl-2 transgene in a mouse model of ischemia/reperfusion injury. *J Gene Med* 2000;2(5):326-33.

[51] Maulik N, Engelman RM, Rousou JA, et al. Ischemic preconditioning reduces apoptosis by upregulating anti-death gene Bcl-2. *Circulation* 1999;100(19 Suppl):II369-75.

[52] Jenkins DP, Pugsley WB, Alkhulaifi AM, et al. Ischemic preconditioning reduces troponin T release in patients undergoing coronary artery bypass surgery. *Heart* 1997; 77(4):314-8.

[53] Perrault LP, Menasche P, Bel A, et al. Ischemic preconditioning in cardiac surgery: a word of caution. *J Thorac CardioVasc Surg* 1996;112(5):1378-86.

[54] Zheng L, Fisher G, Miller RE, et al. Induction of apoptosis in mature T cells by tumour necrosis factor. *Nature* 1995;377(6547):348-51.

[55] Krown KA, Page MT, Nguyen C, et al. Tumor necrosis factor alpha-induced apoptosis in cardiac myocytes. Involvement of the sphingolipid signaling cascade in cardiac cell death. *J Clin Invest* 1996;98(12):2854-65.

[56] Haunstetter A, Izumo S. Future perspectives and potential implications of cardiac myocyte apoptosis. *Cardiovasc Res* 2000;45(3):795-801.

[57] Sheng Z, Knowlton K, Chen J, et al. Cardiotrophin 1 (CT-1) inhibition of cardiac myocyte apoptosis via a mitogen-activated protein kinase-dependent pathway. Divergence from downstream CT-1 signals for myocardial cell hypertrophy. *J Biol Chem* 1997; 272(9):5783-91.

In: Focus on Cell Apoptosis Research
Editor: Daniel W. Cho, pp. 149-173

ISBN: 1-60021-433-9

Chapter VI

APOPTOSIS OF LEUKEMIA CELLS INDUCED BY L-ASCORBIC ACID AND ARSENIC TRIOXIDE: THE EFFECT OF OXIDATIVE STRESS AND GLUTATHIONE HOMEOSTASIS

Seyeon Park
Department of Applied Chemistry, Dongduk Women's University,
Seoul 136-714, Korea

ABSTRACT

L-ascorbic acid (LAA) at concentrations of 0.25–1.0mM induced a dose- and time-dependent inhibition of proliferation in acute myeloid leukemia (AML) cell lines and also in leukemic cells from peripheral blood specimens obtained from three patients with AML. LAA induced oxidation of glutathione to oxidized form (GSSG) and subsequent H_2O_2 accumulation in a concentration-dependent manner, in parallel to induction of apoptosis. Induction of apoptosis in LAA-treated AML cells involved a dose-dependent increase of Bax protein, release of cytochrome C from mitochondria to cytosol, activation of caspase 9 and caspase 3, and cleavage of poly [ADP-ribose] polymerase. LAA treatment of cells induces a dose-dependent phosphorylation of extracellular signalregulated kinases (ERK) and results in activation of its catalytic domain. The small G protein Rafl and MAPKactivated protein kinase 2 are activated by LAA as an upstream and a downstream regulator of ERK, respectively. Although the ERK pathway has been known to activate cell proliferation, pharmacologic inhibition of ERK reduces LAA-dependent apoptosis and growth inhibitory response of acute myeloid leukemia cell lines, suggesting that this signaling cascade positively regulates induction of apoptotic response by LAA.

Arsenic trioxide (As_2O_3) could induce *in vitro* growth inhibition and apoptosis of AML cells. As_2O_3 or LAA suppressed DNA-binding activity of NF-κB composed of p65/p50 heterodimer through preventing the degradation of IκB-α and the nuclear translocation of p65 subsequently as well as interrupting the binding of NF-κB with their consensus sequences. Inhibitory effect of As_2O_3 or LAA on NF-κB DNA activity was dependent upon intracellular glutathione (GSH) and H_2O_2 level, but not superoxide anion.

The addition of LAA to cells resulted in an immediate increase in intracellular total GSH content and GST activity accompanied by the uptake of cysteine. These results suggest a new role of LAA in intracellular sulfur condition and glutathione regulation. Early protein targets of LAA in leukemia cells were identified using a differential proteomics approach in NB4 human leukemia cells treated with LAA. A subunit of protein-disulfide isomerase (a thiol/disulfide exchange catalyst) and immunoglobulin-heavy-chain binding protein (BiP, identical to Hsp70 chaperone) showed quantitative expression profile differences. A myeloid leukemia associated antigen protein (a tropomyosin isoform) showed changes in pI as a result of phosphorylation. The addition of LAA to cells results in an immediate change in the intracellular thiol/disulfide condition and modulates disulfide bond formation in proteins.

INTRODUCTION

Apoptosis has emerged as an important biologic mechanism that contributes to the maintenance of the integrity of multicellular organisms. Impairment of apoptosis has been implicated in any human diseases including cancers [1-3]. For instance, deregulation of apoptosis is regarded as a major factor contributing to the development of chronic lymphocytic leukemia (CLL) and some types of lymphoma, such as follicular B-cell lymphoma with the chromosomal translocation t(14:18), which results in overexpression of the death antagonist Bcl-2 [4-6]. Therefore, understanding the basic mechanisms that underlie apoptosis will help identify new potential targets for treatment of these diseases [1,7,8].

Recently, arsenic trioxide (As_2O_3), an ancient drug used in Traditional Chinese Medicine, attracted wide interest for its ability to induce complete remission in most patients with acute promyelocytic leukemia (APL), mainly through the induction of apoptosis and differentiation [1,9-14]. Moreover, it was reported that clinically achievable concentrations of As_2O_3 can also trigger apoptosis in chronic myelogenous leukemia cells, some multiple myeloma cells, and some solid tumor cells, such as esophageal cancer cancer and neuroblastoma cells, suggesting that the ability of As_2O_3 to induce apoptosis is not limited to APL [1]. There are data suggesting that oxidative stress triggers signal transduction cascade leading to apoptosis by arsenic trioxide [15-18].

Meanwhile, LAA in combination with arsenic trioxide has shown clinical responses in chemorefractory multiple meyloma [15]. There is also increasing evidence that LAA is selectively toxic to some types of tumor cells, functioning as a pro-oxidant, rather than anti-oxidant [19-21]. LAA at concentrations of 10 nM–1mM induced apoptosis in neuroblastoma and melanoma cells [22]. LAA was also shown to be an important modulator for the growth of mouse myeloma cells in an in vitro colony assay [23]. A series of studies have established that the growth of leukemic progenitor cells from patients with acute myeloid leukemia (AML) and myelodysplastic syndromes (MDS) can be profoundly modulated by LAA[24,25]. Intravenous administration of sodium 5,6-benzylidene-l-ascorbate (SBA) to inoperable cancer patients induced a significant reduction in the tumor volume, without any adverse side effects [26]. Further, recent clinical studies indicate that manipulation of LAA levels in vivo can result in clinical benefit for patients with AML and solid tumors [27].

In this chapter, I summarized the common characteristics and differences of the apoptois mechanisms of leukemia cells induced by As_2O_3 and LAA, highlighting the oxidative stress.

Apoptosis and Growth Inhibition in Leukemia Cells after Treatment with L-Ascorbic Acid or Arsenic Trioxide

Treatment of malignant lymphocytic cell lines with LAA (0.25–1 mM) for 24 h led to a marked dose-dependent decrease of cell proliferation [28].The responsive cell lines were human myeloid leukemia cell line HL-60, retinoic acid (RA)-sensitive acute promyelocytic leukemia (APL) cell line NB4 and RA-resistant APL cell line NB4-R1. Different types of leukemia cells such as K562 (chronic myelogenous leukemia cell line) [28] and KG1 (myeloblast cell line) (our unpublished data) were also responsive to LAA. A similar result was obtained with cells containing over 90% blasts from patients with AML (Table 1). In contrast, exposure of three ovarian cancer cell lines including SK-OV-3, OVCAR-3 and 2774 to the same concentrations of LAA had little effect on the proliferation of these cells [28]. The differential sensitivity could be partly associated with differences in LAA uptake when compared to AML cell lines, showing lower uptake efficiency in ovarian cancer cell lines [28].

Our previous results indicated that for three acute myeloid leukemia cell lines (HL-60, NB4 and NB4-R1), 0.5 mM LAA for 24 h induced apoptosis in approximately 16–26% of the treated cells [28]. Induction of apoptosis by LAA in these cell lines demonstrated a dose-dependent pattern. LAA failed to promote apoptosis in ovarian cell lines including SK-OV-3, OVCAR-3 and 2774 [28]. As compared with control cells, LAA treatment resulted in increases in the G1 fraction ranging from 7% (0.25 mM) to 25% (1 mM), with concomitant reduction in the S phase fraction of 3 and 30%, respectively. This indicates the induction of a G1/S arrest by LAA.

As_2O_3 also exerted substantial dose- and time-dependent growth inhibition in malignant lymphocytic cell lines, but different cells exhibited distinct sensitivities to As_2O_3 [1]. In the previous study, the responses of several malignant lymphocytic cell lines and primary acute lymphocytic leukemia (ALL), chronic lymphocytic leukemia (CLL), and lymphoma cells to a wide range of As_2O_3 concentrations (0.1–2 μM) were investigated. As_2O_3 at 0.1–0.25 μM did not substantially inhibit the growth of most cells during 3–5 days of treatment [1]. At 1–2 μM As_2O_3, however, growth inhibition was evident after 1–3 days and became substantial after 3–5 days in almost all cell lines (Table 2). In our experiment, As_2O_3 caused the growth inhibition of HL-60 cells stronger than normal bone marrow (BM) cells [29]. Such a growth inhibitory effect was accompanied by cell apoptosis. The decrease in cell viability was accompanied by the appearance of morphologic characteristics of apoptosis, such as shrinking cytoplasm, condensed chromatin, and nuclear fragmentation with intact cell membrane. Cultures of these cells demonstrated a dose- and time-dependent increase in the number of cells with a sub-G1 DNA content [29].

Although 1-2 μ*M* As_2O_3 inhibited the growth of nearly all cell lines examined (table 2), it did not induce apoptosis in Nalm-6, Raji, and Jurkat cells. It appears to exert a cytostatic effect in these cells [1]. Zhu and coworkers observed a dose- and time-dependent increase in the number of cells with a sub-G1 DNA content in SKW-3 and Molt-4 cells [1]. Also they observed DNA fragmentation in 1 m*M* As_2O_3-treated SKW-3 cells and 2 m*M* As_2O_3-treated Namalwa cells. 2 μ*M* As_2O_3 induced apoptosis in HL-60 [29] NB4 and NB4- R1 (our unpublished data). These results suggested that 2 μ*M* As_2O_3 or less induced apoptosis in most malignant lymphocytic cell lines studied.

The above data suggest that LAA and As_2O_3 treatment had dual effects on malignant lymphocytic cells, i.e., growth arrest and apoptosis.

Table 1. Effects of LAA treatment on cell growth and apoptosis.

LAA	*Cell line*	*% Viability*	*Apoptotic cell %*
(250 μM,	HL-60	70.0 ± 3.0	6.8 ± 3.2
500 μM)		29.6 ± 8.8	24.6 ± 11.0
	NB4	100 ± 11.4	3.1 ± 2.1
		79.4 ± 3.5	15.9 ± 9.0
	NB4-R1	100 ± 4.2	4.4 ± 2.0
		40.5 ± 9.3	19.1 ± 6.9
	NB4-R2	100 ± 3.0	2.6 ± 3.2
		79.1 ± 12.8	21.0 ± 10.1
	KG1	99.6 ± 2.2	5.1 ± 7.8
		78.5 ± 8.4	18.3 ± 3.6
	K562	80.1 ± 1.1	6.3 ± 2.7
		49.5 ± 6.8	21.9 ± 2.9
	U937	61.9 ± 3.6	
		19.5 ± 4.2	
	Normal BM	88 ± 11.7	
		75 ± 7.8	
	Patient with	84.8 ± 10.6	
	AML	69.1 ± 9.6	
	OVCAR	94.8 ± 1.2	6.2 ± 2.1
		91.9 ± 2.3	6.3 ± 3.8
	SK-OV3	100 ± 8.4	0.1 ± 1.6
		98 ± 6.0	0.1 ± 3.0

Data are means ± standard deviations of triplicate experiments. HL-60, human myeloid leukemia; NB4, NB4-R1, NB4-R2, human acute promyelocytic leukemia (APL); KG1 human myeloblast; K562 human chronic myelogenous leukemia; U937, human histiocytic lymphoma; OVCAR, SK-OV3, ovarian cancer.

Table 2. Effects of As_2O_3 treatment on cell growth and apoptosis.

As2O3(1 μM)	*Cell line*	*% Viability*	*Apoptotic cell %*
1day	HL-60	93.4 ± 3.6 [29]	7.2 (control 4.6%)[29]
3day		84.8 [15]	9.0 ± 1.7 (control 4.0%)[15]
	NB4		33 [15] *
		25 [15] *	57.8 ± 7.2 [18]
	KG1		17.5 ± 4.2(48 h)[18]
	K562		6.7 ± 3.1(48 h)[18]
	U937		5.6 ± 3.4 (48 h)[18]
	SKW-3 [1]	87.0 ± 2.9	10 [1] *
		75.3 ± 1.3	25 [1] *
	Molt-4 [1]	96.4 ± 5.1	
		60.6 ± 3.4	
	BJAB [1]	94.0 ± 2.6	
		85.4 ± 1.1	
	Su-DHL-4 [1]	92.0 ± 2.7	10.5 [15] *
		82.0 ± 3.1	12.5 [15] *
	Namalwa [1]	97.4 ± 3.2	3 [1] *
		90.6 ± 4.5	4 [1] *
	Nalm-6 [1]	100.0	
		97.7 ± 2.5	
	Jurkat [1]	95 .8 ± 0.6	
		97.0 ± 1.8	
	Raji [1]	93.2 ± 2.0	
		88.5 ± 3.3	

Data are means ± standard deviations of triplicate experiments. HL-60, human myeloid leukemia; NB4, human acute promyelocytic leukemia (APL); KG1 human myeloblast; K562 human chronic myelogenous leukemia; U937, human histiocytic lymphoma; SKW-3, human T-cell chronic lymphocytic leukemia; Molt-4 and Jurkat, human T-cell acute lymphocytic leukemia; BJAB, B cell lymphoma; su-DHL-4, follicular B-cell lymphoma; Namalwa and Raji, Burkitt's lymphoma; Nalm-6, pre-B cell acute lymphocytic leukemia.

* Data are approximate values determined from the graph.

Correlation of Apoptotic Effect with Redox Status Modulation

It is well known that an important function of GSH is the elimination of reactive oxygen species (ROS) and the reduction of dehydroascorbic acid (DHA) to LAA by GSH-dependent mechanisms in the cell [30]. The correlation of the GSH level with the inhibitory effect of LAA on the proliferation of HL-60, NB4 cells and cells from a patient with AML was investigated [28]. These AML cells were cultured with various concentrations of LAA in the presence or absence of buthione sulfoximine (BSO) or dithiothreitol (DTT) for 24 h, respectively. The addition of 1mM BSO, which inhibited GSH synthesis resulting in lower GSH levels, increased the growth inhibition of HL-60 cells induced by LAA. However, the treatment of 0.2mM DTT, which reduced GSSG back to GSH, decreased the LAA-induced growth inhibition of HL-60 cells [28]. A striking finding is that the inhibitory effect of LAA was completely abrogated by the addition of 500 U/ml catalase, regardless of LAA concentration; whereas the addition of 100 μM NaN_3 synergistically enhanced the effect of LAA [28]. These results were also observed in NB4 cells and in cells from a patient with AML [28]. These results indicated that the removal of H_2O_2 by catalase completely abolished the inhibitory effect of LAA; but the inhibition of endogenous catalase by NaN_3 accelerated it. The inhibitory effect of LAA was not significantly affected by the addition of 1000 U/ml superoxide dismutase (SOD). This result indicated that superoxide anion had no appreciable role in the inhibitory effect of LAA. The results indicated that at least the major portion, if not all, of the growth inhibitory effect of LAA is due to the generation of H_2O_2.

In accordance with cell proliferation assay, LAA-induced apoptosis was affected by the modulation of cellular redox status [28]. The addition of BSO increased the apoptosis in HL-60 cells induced by LAA. However, treatment of DTT decreased the LAA-induced apoptosis of HL-60 cells. On the other hand, treatment with 1 mM BSO alone for 24 hours rarely induced apoptosis in these cells. The results reported that the removal of H_2O_2 by catalase completely abolished the induction of apoptosis by LAA.

Zu and co-workers treated SKW-3, Namalwa, and Jurkat cells simultaneously with 1–2 mM As_2O_3 and 0.2 mM DTT or 1 mM BSO. DTT treatment alone had no substantial effect on mitochondrial membrane potential ($\Delta\psi_m$), cell viability, and the percentage of sub-G1 cells in SKW-3 and Namalwa cells. However, in these cell lines, 0.2 mM DTT blocked the $\Delta\psi_m$ collapse induced by 2 mM. As_2O_3 DTT also blocked the As_2O_3-induced decrease in cell viability, ATP depletion, and the increase in the sub-G1 cell population.

Although 1 mM As_2O_3 had no effect on cell viability and $\Delta\psi_m$ in Namalwa cells, simultaneous treatment with 1 mM As_2O_3 and 1 mM BSO substantially enhanced the magnitude of $\Delta\psi_m$ collapse and apoptosis. This cooperative effect was also evident in Jurkat cells [1] and HL-60 cells [29].

Effects of LAA and Arsenic Trioxide on Intracellular GSH Oxidation State

The glutathione (GSH) redox system is known to modulate the growth-inhibitory effect of arsenicals [15, 31-36]. As^{5+}, cadmium, thalium, selenium, zinc, or mercury did not induce apoptosis at 1 μM in NB4 cells, suggesting an effect of As^{3+} (as in As_2O_3) in this process [14]. The findings that arsenites (40 μM) can induce apoptosis in CHO cells [37], that hydrogen peroxide-resistant CHO cells are less responsive, and that catalase-deficient CHO cells are hypersensitive suggest an important role for hydrogen peroxide as a mediator of arsenic-induced apoptosis [38]. Recently, however, trivalent antimonial compounds (Sb-III), which belongs to the same group of the periodic table (Group V), were investigated as effective apoptosis-triggering compounds [39]. They suggested that antimony induced GSH efflux from THP-macrophages and apoptosis in these cells.

The growth inhibitory and apoptotic effect of As_2O_3 inversely correlates with GSH content of NB4, su-DHL-4, and HL-60 cells [15]. According to the report [15], the GSH content was three times greater in HL-60 cells than in NB4 cells, whereas in su-DHL-4 cells GSH content fell in between these two values. Moreover, Cell growth inhibition and apoptotic effect of As_2O_3 is modulated by experimentally changing the GSH content. Increasing the GSH content by N-acetylcysteine (NAC) treatment [40] in NB4 cells prevented 2 μM As_2O_3 induction of apoptosis [15]. Similarly, lipoic acid completely blocked As_2O_3-induced apoptosis, perhaps by binding As_2O_3 to its vicinal thiol groups and increasing GSH content [41]. Both NAC and lipoic acid treatment inhibited As_2O_3-induced activation of CPP32 and apoptosis. However, NAC did not and lipoic acid only partially inhibited the degradation of PML/RAR-a protein. Thus, the degradation of PML/RAR-a alone does not appear to be sufficient for the As_2O_3 induction of apoptosis.

Dai et al reported that ascorbic acid enhanced As_2O_3 antilymphoma effect from the study using mouse lymphoma P388D1 cells, which demonstrated in vitro ascorbic acid enhancement of As_2O_3 growth inhibition and were implanted into the peritoneum of BDF1 mice [15]. The combined treatment of As_2O_3 (5 mg/kg) and ascorbic acid (500 mg/kg) improved survival time, whereas single-agent treatment at these nontoxic concentrations did not influence survival.

Several groups have shown that LAA augments As_2O_3-mediated apoptosis [15,17]. Our group, however, presented data that LAA alone could modulate malignant cell growth in vitro [28]. Also, LAA alone could effectively induce apoptosis of the AML cell lines HL-60, NB4 and NB4-R1. From our experiments, incubation with LAA for 1 h resulted in the decrease of GSH/GSSG ratio in not only three myeloid leukemia cell lines but also cells from patients with AML, in concentration dependent manners. Interestingly, however, incubation with LAA did not show such decrease of GSH/GSSG ratio in three ovarian cancer cell lines [28], suggesting that the differential sensitivity to LAA-induced apoptosis between cell lines was associated with differences in cellular oxidation states.

EFFECTS OF LAA ON INTRACELLULAR GSH HOMEOSTASIS

From our experiments, LAA treatment changed the GSH oxidation state in cell lines such as HL-60, NB4, NB4-R1 [28] and KG1 (unpublished data). As shown in Figure 1A, incubation with LAA for 1 hr resulted in the decrease of GSH/GSSG ratio in a concentration-dependent manner. However, incubating AML cells with LAA resulted in a marked increase in the concentration of total glutathione relative to the total protein concentration, which reached a maximum after 2 or 3 hr of 163%, 200%, 197% for for HL-60 (Figure 1B), NB4, and KG1 cells, respectively, followed by a gradual decline. The stimulated responses to LAA were dependent on the concentration of LAA.

It follows that if LAA-mediated increase in intracellular GSH oxidation increased intracellular H_2O_2 level, the LAA is likely to induce intracellular oxidative stress via GSH/GSSG ratio changes. We tested the intracellular hydrogen peroxide accumulation physically using H_2DCFDA and flowcytometry analysis. Consistent with the proposed pro-oxidant activity, the LAA-mediated decrease in intracellular GSH/GSSG ratios correlated with intracellular increase in H_2O_2 level as shown (Fig 1C). Additionally, LAA-mediated increase in the total GSH level after around 3 hr of treatment could impair the intracellular H_2O_2 level, explaining that the intracellular ratio of GSH and GSSG correlated functionally with a H_2O_2 accumulation, which results in cellular oxidative stress in HL-60 cells.

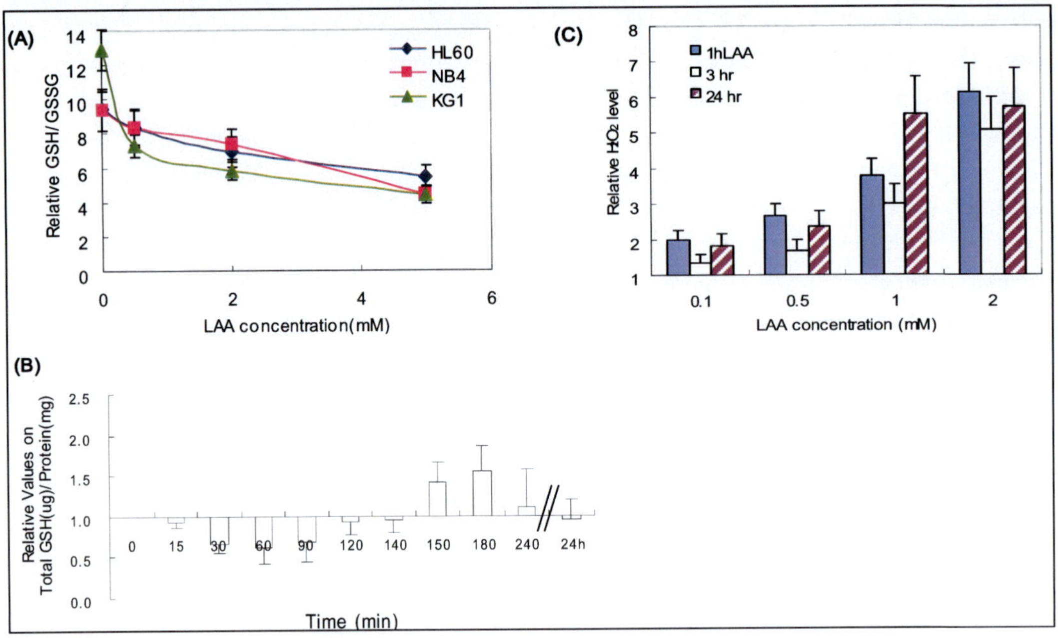

Figure 1. The effect of LAA on the intracellular redox system. (A) The effect of LAA on the GSH/GSSG ratio. HL-60, KG1 and NB4 cells were incubated for 1 hr, and the intracellular GSH and GSSG level were estimated. Values are means ± standard deviations (S.D.) of four experiments. (B) Time course study on the effect of LAA on the concentration of total GSH. HL-60 cells were incubated with 1 mM of LAA, and changes in intracellular GSH were estimated. Values are means ± S.D. of four experiments (C) Analysis of H_2O_2 accumulation in LAA-treated HL-60 cells. HL-60 cells were labeled with H_2DCFDA fluorescent probe for 30 min and then treated with or without LAA for 1 hr, 3 hr and 24 hr. Fluorescence was analyzed by flowcytometry. Data are represented as relative values compared with control.

Effects of LAA on Intracellular-SH Uptake

It is generally believed that cysteine is a rate-limiting precursor for GSH synthesis [42]. We measured cysteine uptake in leukemia cell lines exposed to LAA. The intracellular L-Cys incorporation was measured in intact HL-60 and KG1 cells exposed to LAA by using ^{35}S-labeled-L-Cys in media. Figure 2 illustrates the time-dependent changes in the rate of the uptake of cysteine during culture. It was observed that the rate of uptake, determined under conditions without LAA was very low (at most 119% of baseline by 16 hr). However, it increased greatly showing two peaks after around 1 hr and 3 hr. A glutathione synthesis inhibitor BSO potently inhibited the second peak, suggesting that the incorporation of cysteine corresponding to the second peak is mediated by glutathione synthesis (unpublished data). Our results indicated that LAA induced glutathione synthesis accompanied by intracellular cysteine uptake. Glutathione-S transferases (GSTs) are a family of enzymes that catalyze the conjugation of electrophilic substitute to GSH, and play roles in protecting cells by removing reactive oxygen species and regenerating S-thiolated proteins. Interestingly, intracellular total GSH levels in these AML cells incubated with LAA peaked around 3 hr and declined (Figure 1B), while the increase in [^{35}S]-L-Cys incorporated conformed at 3 hr and kept on going (Figure. 2). This result demonstrated that [^{35}S]-L-Cys once transported into cells through cysteine uptake is incorporated and transferred intracellularly, causing us to hypothesize that the sulfhydryl transfer system is affected by LAA.

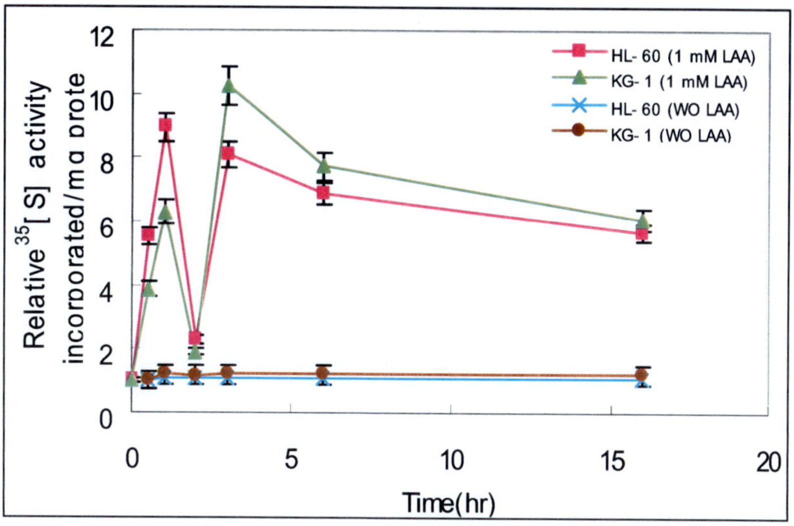

Figure 2. Effect of LAA on the ^{35}[S]-incorporation by cells during culture. Cells were cultured with 5 μCi of ^{35}[S]-cysteine for 30 mins and additionally incubated with or without LAA during the indicated culture period. Then, the cells were washed and the radioactivity of cell extract was measured using scintillation counter. Each point represents a mean value of cumulative data of four separate experiments.

Effect of LAA on GST Enzyme Activity and GST mRNA Synthesis

As expected, LAA also had stimulatory effects on the level of activity of GST in cells (235% increase in HL-60 cells and 159% increase in NB4 cells compared with the control) (Figure. 3). To examine whether the transcriptional regulation of GST genes is responsible for the increase in the GST enzyme activity of the LAA-treated HL-60 cells, RT-PCR was carried out using primers containing non-homologous regions in different genes of GST isoforms (GST-P1, A2, A4, M1, M2). From RT-PCR, the probes were prepared and effects of LAA on the expression of GST-P1, A2, A4, M1, M2 mRNA were estimated by Northern blotting. GST-P1 and A2 mRNA levels increased (unpublished data) but that GST-A4, M1, M2 mRNA levels were not affected by LAA. LAA increased the expression of GST-P1 and A2 mRNA in concentration dependent manners in HL-60 cells. The combined observation points to a stimulatory role for LAA in the expression of GST-P mRNA followed by the concomitant increase in the GST activity.

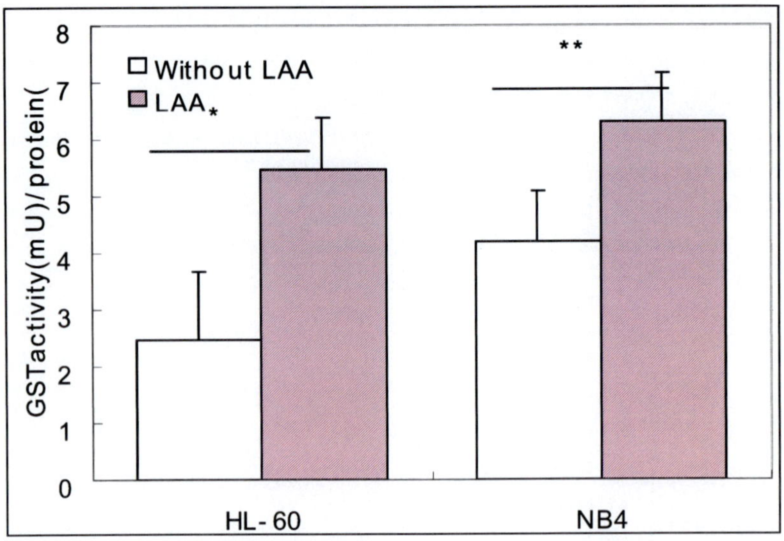

Figure 3. Effect of LAA on the activity of glutathione-S-transferase. HL-60 cells and NB4 cells were treated with 1 mM LAA for 3 hrs, and then GST activity was assayed. Briefly, 1 x 10^6 cells were harvested and washed with PBS, and freeze-thawed for lysis. GST activity was measured at 340 nm in 1-chloro-2,4-dinitrobenzene (CDNB) and GSH for 5 min. Asterisks indicate statistically significant differences(*P<0.01,**P<0.05)

Effects of LAA and Arsenic Trioxide on Intracellular ROS Accumulation

Several previous works indicated that As_2O_3 or LAA-induced apoptosis was modulated by the cellular glutathione redox system, with increased intracellular levels of reduced glutathione (GSH) having an inhibitory effect [1,15,28]. Because of the antioxidant function of GSH, it can be wondered whether modulated GSH level could trigger leukemic cell apoptosis through the generation of ROS. Jing and co-workers showed that the induction of NB4 cell apoptosis and the refractoriness of other leukemic cells to 1 μM As_2O_3 were explained, at least in part, by the accumulation of higher H_2O_2 levels in As_2O_3-treated NB4 as compared with other leukemic cells [18]. Moreover, they showed that, in the presence of glutathione peroxidase (GPx) and catalase inhibitors, it is possible to achieve high apoptotic rates in leukemic cells otherwise refractory to a therapeutic concentration of As_2O_3 (1μM). Although apoptosis was not detected in As_2O_3-treated U937 cells, cell growth was inhibited by As_2O_3 both in NB4 and U937 cells, with IC_{50} of about 0.7 and 1.2 mM at 3 days of treatment, respectively. The growth inhibition was not correlated with arrest in a specific cell cycle phase [18]. These data were consistent with the previous report that As_2O_3 inhibited cell growth in several lymphoma cells by prolongation of the cell cycle without blocking cells in specific phase [1]. Thus, the selective effect of As_2O_3 in NB4 cells was due to apoptosis induction and not to growth inhibition. They suggested that this differential sensitivity to As_2O_3-induced apoptosis in NB4 cells was associated with differences in cellular H_2O_2 levels that were determined.

A causal link between H_2O_2 levels and apoptosis is supported by the coordinate regulation of these events in response to both positive and negative regulation of H_2O_2-catabolizing enzymes.

Consistent with its proposed pro-oxidant activity, the LAA-mediated decrease in GSH/GSSG ratio correlated functionally with an intracellular increase in H_2O_2 level for the four kinds of leukemia cell lines (HL-60, NB4, NB4-R1 and KG1) ([28], Fig1). This differential sensitivity to As_2O_3 or LAA-induced apoptosis in leukemic cells was associated with differences in cellular H_2O_2 levels that were determined by FACS analysis of cells labeled with H_2DCFDA [28,18]. After treatment with 1mM LAA, the marker area percentage representing oxidized DCF increased compared with the untreated controls from 3.7 to 19.7% in HL-60 cells, from 0.1 to 11.9% in NB4 cells, and from 1.0 to 10.4% in NB4-R1 cells within 1 h. LAA-mediated increase in H_2O_2 level was also observed in cells from patients with AML consistent with its GSH oxidation activity in these cells [28]. HL-60, NB4 or NB4-R1 cells had higher oxidized DCF contents than OVCAR-3, 2774 or SK-OV-3 cells 1 h after adding LAA. This demonstrated that the level of H_2O_2 was correlated with the apoptotic activity of LAA. Thus, both the constitutive level of H_2O_2 and the ability of As_2O_3 and LAA to increase the level of H_2O_2 were correlated with the apoptotic activity of As_2O_3 and LAA.

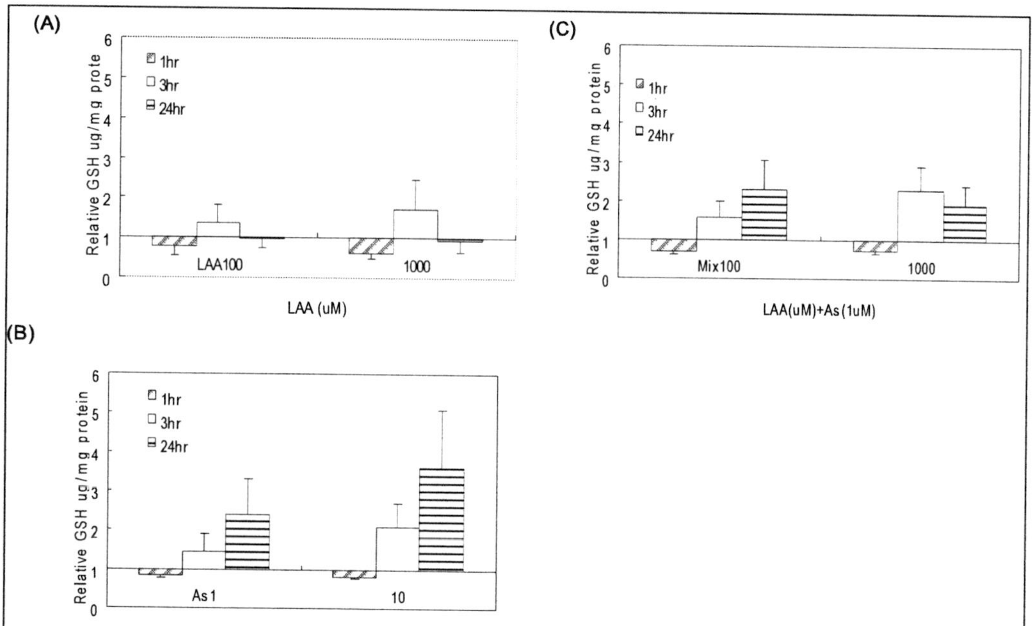

Figure 4. The combined effect of LAA and As_2O_3 on the intracellular redox system. (A) The effect of LAA on the concentration of total GSH. HL-60 cells were incubated with 0.1 and 1 mM of LAA, and changes in intracellular GSH were estimated. Values are means ± S.D. of four experiments (B) The effect of As_2O_3 on the concentration of total GSH. HL-60 cells were incubated with 1 and 10 μM of LAA, and changes in intracellular GSH were estimated. (C) The combined effect of LAA and As_2O_3 on the concentration of total GSH.

However, their characteristic role in the changes of H_2O_2 constitutive level is discriminated. As shown in Figure 1B and 4, LAA dramatically augmented intracellular GSH oxidation and ROS levels within 3 hr and in concentration dependent manners. However, the GSH oxidation and the ROS accumulation do not last for a long period of time. After 3 hr, the total amount of GSH is rather observed to be increased supposedly as a defense mechanism (Figure. 1B). No more ROS accumulation can be inferred from the inclined GSH amount. The dramatic changes of intracellular oxidation state within 3 hr seem to be enough to induce apoptosis, judging from our experiment. It showed that the treatment with 1 mM LAA for only 30 min or 60 min and removal followed by could induce apoptosis in both HL-60 and NB4 cells. In the mean time, As_2O_3 resulted less ROS accumulation than LAA and it was not in concentration dependent manner. However, ROS accumulation increased up to 24 hr with long-lasting effect. It was observed that the treatment of LAA combined with As_2O_3 increases initiative ROS accumulation as well as keeping the effect high up to 24 hr (Figure 5). These results are consistent with the cellular data showing that apoptosis induced by As_2O_3 is evident in three days [1, 15].

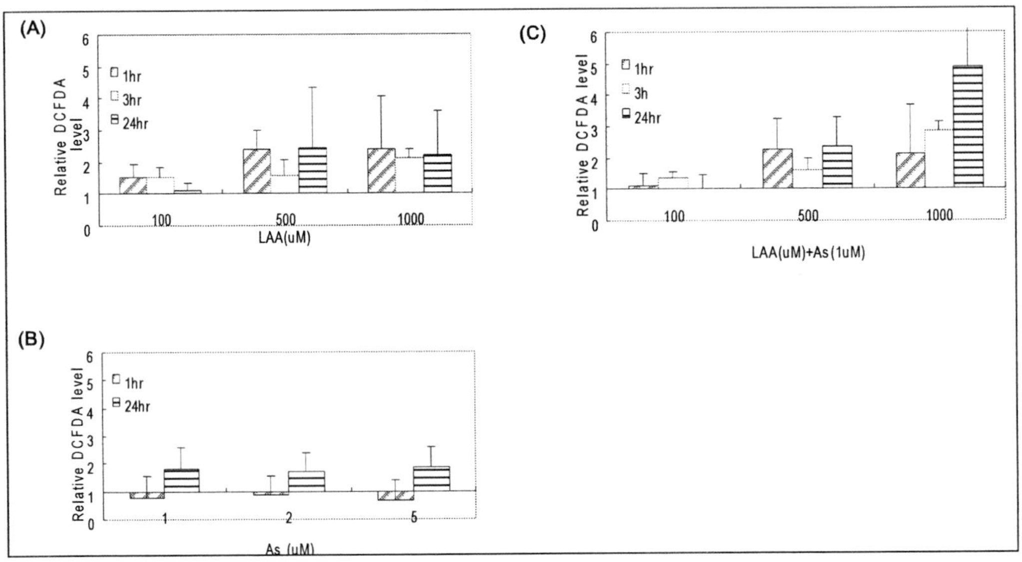

Figure 5. The combined effect of LAA and As_2O_3 on the intracellular ROS. (A) The effect of LAA on the intracellular ROS accumulation. HL-60 cells were labeled with H_2DCFDA fluorescent probe for 30 min and then treated with 0.1, 0.5 and 1 mM of LAA. Fluorescence was analyzed by flowcytometry. Values are means ± S.D. of four experiments (B) The effect of As_2O_3 on the intracellular ROS accumulation. (C) The combined effect of LAA and As_2O_3 on the intracellular ROS accumulation.

Molecular Mechanisms of Apoptosis Induced by LAA or As_2O_3

Lengthened Cell Cycle Time and Decreased Expression of PML Protein

As mentioned above, the cell cycle time was prolonged in 1 mM As_2O_3-treated Namalwa and Raji cells [1]. The cell cycle time was 35.4 hours versus 63.4 hours and 49.9 hours versus 75.3 hours for untreated and 1 mM As_2O_3-treated Namalwa and Raji cells, respectively. No substantial alteration in cell cycle time was found in Jurkat cells upon treatment with 1 mM As_2O_3 (28.1 hours), as compared with untreated cells (32.2 hours). These results suggest that 1 mM As_2O_3 can inhibit proliferation of some malignant lymphocytic cell lines by prolonging the cell cycle instead of arresting cells in a specific phase [1]. Moreover, western blotting revealed that expression of cell cycle-related proteins, including Rb, cyclin D1, CDK4, p16, p53, and c-myc, was not substantially altered in a number of cell lines after treatment with 1 mM As_2O_3 for 12, 24, and 48 hours [1]. However, the PML protein is degraded in NB4 cells treated with 1 mM As_2O_3 [12].

It was reported that As_2O_3 decreased the $\Delta\psi_m$ without altering plasma membrane permeability in SKW-3 and Namalwa cells, in which apoptosis was induced by 1 mM and 2 mM, As_2O_3 respectively [1]. It is known that $\Delta\psi_m$ collapse results in the activation of caspases, which play a central role in the apoptosis-signaling pathway [43,44]. It was found that caspase-3 precursor was cleaved into its active form in 2 mM As_2O_3-treated Namalwa and Raji cells. At the same time, its substrate PARP was also cleaved [1]. However, activation of caspase-3 and degradation of PARP were not observed in As_2O_3-treated BJAB, Molt-4, and SKW-3 cells, all of which underwent substantial apoptosis in response to 1 mM As_2O_3 [1]. Concomitant with the As_2O_3 induction of maximal cytochrome c release, Cpp32 was activated, as shown by the degradation of its precursor (32 kD) and the cleavage of PARP, a Cpp32 substrate. These events were independent of Bcl-2 degradation and were followed by DNA fragmentation [1]. From the result, it is supposed that different cell specific downstream signaling occurs during apoptosis induced by As_2O_3.

Interestingly, it was found that P388 lymphoma cells with multidrug resistance (MDR1) to adriamycin and taxol remain responsive to As_2O_3 [18,45,46]. This suggests that As_2O_3 is not a substrate for the MDR1 drug efflux pathway.

It was demonstrated that apoptosis in As_2O_3-treated NB4 cells proceeds via the classical pathway described for other ROS inductive signals [18,47-50]. Thus, As_2O_3 elicited a rapid decrease in mitochondrial membrane potential, which preceded the release of cytochrome c, Cpp32 activation, DNA fragmentation, and morphologic evidence of apoptosis [18]. However, they did not observe Bcl-2 degradation in NB4 cells treated with 1 to 2 μM As_2O_3 [18]. This is consistent with a report showing that Bcl-2 is not degraded in NB4 cells, even when treated with μM As_2O_3[51]. Moreover, human t(14:18) B-cell lymphoma su-DHL-4 cells overexpressing Bcl-2 are sensitive to 2 μM As_2O_3 without degradation of Bcl-2 [15]. It was suggested that NIH3T342 and HL-60 cells with forced expression of Bcl-2 are more resistant to As_2O_3-induced apoptosis [18]. These findings are consistent with the result that cells overexpressing Bcl-2 are resistant to H_2O_2-induced cell death [52]. As_2O_3- induced apoptosis in NB4 cells was blocked by the caspase inhibitor Z-VAD-FMK [18] suggesting that caspase 9 and caspase 3 were involved in this pathway. This result is consistent with the previous observation that Z-VAD-FMK blocked As_2O_3-induced apoptosis in lymphoma cells [53].

Many previous reports demonstrated that free radical-induced apoptosis may involve the mitochondrial permeability transition pore and is associated with the suppression of Bcl-2 and activation of Bax [1,54]. Treatment of LAA for 24 h in HL-60 cells and for 36 h in NB4 cells repressed the constitutive expression of Bcl-2 while strongly increased the Bax expression via a dose-dependent manner [28]. This result indicated that LAA decreased the ratio of Bcl-2/Bax. It is well known that the decrease of Bcl-2/Bax ratio alters mitochondrial membrane permeability resulting in the release of cytochrome C from mitochondria into the cytosol [55]. Cytochrome C in mitochondria fraction had gradually decreased to a barely detectable level at 0.5mM and then completely disappeared at 1mM of LAA. On the contrary, cytochrome C in cytosolic fraction was increased. Caspase-9 is known to be activated by proteolytic cleavage when interacted with dATP, Apaf-1, and extra mitochondrial cytochrome C [56]. Caspase-9 can activate caspase-3 by proteolytic cleavage and the activated caspase-3 may then cleave vital cellular proteins such as PARP or activate additional caspases by proteolytic cleavage. Caspase-9 and caspase-3 were both activated through the cleavage of the inactive form of procaspase-9 and procaspase-3 via a dose-dependent manner, after being treated with LAA for 24 h in HL-60 cells and for 36 h in NB4 cells [28]. We also found that constitutive PARP was cleaved by the treatment of LAA. PARP is an enzyme involved in DNA repair and genomic integrity. 116 kDa of PARP is proteolytically cleaved into 86 kDa and 23 kDa fragments during apoptosis by caspase-3 which reduces PARP's enzymatic activity, thereby inhibiting DNA repair. These results indicated that LAA caused apoptosis by way of inducing a decrease in Bcl-2/Bax ratio, release of cytochrome C from mitochondria to cytosol, activation of caspase-9 and caspase-3, and cleavage of PARP.

It has been demonstrated that, in acute promyelocytic leukemia cells, As_2O_3 lowers bcl-2 levels and reduces NF-κB translocation to the nucleus, whereas it also induces collapse of mitochondrial transmembrane potential, resulting in cytochrome *c* release and activation of caspase-3 [57,58]. All these events appear to contribute to the induction of apoptosis, and the upstream regulatory mechanisms of such pro-apoptotic signals are recently suggested.

Recent studies showed that all-trans-retinoic acid (RA) or As_2O_3 induce activation of the p38 MAP kinase pathway, and that pharmacological inhibition of p38 augments the antiproliferative and differentiating effects of RA on NB4 cells[59,60]. These studies established that the small G-protein Rac1 and the MAPKAP kinase 2 are both RA-dependent and arsenic-dependent upstream and downstream effectors, respectively, of p38 MAP kinase. Furthermore, they indicated that treatment with a p38 inhibitor enhanced RA-dependent induction of cell proliferation and arsenic-dependent growth inhibitory responses. In our previous study, we determined that the treatment with LAA induced activation of the ERK MAP kinase pathway in HL60, NB4, and NB4-R1 acute myeloid leukemia cell lines [61]. Considering that LAA induced activation of ERK occurs both in cells that are resistant and sensitive to differentiation by RA, the upstream signals that mediate LAA-dependent activation of ERK are separated from the ones mediating RA-dependent signals. From the above results, ERK rather than p38 is activated by LAA in acute myeloid leukemia cell lines such as HL-60, NB4 and NB4-R1, indicating that it also involved in the induction of LAA-response in these cells.

It is well known that Raf1 is an upstream regulator of the ERK MAP kinase in response to stress [62]. Ras is a member of family of small GTPases and is a key mediator of intracellular signaling cascades, including the Raf pathway [63]. The Raf-dependent signaling pathways have been shown to regulate the ERKs [62]. ERK1 and ERK2 were among the first substrates of Raf to be discovered and have proven to be ubiquitous and versatile mediators of Raf signal transduction [64-67]. LAA treatment resulted in activation of Raf1 in three cell lines including HL-60, NB4 and NB4-R1 [61] providing evidence for involvement of Raf1 kinase in a LAA-activated signal pathway. These results suggest that Raf1 is an upstream regulator of the LAA-mediated activation of ERK and that the early upstream signal induced by LAA is different from an As_2O_3-activated upstream signal utilizing Rac1 and subsequent p38 MAP kinase activation.

The pre-treatment with the chemical antioxidant, N-acetyl cysteine attenuated Raf-1 activation, suggesting that H_2O_2 may be required for the signaling pathway leading to Raf-1 activation [61]. In addition, genistein, a tyrosine kinase inhibitor, reduced the activation of Raf-1 kinase by LAA [61]. It is assumed that tyrosine phosphorylation is required for the activation of Raf-1.

Previous work has shown that MAPK-activated protein kinase 2 (MAPKAP kinase 2) is a downstream effector of p38, which mediates intracellular signals in response to growth factors, hormones and neurotransmitters [68]. In our study [61], however, LAA-induced activation of ERK promotes downstream activation of MAPKAP kinase 2, in HL-60, NB4 or NB4-R1 cells. Lysates from LAA-treated cells were immunoprecipitated with an anti-MAPKAP kinase 2 antibody, and applied to an in vitro kinase assay using peptide as exogenous substrates. LAA treatment of cells activated the MAPKAP kinase 2, evidenced by the phosphorylation of substrate peptide in the kinase assay [61]. In addition, treatment of cells with the ERK-specific inhibitor PD98059 blocked the LAA-dependent activation of MAPKAP kinase 2 [61] indicating the downstream engagement of MAPKAP kinase 2 in an LAA-activated signal transduction pathway.Although LAA slightly activated p38 by Western blot, SB203580 partially reduced the activation of MAPKAP kinase 2 [61]. Judging from our results, LAA did not induce p38 activation. Interestingly, however, MAPKAP kinase 2 was involved as a downstream modulator for both p38 and ERK signal pathway. Despite the fact that MAPKAP kinase 2 was among the specific substrates of p38 identified, its mechanism of action seems to be also regulated by ERK. One ERK-catalyzed phosphorylation site was identified in the activation loop of the C-terminal kinase of MAPKAP kinase 2 [69]. In addition, it has been observed that ERK alone induced no activation of MAPKAP kinase 2, but enhanced activation of MAPKAP kinase 2 by 3-phosphoinositide-dependent protein kinase-1 (PDK1) [68]. Thus, full activation of the MAPKAP kinase 2 may require the collaboration of protein kinases such as p38, PDK1 and ERK. Altogether, these studies suggest that treatment of cells with LAA results in sequential activation of a Raf1, ERK1/2 and MAPKAP kinase 2 pathway, whose activation is probably upstream of cell apoptosis induced by LAA. In our experiments to define the functional role of this cascade, we found that inhibition of this pathway using ERK-specific inhibitor PD98059 suppressed LAA-dependent induction of apoptosis and inhibition of cell growth [61]. Thus, it appears that important signals for regulation of LAA-dependent responses are mediated by the ERK1 and/or 2 isoforms. Consistent with this, overexpression of a dominant negative ERK mutant inhibited LAA-dependent cytotoxicity in HL-60 cells [61].

Overexpression of Raf proteins is associated with proliferation in cells, including human hematopoietic cells or human erythroid progenitor cells [70-72]. However, overexpression of activated Raf proteins is associated with cell growth arrests in some cell lines [73-75]. It is not clear why overexpression of the Raf gene can lead to such conflicting results, but it has been suggested that the opposite outcomes may be determined by the amount or activity of the Raf oncoprotein [70,76]. Under LAA exposure, strong Raf signaling results in the activation of downstream kinases including ERK and MAPKAP kinase 2, which may provide another evidence for an antiproliferative effect of not only Raf but also ERK. Our data using cells obtained freshly from three patients with AML were consistent with the previous data using cell lines, showing that LAA suppressed leukemic progenitor cell growth with this suppressive effect abrogated by an ERK inhibitor [61]. This indicated the involvement of ERK in the LAA suppressive effect in patient specimens as well as cell lines. Our current view of the apoptotic pathway induced by LAA, as supported by the abovenoted data, is depicted diagrammatically in Figure 6. Our interim conclusion is that LAA induces the apoptosis of HL-60 cells, augmented by As_2O_3, through modulation of redox status, definitely, depletion of GSH by intracellular dehydroascorbic acid reduction and consequent hydrogen peroxide accumulation. And through MAPK modulation, and induction of apoptotic cascade including Bcl2/bax, cytochrome C, caspase 9, caspase 3 and PARP.

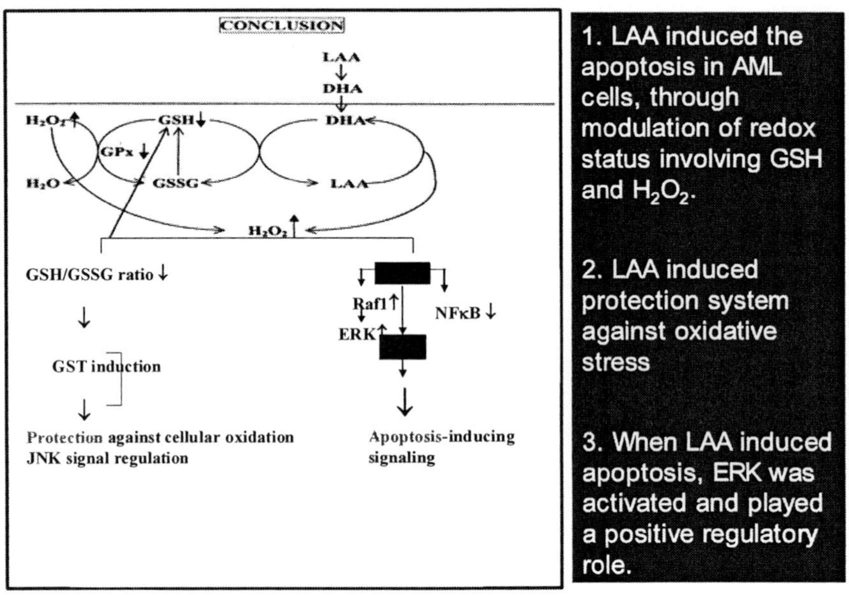

Figure 6. Diagrammatic representation of the proposed signaling pathway induced by LAA in HL-60 cells.

Inhibition of NF-κB Activation

As mentioned above, it has been demonstrated that, in acute promyelocytic leukemia cells, arsenic trioxide reduces NF-κB translocation to the nucleus [57,58]. Nuclear factor kappa-B (NF-κB) has been known to regulate expression of many genes in modulating cellular proliferation, inflammatory responses, and apoptosis [77,78]. NF-κB exists mainly as a hetero- or homo-dimer consisting of subunits of Rel family such as p50, p52, c-Rel, p65 (RelA), or RelB. NF-κB is normally sequestered in the cytoplasm as an inactive complex by binding to an inhibitory protein, IκB. IκB tightly modulates NF-κB activity by preventing nuclear translocation of NF-κB and inhibiting DNA binding. When cells are exposed to various extracellular stimuli, IkB is rapidly degraded by proteosomes following its phosphorylation and subsequent ubiquiti-nation. Hence, free NF-κB translocates to the nucleus, where it binds to its binding sites in the promoter of target genes, thereby controlling their expression [79,80].

Recently, many drugs with the ability to inhibit the cyclooxygenase (COX) enzymes have been shown to prevent or delay development of certain tumors. COX-1 is constitutively expressed to fulfill its beneficial housekeeping roles in most human tissues; whereas cyclooxygenase-2 (COX-2), with multiple binding sites on its promoter for various transcription factors including NF-κB, is not constitutively expressed in most normal tissues but it is rapidly induced by cytokines, tumor promoters, growth factor, and oncogenes [81]. The expression of COX-2 has been shown to be abnormally or overexpressed in various human tumors, including breast, head and neck, lung, pancreatic, and gastric cancers [82-86].

Unstimulated HL-60 cells had a high basal level of NF-κB activity, which was downregulated by treatment of As_2O_3, in a concentration- and time-dependent manner [29]. The NF-κB complex inactivated by As_2O_3 was indeed NF-κB existing as a heterodimer of p50 and p65 subunits. As_2O_3 inhibited NF-κB activation by preventing the degradation of IκB-α and the translocation of p65 subunit as well as inhibiting the binding between NF-κB and DNA [29]. These results are related to the data that inhibition of NF-κB strongly increases the apoptosis rate and slows down cell proliferation in Hodgkin/Reed-Sternberg cells [87]. Also, these results are important considering the data that up to 25% of hematopoietic malignancies in adults contain rearrangements or amplifications of members of the NF-κB/IκB gene families [88].

The high basal level of NF-κB activity in unstimulated HL-60 cells was almost completely downregulated by treatment of LAA, with or without 1 mM As_2O_3, in a concentration- and time-dependent manner [89]. LAA Inhibited NF-κB activation by preventing the degradation of IkB-α and the translocation of p65 Subunit but did not directly disrupt binding of NF-κB to DNA [89].

Our results indicated that inhibitory effect of As_2O_3 and LAA on NF-κB activity was dependant on GSH level confirming that intracellular GSH played a suppressive role in the inhibitory effect of As_2O_3 and LAA on NF-κB activity. Also, both As_2O_3- and LAA–induced inactivation of NF-κB activity is correlated with H_2O_2 level [29,89]. As_2O_3 and LAA inhibit expression of COX-2 and it appears to be through suppression of NF-κB DNA binding activity [29,89].

The property of As_2O_3 that disrupts the DNA binding and prevents the degradation of IkB-α and the translocation of the p65 subunit, appears to be via binding to SH groups in the DNA-binding region of NF-κB [29]. These results support the concept that intracellular GSH level could be associated with the sensitivity of cells to As_2O_3-induced apoptosis. Moreover, As_2O_3 inhibits glutatathione peroxidase activity, important in elimination of H_2O_2, and thus increases cellular H_2O_2 content in NB4 cells [90]. However, it was unclear whether LAA could bind to SH groups in the DNA-binding protein.

Early Molecular Targets Modified by L-Ascorbic Acid

To understand the molecular basis of LAA-induced oxidative stress, it is important to identify the expression or function of proteins which are modified by LAA. We analyzed molecular modifications induced by LAA in NB4 leukemia cells using a proteomic approach. This methodology provides important qualitative information on post-translational modifications to each protein and quantitative data on protein expressions in response to a particular stimulus. This information is particularly important because it provides data on early cellular events, such as, on the stimulus and signaling cascades triggered independently of protein neosynthesis.

A 30 min incubation of NB4 cells with LAA resulted in the uptake of LAA in a concentration-dependent manner (in the 0–2 mM range) [28,91]. In accordance with its proposed pro-oxidant activity, LAA-mediated reduction in the GSH/GSSG ratio correlated functionally with an intracellular H_2O_2 increase in the NB4 cell line.

The immediate effects of cell stimuli are associated with protein post-translational modifications, such as, phosphorylation, glutathionylation and cysteine oxidation. To study these early modifications, NB4 human leukemia cells were treated with 0.5 mM LAA and then collected for two-dimensional electrophoresis analysis. Approximate 240 different spots that were focused in the 4-7 pH range were analyzed per sample. After exposing cells to LAA, we observed; one new spot, three intensified spots, and five attenuated spots [91]. Each of these spots were excised, digested with trypsin and analyzed by MALDI–TOF. Peptide mass fingerprint analysis and nonredundant sequence database matching allowed the unambiguous identification of all of the analyzed species [91]. From the result, tropomyosin was identified as a marker of the oxidative effect of LAA in NB4 cells. The spot corresponding to tropomyosin was positioned at a pI value of approximately 5.0. This spot was attenuated in LAA-treated cells, in addition, a new spot was detected at pI, 4.9 of almost the same molecular weight [91]. This new spot was also identified as tropomyosin, which suggested that some chemical modification had affected its pI value. This result is consistent with the previous data that the extracellular signal-regulated kinase (ERK) mediated phosphorylation of tropomyosin-1 promotes cytoskeleton remodeling in response to oxidative stress [92]. The acidic shift of the spot with a pI 5.0 (T) to the phosphorylated tropomyosin spot by the treatment with LAA was found to be abrogated by the co-treatment with PD98059 [91], demonstrating that the phosphorylation of tropomyosin was responsible for the observed acidic shift.

The significance of this is unclear but may be related to differences in the regulation of the actomyosin contractile system in non-muscle cells as compared with that present in muscle cells. One additional point worth noting is that the proteins which specifically reacted with sera from chronic myeloid leukemia patients included structural proteins, such as, beta-tubulin and tropomyosin isoforms [31]. Although the function of these proteins in myeloid leukemia needs further investigation, tropomyosin may have value as a leukemia-associated antigen and as a molecular target in antigen-based immunotherapy. From this point of view, it is important that LAA causes a tropomyosin isoform to be modified during the immediate early response.

Another protein identified from the proteomic analysis is a thiol/disulfide exchange catalyst as a marker of the effect of LAA in NB4 cells [91]. Protein disulfide isomerase (PDI) belongs to the superfamily of thiol/disulfide exchange catalyst which acts as protein-thiol oxidoreductase enzymes and shares sequence similarities with thioredoxin [93]. PDI is composed of four domains which have similarities with thioredoxin folds (a-b-b'-a') [94]. In our study [91], the intensity of spot corresponding to the PDI b subunit was weakened in LAA-treated cells versus control cells. These results demonstrate that thiol/disulfide exchange proteins are regulated in NB4 cells after LAA exposure. This is not surprising because our study also shows that intracellular GSH/GSSG exchange occurs shortly after LAA exposure ([28], Figure 1). Therefore, we speculate that the biochemical pathway leading to thiol/disulfide redox regulation could be triggered by LAA.

Interestingly, of the proteins with expressions changed by LAA treatment, immunoglobulin-heavy-chain binding protein (BiP, identical to Hsp70 chaperone) [91] like PDI was also a multi-domain chaperone. BiP was associated with the α-subunit of prolyl 4-hydroxylase (P4-H) by a disulfide bond [90]. P4-H consists of two distinct polypeptides, the catalytically more important α-subunit and the β-subunit, which is identical to the multifunctional enzyme, PDI [90]. Thus, BiP was associated with α-subunit of P4-H, which is another partner of PDI. It was determined that the interaction of PDI with its substrates is due to a change in disulfide bonds, indicating that intra-chain disulfide bonds between domains and substrates had been reduced [91]. Taken together, these results suggest that LAA oxidizes intracellular reduced glutathione and modulates disulfide bond formation in proteins.

REFERENCES

[1] Zhu, X. H., Shen, Y. L., Jing, Y. K., Cai, X., Jia, P. M., Huang, Y., Tang, W., Shi, G. Y., Sun, Y. P., Dai, J., Wang, Z. Y., Chen, S. J., Zhang, T. D., Waxman, S., Chen, Z., Chen, G. Q. (1999). Apoptosis and Growth Inhibition in Malignant Lymphocytes After Treatment With Arsenic Trioxide at Clinically Achievable Concentrations. *J Natl Cancer Inst, 91*, 772-778.

[2] Rcw, D. A. (1988). Cell and molecular mechanisms of pathogenesis and treatment of cancer. *Postgrad Med J, 74*, 77-88.

[3] Steller, H. (1995). Mechanisms and geges of cellular suicide. *Science, 267*, 1445-1449.

[4] Campana, D., Coustan-Smith, E., Manabe, A., Buschle, M., Raimondi, S. C., Behm, F. G., et al. (1993). Prolonged survival of B-lineage acute lymphoblastic leukemia cells is accompanied by overexpression of bcl-2 protein. *Blood, 81*, 1025-1031.

[5] Coustan-smith, E., Kitanaka, A., Pui, C. H., Nc-Ninch, L., Evans, W. E., Raimindi, S. C., et al. (1996). Clinical relevance of BCL-2 overexpression in childhood acute lymphoblastic leukemia. *Blood, 87*, 1140-1146.

[6] Kitada, S., Andersen, J., Akar, S., Zapata, J. M., Takayama, S., Krajewski, S., et al. (1998). Expression of apoptosis-regulating proteins in chronic lymphocytic leukemia: correlations with *in vitro* and *in vivo* chemoresponses. *Blood, 91*, 3379-3389.

[7] Fisher, D. E. (1994). Apoptosis in cancer therapy: crossing the threshold. *Cell, 78*, 539-542.

[8] Staunton, M. J., Gaffney, E. F. (1998). Apoptosis: basic concepts and potential significance in human cancer. *Arch Pathol Lab Med, 122*, 310-319.

[9] Sun, H. D., Ma, L., Hu, X. C., Zhang, T. D. (1992). Ai-Lin I treated 32 cases of acute promyelocytic leukemia. *Chin J Integrat Chinese Western Med, 12*, 170-172.

[10] Zhang, P., Wang, S. Y., Hu, X. H. (1996). Arsenic trioxide treated 72 cases of acute promyelocytic leukemia. *Chin J Hematol. 17*, 58-61.

[11] Mervis, J. (1996). Ancient remedy performs new tricks[news]. *Science, 273*, 578.

[12] Chen, G. Q., Zhu, J., Shi, X. G., Ni, J. H., Zhong, H.J., Si, G.Y., Jin, X. L., Tang, W., Li, X. S., Xong, S. M., Shen, Z. X., Sun, G. L., Ma, J., Zhang, P., Zhang, T. D., Gazin, C., Naoe, T., Chen, S. J., Wang, Z. Y., Chen, Z. (1996). *In vitro* studies on cellular and molecular mechanisms of arsenic trioxide (As_2O_3) in the treatment of acute promyelocytic leukemia: As_2O_3 induces NB4 cell apoptosis with downregulation of Bcl-2 expression and modulation of PML-RAR alpha/PML proteins. *Blood, 88*, 1052-1061.

[13] Chen, G. Q., Shi, X. G., Tang, W., Xiong, S. M., Zhu, J., Cai, X., et al. (1997). Use of arsenic trioxide (As_2O_3) in the treatment of acute promyelocytic leukemia (APL): I. As_2O_3 exerts dose-dependent dual effects on APL cells. *Blood, 89*, 3345-3353.

[14] Kitamura, K., Yoshida, H., Ohno, R., Naoe, T. (1997). Toxic effects of arsenic (As^{3+}) and other metal ions on acute promyelocytic leukemia cells. *Int J Hematol, 65*, 179-185.

[15] Dai, J., Weinberg, R. S., Waxman, S., Jing, Y. (1999). Malignant cells can be sensitized to undergo growth inhibition and apoptosis by arsenic trioxide through modulation of the glutathione redox system. *Blood, 93*, 268-277.

[16] Bahlis, N. J., Jordan-Mcmurry, I., Grad, J. M., Reis, I., Neel, J., Kharfan-Dabaja, M. et al. (2001). Phase I results from a phase I/II study of arsenic trioxide (As_2O_3) and ascorbic acid (AA) in relapsed and chemorefractory multiple myeloma. *Blood, 98*, 375a.

[17] Grad, J. M., Bahlis, N. J., Reis, I., Oshiro, M. M., Dalton, W. S., Boise, L. H. (2001). Ascorbic acid enhances arsenic trioxide-induced cytotoxicity in multiple myeloma cells. *Blood, 98*, 805-813.

[18] Jing, Y., Dai, J., Chalmers-Redman, R. M. E., Tatton, W. G., Waxman, S.(1999). Arsenic trioxide selectively induces acute promyelocytic leukemia cell apoptosis via a hydrogen peroxide-dependent pathway. *Blood, 94*, 2102-2111.

[19] Bram, S., Froussard, P., Guichard, M. (1980). Ascorbic acid preferential toxicity for malignant melanoma cells. *Nature, 284*, 629-631.

[20] Bruchelt, G., Baader, L., Reith, A. G., Holger, N. L., Gebhardt, S., Niethammer, D. (1993). Rationale for the use of ascorbic acid in neuroblastoma therapy. In M. Schwab (Ed.), *Human neuroblastoma* (pp. 34-40). Harwood Academic Publishers.

[21] Fujinaga, S., Sakagami, H., Kuribayashi, N., Takahashi, H., Amano, Y., Sakagami, T. et al. (1994). Possible role of hydrogen peroxide in apoptosis induction by ascorbic acid in human myelogenous leukemic cell lines. Showa University *Medical Science, 6*, 135-144.

[22] De Laurenzi, V., Melino, G., Savini, I., Annicchiarico-Petruzzelli, M., Finazzi-Agro, A., Avigliano, L. (1995). Cell death by oxidative stress and ascorbic acid regeneration in human neuroectodermal cell lines. *Eur J Cancer, 31*, 463-466.

[23] Park, C. H., Bergsagel, D. E., McCulloch, E. A. (1971). Ascorbic acid: a culture requirement for colony formation by mouse plasmacytoma cells. *Science, 174*, 720-722.

[24] Park, C. H. (1985). The biological nature of the effect of ascorbic acids on the growth of human leukemic cells. *Cancer Res, 45*, 3969-3973.

[25] Park, C. H., Kimler, B. F., Bodensteiner, D., Lynch, S. R., Hassanein, R. S. (1992). In vitro growth modulation by L-ascorbic acid of colony-forming cells from bone marrow of patients with myelodysplastic syndromes. *Cancer Res, 52*, 4458-4466.

[26] Sakagami, H., Asano, K., Fukuchi, K., Gomi, K., Ota, H., Kazama, K. et al. (1991). Induction of tumor degeneration by sodium benzylideneascorbate. *Anticancer Res, 11*, 1533-1538.

[27] Park, C. H., Kim, W. S., Park, C., Lee, M. H., Boo, Y. C., Yoon, S. S. (1999). Clinical disease suppression and reduction in acute myeloid leukemia and solid tumors by very high dose of L-ascorbic acid: A new concept and in search of molecular targets. *Clin Cancer Res, 5 (SUPPL.)*, 3784s

[28] Park, S., Han, S. S., Park, C. H., Hahm, E. R., Lee, S. J., Park, H. K., Lee, S. H., Kim, W. S., Jung, C. W., Park, K., Riordan, H. D., Kimler, B. F., Kim, K., Lee, J. H. (2004). L-Ascorbic acid induces apoptosis in acute myeloid leukemia cells via hydrogen peroxide-mediated mechanisms. *Int J Biochem Cell Biol, 36*, 2180-2195.

[29] Han, S. S., Kim, K., Hahm, E. R., Park, C. H., Kimler, B. F., Lee, S. J., Lee, S. H., Kim, W. S., Jung, C. W., Park, K., Kim, J., Yoon, S. S., Lee, J. H., Park, S. (2005). Arsenic trioxide represses constitutive activation of NF-kappaB and COX-2 expression in human acute myeloid leukemia, HL-60. *J Cell Biochem, 94*, 695-707.

[30] Meister, A. (1994). Glutathione-ascorbic acid antioxidant system in animals. *J Biol Chem. 269*, 9397-9400.

[31] Zou, L., Wu, Y., Pei, L., Zhong, D., Gen, M., Zhao, T., Wu, J., Ni, B., Mou, Z., Han, J., Chen, Y., Zhi, Y. (2005). Identification of leukemia-associated antigens in chronic myeloid leukemia by proteomic analysis. *Leuk Res, 29*, 1387-1391.

[32] Lee, T.C., Wei, M. L., Chang, W. J., Ho, I. C., Lo, J. F., Jan. K. Y., Huang, H. (1989). Elevation of glutathione levels and glutathione S-transferase activity in arsenic-resistant Chinese hamster ovary cells. *In Vitro Cell Dev Biol, 25*, 442.

[33] Lee, T. C., Ko, J. L., Jan, K. Y. (1989). Differential cytotoxicity of sodium arsenite in human fibroblasts and Chinese hamster ovary cells. *Toxicology, 56*, 289.

[34] Ochi, T., Kaise, T., Oya-Ohta, Y. (1994). Glutathione plays different roles in the induction of the cytotoxic effects of inorganic and organic arsenic compounds in cultured BALB/c 3T3 cells. *Experientia, 50*, 115.

[35] Ochi, T., Nakajima, F., Sakurai, T., Kaise, T., Oya-Ohta, Y. (1996). Dimethylarsinic acid causes apoptosis in HL-60 cells via interaction with glutathione. *Arch Toxicol, 70*, 815.

[36] Scott, N., Hatlelid, K. M., MacKenzie, N. E., Carter, D. E. (1993). Reactions of arsenic(III) and arsenic(V) species with glutathione. *Chem Res Toxicol, 6*, 102.

[37] Wang, T. S., Kuo, C. F., Jan, K. Y., Huang, H. (1996). Arsenite induces apoptosis in chinese hamster ovary cells by generation of reactive oxygen species. *J Cell Physiol, 169*, 256.

[38] Cantoni, O., Hussain, S., Guidarelli, A., Cattabeni, F. (1994). Cross-resistance to heavy metals in hydrogen peroxide-resistant CHO cell varients. *Mutat Res, 324*, 1.

[39] Wyllie, S., Fairlamb, A. H. (2006). Differential toxicity of antimonial compounds and their effects on glutathione homeostasis in a human leukaemia monocyte cell line. *Biochem Pharmacol, 71(3)*, 257-267.

[40] Droge, W., Eck, H. P., Mihm, S. (1992). HIV-induced cysteine deficiency and T-cell dysfunctionA rationale for treatment with N-acetylcysteine. *Immunol Today, 13*, 211.

[41] Han, D., Handelman, G., Marcocci, L., Sen, C. K., Roy, S., Kobuchi, H., Tritschler, H. J., Flohe, L., Packer, L. (1997). Lipoic acid increases de novo synthesis of cellular glutathione by improving cystine utilization. *Biofactors, 6*, 321.

[42] Watanabe, H., Bannai, S. (1987). Induction of cystine transport activity in mouse peritoneal macrophages. *J Exp Med, 165*, 628-640.

[43] Kumar, S. (1995). ICE-like protease in apoptosis. *Trends Biochem Sci, 20*, 198-204.

[44] Takahashi, A., Earnshaw, W. C. (1996). ICE-related proteases in apoptosis. *Curr Opin Genet Dev, 6*, 50-55.

[45] Wang, H. F., Lee, T. C. (1993). Glutathione S-transferase pi facilitates the excretion of arsenic from arsenic-resistant Chinese hamster ovary cells. *Biochem Biophys Res Commun, 192*, 1093.

[46] Rappa, G., Lorico, A., Flavell, R. A., Sartorelli, A. C. (1997). Evidence that the multidrug resistance protein (MRP) functions as a co-transporter of glutathione and natural product toxins. *Cancer Res, 57*, 5232.

[47] Marchetti, P., Decaudin, D., Macho, A., Zamzami, N., Hirsch, T., Susin, S. A., Kroemer, G. (1997). Redox regulation of apoptosis: Impact of thiol oxidation status on mitochondrial function. *Eur J Immunol, 27*, 289.

[48] Satoh, T., Enokido, Y., Aoshima, H., Uchiyama, Y., Hatanaka, H. (1997). Changes in mitochondrial membrane potential during oxidative stress-induced apoptosis in PC12 cells. *J Neurosci Res, 50*, 413.

[49] Yang, J. C., Cortopassi, G. A. (1998). Induction of the mitochondrial permeability transition causes release of the apoptogenic factor cytochrome c. *Free Radic Biol Med, 24*, 624.

[50] Stridh, H., Kimland, M., Jones, D. P., Orrenius, S., Hampton, M. B. (1998). Cytochrome c release and caspase activation in hydrogen peroxide- and tributyltin-induced apoptosis. *FEBS Lett, 429*, 351.

[51] Ma, D. C., Sun, Y. H., Chang, K. Z., Ma, X. F., Huang, S. L., Bai, Y. H., Kang, J., Liu, Y. G., Chu, J. J. (1998). Selective induction of apoptosis of NB4 cells from G2+M phase bysodium arsenite at lower doses. *Eur J Haematol, 61*, 27.

[52] Hockenbery, D. M., Oltvai, Z. N., Yin, X. M., Milliman, C. L., Korsmeyer, S. J. (1993). Bcl-2 functions in an antioxidant pathway to prevent apoptosis. *Cell, 75*, 241.

[53] Akao, Y., Mizoguchi, H., Kojima, S., Naoe, T., Ohishi, N., Yagi, K. (1998). Arsenic induces apoptosis in B-cell leukaemic cell lines in vitro: Activation of caspases and down-regulation of Bcl-2 protein. *Br J Haematol, 102*, 1055.

[54] Kroemer, G., Zamzami, N., Susin, S. A. (1997). Mitochondrial control of apoptosis. *Immunol Today, 18*, 44-51.

[55] Kluck, R. M., Martin, S. J., Hoffman, B. M., Zhou, J. S., Green, D. R., Newmeyer, D. D. (1997). Cytochrome c activation of CPP32-like proteolysis plays a critical role in a Xenopus cell-free apoptosis system. *EMBO Journal, 16 (15)*, 4639-4649.

[56] Li, P., Nijhawan, D., Budihardjo, I., Srinivasula, S. M., Ahmad, M., Alnemri, E. S., Wang, X. (1997). Cytochrome c and dATP-dependent formation of Apaf-1/caspase-9 complex initiates an apoptotic protease cascade. *Cell, 91*, 479-489.

[57] Wang, Z. G., Rivi, R., Delva, L., Konig, A., Scheinberg, D. A., Gambacorti-Passerini, C., Gabrilove, J. L., Warrell, R. P., Jr., and Pandolfi, P. P. (1998). Arsenic Trioxide and Melarsoprol Induce Programmed Cell Death in Myeloid Leukemia Cell Lines and Function in a PML and PML-RAR Independent Manner. *Blood, 92*, 1497-1504.

[58] Mahieux, R., Pise-Masison, C., Gessain, A., Brady, J. N., Olivier, R., Perret, E., Misteli, T., and Nicot, C. (2001). Arsenic trioxide induces apoptosis in human T-cell leukemia virus type 1- and type 2-infected cells by a caspase-3-dependent mechanism involving Bcl-2 cleavage. *Blood, 98*, 3762-3769.

[59] Alsayed, Y., Uddin, S., Mahmud, N., Lekmine, F., Kalvakolanu, D. V., Minucci, S., Bokoch, G., Platanias, L. C. (2001). Activation of Rac1 and the p38 Mitogen-activated Protein Kinase Pathway in Response to All-trans-retinoic Acid. *J Biol Chem, 276*, 4012-4019.

[60] Verma, A., Mohindru, M., Deb, D. K., Sassano, A., Kambhampati, S., Ravandi, F., Minucci, S., Kalvakolanu, D. V., Platanias, L. C. (2002). Activation of Rac1 and the p38 mitogen-activated protein kinase pathway in response to arsenic trioxide. *J Biol Chem, 277*, 44988-44995.

[61] Park, S., Park, C. H., Hahm, E. R., Kim, K., Kimler, B. F., Lee, S. J., Park, H. K., Lee, S. H., Kim, W. S., Jung, C. W., Park, K., Riordan, H. D., Lee, J. H. (2005). Activation of Raf1 and the ERK pathway in response to l-ascorbic acid in acute myeloid leukemia cells. *Cell Signal, 17*, 111-119.

[62] English, J., Pearson, G., Wilsbacher, J., Swantek, J., Karandikar, M., Xu, S., Cobb, M. H. (1999). New insights into the control of MAP kinase pathways. *Exp Cell Res, 253*, 255-270.

[63] Malumbres, M., Pellicer, A. (1998). RAS pathways to cell cycle control and cell transformation. *Front Biosci, 3*, 887-912.

[64] Jones, S. W., Erikson, E., Blenis, J., Maller, J. L., Erikson, R. L. (1988). A Xenopus ribosomal protein S6 kinase has two apparent kinase domains that are each similar to distinct protein kinases. *Proc Natl Acad Sci U S A, 85*, 3377-3381.

[65] Sturgill, T. W., Ray, L. B., Erikson, E., Maller, J. L. (1988). Insulin-stimulated MAP-2 kinase phosphorylates and activates ribosomal protein S6 kinase II. *Nature, 334*, 715-718.

[66] Alcorta, D. A., Crews, C. M., Sweet, L. J., Bankston, L., Jones, S. W., Erikson, R. L. (1989). Sequence and expression of chicken and mouse rsk: Homologs of Xenopus laevis ribosomal S6 kinase. *Mol Cell Biol, 9*, 3850-3859.

[67] Zhao, A., Li, J., Lu, Y. (1995). Expression of Ras and erbB in benign and malignant lesions of the stomach. *Chinese Journal of Oncology, 17*, 336-339.

[68] Frodin, M., Gammeltoft, S. (1999). Role and regulation of 90 kDa ribosomal S6 kinase (RSK) in signal transduction. *Mol Cell Endocrinol, 151*, 65-77.

[69] Sutherland, C., Campbell, D. G., Cohen, P. (1993). Identification of insulin-stimulated protein kinase-1 as the rabbit equivalent of rsk(mo)-2. Identification of two threonines phosphorylated during activation by mitogen-activated protein kinase. *Eur J Biochem, 212*, 581-588.

[70] Lee J. T., McCubery, J. A. (2002). The Raf/MEK/ERK signal transduction cascade as a target for chemotherapeutic intervention in leukemia. *Leukemia, 16*, 486-507.

[71] Hochhaus, A., McCubrey, J., Killmann, N. M. B. (2002). Molecular targets for therapy, signal transduction inhibition and imatinib. *Leukemia, 16*, 1205-1206.

[72] McGlynn, A. P., Padua, R. A., Burnett, A. K., Darley, R. L. (2000). Alternative effects of RAS and RAF oncogenes on the proliferation and apoptosis of factor-dependent FDC-P1 cells. *Leuk Res, 24*, 47-54.

[73] Cioffi, C. L., Garay, M., Johnston, J. F., McGraw, K., Boggs, R. T., Hreniuk, D., Monia, B. P. (1997). Selective inhibition of A-Raf and C-Raf mRNA expression by antisense oligodeoxynucleotides in rat vascular smooth muscle cells: Role of A-Raf and C-Raf in serum-induced proliferation. *Mol Pharmacol, 51*, 383-389.

[74] Ravi, R. K., Weber, E., McMahon, M., Williams, J. R., Baylin, S., Mal, A., Harter, M. L., Dillehay, L. E., Claudio, P. P., Giordano, A., Nelkin, B. D., Mabry, M. (1998). Activated Raf-1 causes growth arrest in human small cell lung cancer cells. *J Clin Invest, 101*, 153-159.

[75] Lloyd, A. C., Obermüller, F., Staddon, S., Barth, C. F., McMahon, M., Land, H. (1997). Cooperating oncogenes converge to regulate cyclin/cdk complexes. *Genes Dev, 11*, 663-677.

[76] Woods, D., Parry, D., Cherwinski, H., Bosch, E., Lees, E., McMahon, M. (1997) Raf-induced proliferation or cell cycle arrest is determined by the level of Raf activity with arrest mediated by p21(Cip1). *Mol Cell Biol, 17*, 5598-5611.

[77] Sen, P. K., Packer, L. (1996). Antioxidant and redox regulation of gene transcription. *FASEB J, 10*, 709-720.

[78] Barnes, P. J., Karin, M. (1997). Nuclear factor-kappaB: a pivotal transcription factor in chronic inflammatory diseases. *N Engl J Med, 336*, 1066-1071.

[79] Beg, A. A., Baltimore, D. (1996). An essential role for NF-kappaB in preventing TNF-alpha-induced cell death. *Science, 274*, 782-784.

[80] Grimm, S., Bauer, M. K., Baeuerle, P. A., Schulze-Osthoff, K. (1996). Bcl-2 down-regulates the activity of transcription factor NF-kappaB induced upon apoptosis. *J Cell Biol, 134*, 13-23.

[81] Prescott, S. M., Fitzpatrick, F. A. (2000). Cyclooxygenase-2 and carcinogenesis. *Biochim Biophys Acta, 1470*, M69-78.

[82] Subbaramaiah, K., Telang, N., Ramonetti, J. T., Araki, R., DeVito, B., Weksler, B. B., Dannenberg, A. J. (1996). Transcription of cyclooxygenase-2 is enhanced in transformed mammary epithelial cells. *Cancer Res, 56*, 4424-4429.

[83] Ristimaki, A., Honkanen, N., Jankala, H., Sipponen, P., Harkonen, M. (1997). Expression of cyclooxygenase-2 in human gastric carcinoma. *Cancer Res, 57*, 1276-1280.

[84] Shen, Z. X., Chen, G. Q., Ni, J. H., Li, X. S., Xiong, S. M., Qiu, Q. Y., Zhu, J., Tang, W., Sun, G. L., Yang, K. Q., Chen, Y., Zhou, L., Fang, Z. W., Wang, Y. T., Ma, J., Zhang, P., Zhang, T. D., Chen, S. J., Chen, Z., Wang, Z. Y. (1997). Use of arsenic trioxide (As_2O_3) in the treatment of acute promyelocytic leukemia (APL): II. Clinical efficacy and pharmacokinetics in relapsed patients. *Blood, 89*, 3354-3360.

[85] Chan, G., Boyle, J. O., Yang, E. K., Zhang, F., Sacks, P. G., Shah, J. P., Edelstein, D., Soslow, R. A., Koki, A. T., Woerner, B. M., Masferrer, J. L., Dannenberg, A. J. (1999). Cyclooxygenase-2 expression is up-regulated in squamous cell carcinoma of the head and neck. *Cancer Res, 59*, 991-994.

[86] Tucker, O. N., Dannenberg, A. J., Yang, E. K., Zhang, F., Teng, L., Daly, J. M., Soslow, R. A., Masferrer, J. L., Woerner, B. M., Koki, A. T., Fahey, T. J. 3rd. (1999). Cyclooxygenase-2 expression is up-regulated in human pancreatic cancer. *Cancer Res, 59*, 987-990.

[87] Bargou, R. C., Emmerich, F., Krappmann, D., Bommert, K., Mapara, M. Y., Arnold, W., Royer, H. D., Grinstein, E., Greiner, A., Scheidereit, C., Dorken, B. (1997). Constitutive nuclear factor-kappaB-RelA activation is required for proliferation and survival of Hodgkin's disease tumor cells. *J Clin Invest, 100*, 2961-2969.

[88] Luque, I., Gelinas, C. (1997). Rel/NF-kappa B and I kappa B factors in oncogenesis. *Semin Cancer Biol, 8*, 103-111.

[89] Han, S. S., Kim, K., Hahm, E. R., Lee, S. J., Surh, Y. J., Park, H. K., Kim, W. S., Jung, C. W., Lee, M. H., Park, K., Yang , J. H., Yoon, S. S., Riordan, N. H., Riordan, H. D., Kimler, B. F., Park, C. H., Lee, J. H., Park, S. (2004). L-ascorbic acid represses constitutive activation of NF-kappaB and COX-2 expression in human acute myeloid leukemia, HL-60. *J Cell Biochem, 93*, 257-270.

[90] John, D. C., Bulleid, N. J. (1996). Intracellular dissociation and reassembly of prolyl 4-hydroxylase: the alpha-subunits associated with the immunoglobulin-heavy-chain binding protein (BiP) allowing reassembly with the beta-subunit. *Biochem J, 317*, 659-665.

[91] Park, S., Lee, J., Yeom, C. H. (2006). A proteomic approach to the identification of early molecular targets changed by L-ascorbic acid in NB4 human leukemia cells. *J Cell Biochem,* published online Jul 5 2006, DOI:10, 1002/jcb20971.

[92] Houle, F., Rousseau, S., Morrice, N., Luc, M., Mongrain, S., Turner, C. E., Tanaka, S., Moreau, P., Huot, J. (2003). Extracellular signal-regulated kinase mediates phosphorylation of tropomyosin-1 to promote cytoskeleton remodeling in response to oxidative stress: impact on membrane blebbing. *Mol Biol Cell, 14*, 1418-1432.

[93] Noiva, R. (1999). Protein disulfide isomerase: the multifunctional redox chaperone of the endoplasmic reticulum. *Semin Cell Dev Biol, 10*, 481-493.

[94] Kemmink, J., Darby, N. J., Dijkstra, K., Nilges, M., Creighton, T. E. (1997). The folding catalyst protein disulfide isomerase is constructed of active and inactive thioredoxin modules. *Curr Biol, 7*, 239-245.

In: Focus on Cell Apoptosis Research
Editor: Daniel W. Cho, pp. 175-202

ISBN: 1-60021-433-9

Chapter VII

"APOPTOSIS VERSUS SURVIVAL" IN CHRONIC LYMPHOCYTIC LEUKEMIA B CELLS: ROLE OF DEATH RECEPTORS AND OF CYTOKINES

Daniela de Totero** and Silvano Ferrini
National Institute for Cancer Research, Genoa, Italy

ABSTRACT

Chronic lymphocytic leukemia (CLL) of B cells is the most common type of leukemia in the Western world and is characterized by the abnormal expansion of CD5+ mature B lymphocytes. The study of the pathways leading to apoptosis resistance in these cells is critical both in understanding the disease process and in devising novel therapeutical strategies. In CLL the accumulation of monoclonal mature B-cells may result from an extended lifespan of B cells, their increased capacity to proliferate, and/or defective apoptosis. Growth of CLL B cells might be a direct consequence of defects of genes regulating apoptosis, but it could be also indirectly affected by abnormal interactions between leukemic cells and the surrounding bystander non-tumoral cells. On CLL B cells levels of expression of various molecules representing major components of apoptotic pathways, such as Bcl-2/Bax, Bid, or the DISC complex (FAS, FADD, Caspase-8) have a key role in regulating the ability of neoplastic cells to undergo apoptosis. In addition, it has been demonstrated that particular cytokines, and specific ligand-receptor interactions, can sustain overexpression of molecules regulating apoptosis, as in the case of IL-4 for Bcl-2 and of CD40-CD40L for FAS and Bid expression. These observations may thus suggest that in CLL, imbalanced production of IL-4 could prolong survival through increased Bcl-2 expression, while CD40 triggering,

* Correspondence concerning this article shoud be addressed to: Dr. Daniela de Totero, Immunological Therapies Unit, National Institute for Cancer Research, Largo R. Benzi n.10; 16132 Genoa, Italy; e-mail: daniela.detotero@istge.it

of leukemic B cells, might sensitize neoplastic B cells to apoptosis through upregulation of FAS and of Bid. However the apoptotic pathway that is triggered by FAS/ FAS ligand (FASL) interaction may be not functional in CLL, and this despite apparently normal levels of expression of the molecules composing the DISC complex both in resting and in CD40-activated cells.

We further described in CLL, expansion of a CD3+/CD8+ T cell subset with unique phenotypic features of Tc2, capable of releasing IL-4, that may support CLL survival. Moreover also IL-2 and IL-15, sharing the common γ chain and the IL-2Rβ chain capable of transducing signals, appear to induce proliferation of CLL B cells. We recently demonstrated that IL-21, utilizing the common γ chain as IL-2 and IL-15, induces, instead, apoptosis of CLL B cells, especially after CD40 triggering, which upregulates IL-21 receptor expression in all the cases studied. In this review we will discuss how CD40 triggering of CLL B cells may modulate expression of cytokine receptors and of molecules implicated in apoptosis of CLL B cells and the mechanisms that regulate these phenomena.

Keywords: apoptosis, cytokines, receptors

INTRODUCTION

Apoptosis or programmed cell death (PCD) is a physiological mechanism regulating cellular selection, differentiation and playing critical roles in tissue homeostasis. On a morphologic point of view apoptosis is characterized by cell shrinkage, membrane remodelling and blebbing, chromatin condensation and DNA and cellular fragmentation into apoptotic bodies. Signals triggering cellular apoptosis may derive by external stimuli such as interactions between specific ligands and surface death receptors or through specific cell sensors residing in the cell nucleus and cytoplasm [1]. In lymphoproliferative disorders defective regulation of signals mediating cellular death lead to uncontrolled cell accumulation and, in chronic lymphocytic leukemia (CLL), deregulation of one or more steps of the apoptosis signaling pathway may decisively influence disease progression. Survival or apoptosis thus represent final outcomes of a relative balance of expression of several proteins controlling cell's fate. Several molecules are operating in this system including TNF receptors and ligands, the extended Bcl-2 family, myc, p53 and p73 isoforms [2]. Moreover, in normal B cells, the balance between the dinamically opposing functions mediated by CD40 and FAS receptor (CD95) (anti- and pro-apoptotic, respectively) is critical in driving B cell selection, maturation and death. In addition, it is becoming clear that interactions between malignant cells and the microenviroment play a fundamental role in the onset and progression of the disease. Cytokines produced by CLL B cells modulate the enviroment where they accumulate and, in turn, the behavior of leukemic cells is influenced by cytokines produced by surrounding non-tumoral cells. It has been demonstrated that "in vitro" IL-4, IFNα and IFNγ inhibit apoptosis of CLL B cells through preservation or up-regulation of Bcl-2 expression [3-6].

Apoptosis is caused by the activation of intracellular proteases known as caspases and two major pathways for triggering caspases activation have been elucidated in detail (extrinsic and intrinsic). The extrinsic pathway relies on tumor necrosis factor (TNF) family death receptors for triggering apoptosis, while the intrinsic pathway centers on mitochondria as initiators of cell death [1]. We and others have however demonstrated a block of the apoptotic pathway triggered by CD95-CD95L (FAS-FASL) interaction in CLL B cells, thus suggesting that prolonged survival and accumulation of neoplastic CLL cells may result from defects in apoptotic programs [7-9]. The complex network of pro-apoptotic and anti-apoptotic molecules interacting in CLL leading to an imbalance versus anti-apoptotic signals that promote overgrowth of the neoplastic B cell clone will be further described and discussed.

TNF Family Ligands and Receptors in CLL

TNFα has been initially regarded as a promising cytokine in cancer treatment in view of its ability of selectively kill tumor but not normal cells. Unfortunately the proinflammatory and cardiovascular actions of TNFα precluded its systemic administration, although TNFα is approved for the treatment of sarcomas of the limbs by the isolated limb perfusion, which limits systemic toxicities [10]. Intriguingly, TNFα has also been described as inducing proliferation of CLL and potentially playing a role in disease progression [11-13]. An autocrine secretion of TNFα from leukemic B cells has been reported and some recent data showed that TNFα produced by resting or activated CD8+ T cells, is higher in CLL than in healthy controls, in particular in later stages of disease [14,15]. In spite of these observations, the actual knowledge of the different pathways activated by TNFα and by TNF-related ligands, such as TRAIL and FASL, may allow to exploit novel therapeutic strategies. Binding of TNFα to its cellular receptor, TNFRI triggers two parallel signaling pathways, which diverge at the adapter protein TNF receptor-associated death domain (TRADD). One of these pathways results in the recruitment of FAS-associated death domain (FADD) and in the activation of the apoptotic cascade initiated by caspase-8. The other parallel pathway triggers activation of nuclear factor kB (NF-kB) family transcription factors and NF-kB influences the expression of many target genes involved in immune regulation, in host defenses but also of genes that suppress apoptosis. TNF family cytokines, such as FASL (CD95L) and TRAIL, thus trigger activation of the caspase pathway without concomitant induction of NF-kB. On ligand binding, TNF family receptors containing cytosolic death domain (DD) such as FADD and TRADD, cluster in membrane and bind to specific pro-caspases through a death effector domain (DED) forming the "death-inducing signaling complex" (DISC) and finally resulting in activation of caspases, the principal executioners of cellular death [16].

We studied the expression of FAS (CD95) and of TNFRI/II (CD120a/b) on CLL B cells, both at resting and at activated conditions, and found that FAS is not constitutively expressed while TNFRI/II are present at low levels. However, all the three receptors were strongly upregulated after CD40 triggering [8,17], as indicated in figure 1. Within a group of 20 CLL cases studied, FAS expression was faintly detectable only in one patient, but, after CD40 activation, it was strongly up-regulated in all the cases. The observation of a constitutive expression of TNFRII (CD120b), lacking the death domain and predominantely involved in

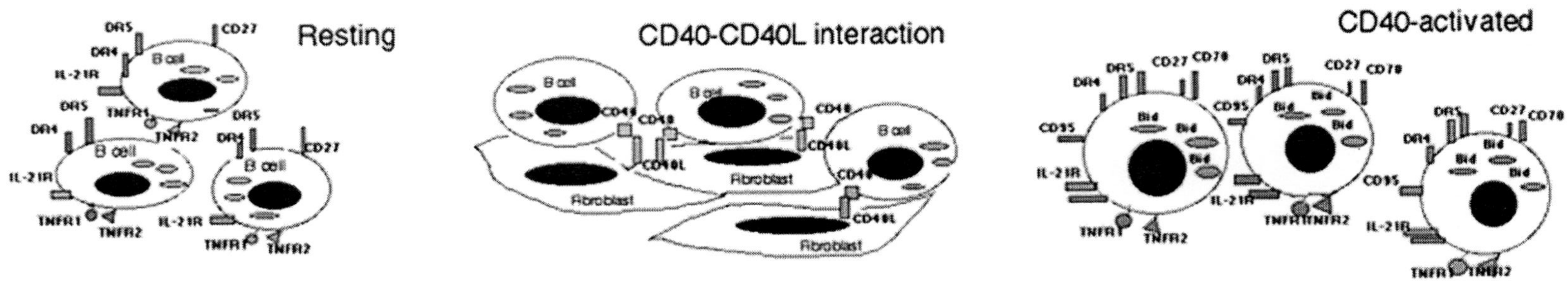

Figure 1. Modulation of TNF ligands and receptors superfamily and of IL-21R after interaction between CD40+ CLL B cells and CD40L-transfected mouse fibroblasts. In CD40-activated CLL-B cells TNFR1, 2, DR5 and IL-21R are upregulated; CD95 and CD70 are "de novo" expressed and CD27 is downregulated. Also Bid expression is "de novo" induced by CD40 activation.

cell survival and differentiation, suggested a role of this receptor in mediating proliferation of the leukemic clone to autocrine TNFα [18]. Residual CLL T-cells may represent, in fact, a source of TNFα and its intracytoplasmic expression, higher in T- than in B-cells, is significantly increased in patients with more advanced disease (stage III/IV) [19]. In addition, the lack of FAS on quiescent B cells and the finding that, even if upregulated after CD40 triggering, it is still unable of inducing apoptosis after binding of an agonist monoclonal antibody (moAb) [7,8,20] further suggested that an altered FAS pathway might contribute to the relentless accumulation of malignant cells.

In order to better clarify whether apoptosis induced by FAS/FASL and TNFα/CD120a can be limited by a deregulated expression of molecules involved in the formation of the DISC (FAS, FADD, FLICE), we analyzed the DISC complex both at mRNA and at protein levels in CLL B cells. In agreement with previous cytofluorimetric analysis we confirmed lack of FAS mRNA in resting CLL B cells and significant increase in CD40-activated cells [8]. Moreover, FLICE (pro-caspase-8), FADD and TRADD mRNAs were expressed in both, resting and activated conditions in all the cases studied. Analyses at the protein level by Western Blot further confirmed that both FLICE and FADD are constitutively expressed and slightly upregulated after CD40-activation [8]. Our findings thus suggest that leukemic CLL cells possess a complete machinery required to carry out the FAS-triggered apoptotic program.

Activation of caspase-8 is the initiating event of the extrinsic pathway, which can be triggered by FAS, TRAIL and TNFRI receptor. However, levels of FLICE-inhibitory proteins (FLIPs), represented by truncated forms of FLICE lacking the catalytic subunit [21], are present and strongly upregulated in CLL B cells after CD40/CD40L interaction [20], thus suggesting their potential interference with activation of this death pathway. Kipps and coworkers however reported that CD40-activated CLL B cells, initially resistant to FAS-mediated apoptosis, become sensitive after prolonged "in vitro" co-culture with FASL-transfected cells (5 days or more), possibly in relation with a decline of the expression of two FLIPs' isoforms (long and short) [22]. As reported by the same authors FLIP downregulation by antisense oligonucleotides or pharmacological agents was, however, insufficient to induce FAS-mediated apoptosis, thus indicating that other factors may contribute to apoptosis resistance in the initial period after CD40 triggering. Moreover CLL B cells, like many other tumors, express high levels of the inhibitors of apoptosis protein family member X-linked (XIAP) [20]. IAPs can negatively regulate apoptosis by inhibiting caspase activity directly and they might thus contribute to FAS-apoptosis resistance of CD40-activated CLL B cells. CD40 triggering of CLL B cells, however, induces a pro-apoptotic protein called the B-cell leukemia 2 homology 3 (BH3)-interacting domain death agonist (Bid), which is capable of connecting the death receptor with the mitochondrial apoptosis pathway. Following CD95 ligation Bid is degraded in a small truncated form (tBid), which may rapidly traffick to the mitochondria and trigger the release of Smac/DIABLO. In turn, Smac/DIABLO may inhibit IAPs allowing sensitization of CLL to FAS-mediated apoptosis. In agreement with this hypotesis XIAP-inhibitors sensitize CD40-activated chronic lymphocytic leukemia cells to CD95-mediated apoptosis [23]. The events characterizing the apoptotic pathways mediated by death receptors of the TNF-related apoptosis-inducing ligand (TRAIL) are very similar to those decribed for FAS and require FADD and FLICE in the TRAIL-induced formation of the DISC. However ligation of TRAIL receptors also can induce activation of caspases other than FLICE, which in turn can induce cleavage of Bid for initiation of apoptosis via the intrinsic pathway. CLL B cells at resting conditions appeared resistant to TRAIL-induced apoptosis “in vitro”, even when exogenous human recombinant TRAIL was added at high concentration [24]. Resistance to TRAIL-induced apoptosis in resting CLL B cells may be due partly to low surface expression of the death receptors (TRAIL-R1/DR4 and TRAIL-R2/DR5) [24]. However CD40 stimulation of CLL B cells was capable of up-regulating TRAIL-R2 and not TRAIL-R1 [25], in addition to the above described expression of the pro-apoptotic molecule Bid [25,26]. After CD40-activation CLL B cells, expressing both FAS and DR5, become thus sensitive to FAS- and TRAIL-induced apoptosis. Simultaneous expression of more than one death receptor, i.e. FAS and DR5, and the simultaneous engagement of these two by corrisponding ligands (FASL, TRAIL), further resulted more effective in inducing apoptosis than a single receptor-ligand combination [26].

From these data, it appears evident how the CD40/CD40L interaction system is critical in regulating survival or apoptosis of leukemic CLL cells, through up- or downregulation of various molecules each other functionally interconnected. The CD40 antigen, another member of the TNFR superfamily, is constitutively expressed on normal B cells and CD40/CD40L interaction has a key role in T-B cell dialogue. In normal lymph node germinal centres, where follicular centre B cells express FAS and CD40 receptors, CD40L and FASL have opposing biological functions: FASL induces apoptosis and CD40L rescues the cells from apoptosis. The balance in these opposing actions appears important for B-cell selection and normal B-cell development [27]. Stimulation of CD40 by either the CD40 ligand or agonistic anti-CD40 antibodies induces a potent activation signal in mature B cells [28,29] leading to expression of activation antigens [30,31], tyrosine kinase activity [32], Ig class switching [33,34] and increased DNA synthesis [35]. Defective CD40L (CD154) expression can result in severe immunodeficiency. In fact, individuals with genetic defects in the X-linked gene encoding CD154 have recurrent bacterial infections secondary to their inability to perform Ig class-switch and to generate antigen-specific secondary antibody [36-39]. Immune defects are often also noted in CLL. Such defects might be due to a deregulated expression of the CD40L or of the CD28 antigen on residual T-cells [40,41], as well as the weak expression of CD80/CD86 on B cells [42], which altogether may limit the T/B cross-talk. Several reports have previously shown that CLL B cells, despite expressing high levels of MHC class I/II, are ineffective antigen presenting cells (APCs), unable to stimulate even normal allogeneic T cells in mixed lymphocyte reaction (MLR). However, in vitro activated T cells can induce CLL B cells to become effective APCs [43]. This effect is largely mediated by the CD40L, transiently expressed by activated CD4+ T cells. In fact CLL B cells (CD40+), after CD40L ligation, upregulate accessory surface molecules important in cognate co-stimulatory cell-cell interactions, such as CD54, CD80, CD86, CD70, thus allowing these cells to function as APCs and, conceivably, to activate immune responses against tumor associated antigens. For this reason, transgenic expression of CD40L by tumor cells has been exploited in CLL. A replication-defective adenovirus vector was employed to transduce CLL B cells to express the gene encoding CD40L (Ad-CD154). Upon expression of CD40L, CD40 was cross-linked on transduced cells as well as bystander cells, thereby inducing them to express co-stimulatory antigens, in the attempt to induce a T-cell immune response against leukemia cells [44]. A phase I clinical trial was further performed employing this strategy and CLL patients receiving intravenous infusion of autologous leukemia cells transfected with Ad-CD154 experienced acute reductions in leukemia cell counts and in the size of lymph node and spleen [45]. Moreover the numbers of circulating T cells specific for autologous leukemic cells increased, and elevated levels of IFNγ, IL-12, IL-6 were detected in the sera of patients [46]. Some studies appeared, however, in contradiction with these findings. CD40 stimulation, in fact, may have also an anti-apoptotic effect and may further impair the dysregulated response of CLL to apoptotic stimuli. Knowledge about CD40 stimulation in CLL and its ability to modulate a complex array of molecules involved in survival or apoptosis has been however improved [26] and some reports demonstrated that CD40-activated CLL B cells are more susceptible to cytotoxic T-lymphocytes (CTLs) [20,26] or to cytotoxic drugs [8]. CTLs from allogeneic normal donors were, in fact, capable of inducing apoptosis of both resting and CD40-activated CLL B cells trough a FAS-indipendent granule-exocytosis pathway [20]. Moreover CD40-CLL gene therapy "in vivo" induced tyrosine phosphorylation and functional activation of signal transducer and activator of transcription (STAT) 1 protein [46] through a soluble mediators, further supporting the critical role played by activated T cells in clinical and immunological remission.

The role of two other members of the TNF superfamily, named BAFF and APRIL, that potently induce survival of normal B-lymphocytes, was recently studied in CLL and revealed a potential role in the onset of the disease [47]. BAFF and APRIL bind with high affinity two members of the TNF-receptor (TNF-R) superfamily: BCMA and TAC1. BCMA signals through TNF-R-associated factor-1 (TRAF)-1, TRAF-2, and TRAF-3 and may result in activation of NF-kB, Elk-1, JNK, and p38 [48,49]. The intracellular domain of TAC1 further interacts with TRAF-2, TRAF-5 and TRAF-6 thus activating NF-kB and JNK. In addition, BAFF binds to a third receptor specific for BAFF, BAFF-R or BR3 [50]. CLL B cells express BAFF and APRIL as well as the primary BAFF-R and the two other receptors BCMA1 and TAC1 [51]. Moreover addition of exogenous BAFF and APRIL to the leukemic cells was capable of stimulating NF-kB activation and of protecting them from spontaneous and flavopiridol-induced apoptosis. On the contrary the addition of soluble BCMA-Fc or of anti-BAFF and anti-APRIL neutralizing antibodies resulted in apoptosis induction. Altogether these findings suggest that CLL B cells coud be rescued from entering apoptosis through an autocrine survival pathway mediated by BAFF and APRIL [52,53]. In addition to this autocrine loop, however, the possibility of a paracrine pathway has been suggested in CLL by the presence of a particular subset of cells, Nurse-like cells, which show significant high levels of BAFF and APRIL expression and appear to promote survival of leukemic B cells [54].

Fludarabine-Induced Apoptosis of CLL B Cells is Increased after CD40 Triggering

In view of the ability of CD40 signalling to sensitize CLL B cells to apoptosis, we tested the possibilty that CD40-activated cells might become more sensitive to the apoptotic effects of fludarabine (9-β-D-arabinofuranosyl-2-fluoroadenine, F-ara-A) treatment "in vitro" and studied the apoptotic pathway involved. To this end, purified CLL B cells were first co-cultured with adherent mouse fibroblasts transfected with human CD40L, and then harvested and re-cultured with fludarabine for 72h more [8]. Apoptosis was determined in cytofluorographic analysis by three different methods of cell staining, which allowed us to analyze cell cycle (Propidium Iodide= PI), early and late apoptosis (Annexin-V-FITC/Propidium Iodide= ANN-V/PI), and membrane mitochondrial depolarization (3, 3'dihexylocarbocyanine iodide= DiOC6). Altogether the data obtained by these three methods showed a higher sensitivity of CD40-activated cells, than resting cells, to fludarabine. Within a group of 20 CLL cases studied, in fact, there was a clear enhancement of apoptosis induction after CD40 triggering ($p< 0.0001$ $n=20$) [8]. Furthermore, PI staining showed, after CD40 pre-activation, a sharper subdiploid peak that suggested a more rapid kinetics of fludarabine-induced apoptosis. Accordingly, double staining by Annexin-V/PI demonstrated that the percentage of cells already in late apoptosis (AnnexinV+/PI+) was higher in fludarabine-treated and activated cells than in cells treated at resting conditions. The staining of CLL B cells with DiOC6, a green cell-permeable fluorochrome that is concentrated in the charged mitochondria of living cells but not in the depolarized mitochondria of apoptotic cells, further indicated that fludarabine, both in resting and in activated cells, induced a drop of membrane mitochondrial potential ($\Delta\Psi_m$), an event associated with cells undergoing apoptosis [8].

Bcl-2 is another molecule overexpressed on leukemic B cells in CLL, contributing to disease progression and resistance to apoptosis [55]. Evaluation of Bcl-2 expression and of its modulation in response to various stimuli or pharmacological treatment can be helpful in determining cell sensitivity to apoptosis induction. Bcl-2 belongs to an extended family of genes (Bax, Bcl-x_L, Bcl-x_S) each other interrelated in controlling apoptosis [56]. With exception of < 5% of CLL cases in which the Bcl-2 gene is juxtaposed to Ig loci [57], the mechanisms involved in deregulated expression of Bcl-2 are currently unclear. However the propensity of CLL B cells to undergo apoptosis "in vitro" have previously suggested that the fine regulation of apoptosis "in vivo" can be influenced by cell-cell interactions or by increased amounts of apoptosis-protecting cytokines, such as IL-4, IFNα, IFNγ, preserving high Bcl-2 expression. Furthermore, recent observations indicated a mechanism of regulation of Bcl-2 expression in hematopoietic cancer cells consisting of posttranscriptional downregulation by two microRNAs (miR-15 and miR-16) [58]. These two miRs are natural antisense Bcl-2 interactors that downregulate Bcl-2 expression, but over than 65% of CLL present deletion or translocation in miR15a and miR16-1 [58]. This finding can offer a potential explanation of deregulated Bcl-2 expression in this disease, which can be associated with CLL prognosis and progression [59].

Through analysis of Bcl-2 expression, in cells at basal conditions and after CD40 triggering, we further showed that Bcl-2 was downmodulated at a more significant level in CD40-activated than in resting cells after fludarabine treatment "in vitro" ($p<0.0001$ and $p=0.0003$ respectively) [8], thus suggesting an involvement of Bcl-2 in fludarabine-triggered apoptosis. In agreement with our data Berrebi et al. described an increased fludarabine sensitivity of CLL B cells after pokeweed mitogen activation and Grdisa et al. showed that CD40 ligation does not affect fludarabine activity [60,61]. On the contrary two previous reports described that sensitivity of CLL B cells to fludarabine was decreased by CD40 activation [62,20]. These results, apparently in contrast with our data, could be due to different experimental conditions employed. In the first work, in fact, CD40 activation was carried out by an agonistic anti-CD40 moAb, while we activated CLL B cells through co-cultures with transfected CD40L cells that are more efficient in priming leukemic cells. In the second report, further, the authors added fludarabine during co-culture of CLL B cells with CD40L-transfected cells. We, instead, performed a previous pre-activation with transfected cells and then treated CLL B cells with fludarabine, in a system free of contaminating transfected cells. Recently Kipps and coworkers [63] confirmed and extended our previous findings further demonstrating an increased sensitivity of CD40-activated CLL cells to fludarabine, in particular of those cases lacking p53 that develop a resistance to chemotherapy. This study further highlight that CD40 stimulation sensitize leukemia cells to apoptosis via a c-Abl-dependent activation of p73 and mitigate the resistance of p53-deficient CLL to anti-cancer drug therapy [63].

We further examined the pathway leading to apoptosis [8] and demonstrated that caspases-6 and -1 are the major caspases driving to apoptosis CLL B cells fludarabine-treated at resting conditions, while caspases-3, -9 and-8 appeared activated only in a limited number of cases and caspases-2 and-5 were never involved. The limited involvement of caspase-8 and-3 can be related to a possibile block of the FAS-apoptotic pathway that we described. However, it is also conceivable that cytotoxic drugs, such as fludarabine, induce apoptosis through a pathway that is indipendent of FAS/ TRAIL triggering. Our observation that the use of a blocking anti-FAS moAb (ZB4) failed to inhibit fludarabine-induced apoptosis of CD40-activated cells further confirmed that FAS receptor is not involved in fludarabine-mediated apoptosis. Interestingly, however, following CD40-activation fludarabine-induced apoptosis of CLL B cells is no longer blocked by caspase inhibitors, thus suggesting that others and multiple factors might be relevant to induce death, or on the contrary survival.

Among these factors, cytokines have a prominent role. We found that addition of anti-TNFα or anti-IFNγ moAbs were capable of inhibiting fludarabine-induced apoptosis of CD40-activated cells [8]. In agreement TNFα and IFNγ secretion by CLL B cells and TNFRI/II were strongly upregulated only after CD40 triggering [8]. These data suggest that autocrine secretion of specific cytokines, such as TNFα and IFNγ, may contribute to sensitize CD40-activated CLL B cells to fludarabine-induced apoptosis. It has been widely demonstrated that TNFRI is able to mediate apoptosis following association between its cytoplasmic death domain (DD) and the adapter proteins TRADD and FADD. TNFRII, instead, belongs to the subgroup of TNFR molecules, such as CD40 and CD30, that do not contain a death domain and are primarily involved in cell survival, growth and differentiation through induction of specific gene transcription. However, it now appears clear that under certain conditions these molecules can also induce cell death. In different cellular systems, Grell et al., in fact, reported that the cytotoxic effects induced by TNFRII, CD40 and CD30 are mediated by endogenous production of TNFα and autotropic or paratropic activation of TNFRI, and that stimulation of TNFRII and CD40 synergistically enhances TNFRI-induced cytotoxicity [64]. Yang et al., in addition, suggested that both TNFRI (p55) and TNFRII (p75) are able to transmit apoptotic signals [65]. It is further interesting to note that exogenous TNFα is capable of up-regulating CD20 expression on CLL B cell [66], thus suggesting that TNFα might improve Rituximab (anti-CD20 moAb) therapy. Taken together these data further highlight how survival or death of CLL B cells can be regulated by a complex network of interactions between leukemic cells and microenviroment and by the autocrine or paracrine production of particular cytokines.

Pro- and Anti-Apoptotic Cytokines in CLL

In CLL the majority of circulating cells appear to be nondividing and it has been suggested that a prolonged life span and decreased cell death can be responsible for the accumulation of the leukemic cells. CLL B cells cultured "in vitro" become, however, more prone to undergo spontaneous apoptosis, further suggesting that survival factors might be present "in vivo". Various cytokines have been proposed as anti-apoptotic factors in CLL and these cytokines may also stimulate the proliferation of leukemic cells or sustain the overexpression of anti-apoptotic molecules, such as Bcl-2. On the opposite other cytokines may enhance the apoptotic rate of CLL B cells and may counteract the activity of anti-apoptotic cytokines.

The cytokines IL-2, -4, -7, -9, -15 and –21 share the usage of the common cytokine receptor γ-chain (γ_c) and play a crucial role in the regulation of lymphoid cell development and function. The γ_c is an essential element of the receptors of these cytokines and is involved in a common pathway of signal transduction. The γ_c associates with cytokine-specific receptor chains that confer binding specificities and may provide additional signaling properties. Therefore cytokines of this family display specific biological activities, but also a noticeable degree of functional redundancy [67,68]. The genetic defects of the γ_c [69] or of the tyrosine-kinase JAK3 [70,71], which is strictly required for signaling via γ_c, results in a severe combined immunodeficiency (SCID) with profound T, B and NK cell defects.

Among the γ_c cytokine family, IL-2, and its receptors might have a critical role in the mechanisms controlling the growth of neoplastic B cells [72]. IL-2 is produced by Th cells upon activation, and has been primarily involved in the immune response by stimulating activated T, B and NK cell proliferation and functional differentiation. More recently, IL-2 has also been shown to induce feed-back mechanisms of immune-regulation, involving activation-induced cell death of T cells and stimulation of CD4+CD25+ regulatory T cell function [73]. The IL-2 receptor (IL-2R) complex is composed of the IL-2Rα (CD25, p55), and β (CD122, p75) chains, which combine to the γ_c (CD132, p64) to form a high affinity ($K_D=10^{-11}$M) trimeric complex, expressed by normal lymphoid cells only upon activation [74,75]. Lymphoid cells may also express an intermediate affinity ($K_D=10^{-10}$M) IL-2Rβ/γ dimeric complex. Both the high and the intemediate affinity receptors are functional and can be triggered by IL-2 to signal trough the janus kinase 1 (JAK1) and JAK3 tyrosine kinases (TK), which are recruited by the IL-2Rβ and γ_c, respectively. JAK1 and JAK3 autophosphorylate and generate docking sites for STAT-1 and –3 and for STAT-5a and –5b, respectively. IL-2Rα chain, with low affinity for IL-2 ($K_D=10^{-8}$M), on its own is instead unable to mediate signal transduction. Leukemic CLL B cells, analyzed by cytofluorimetric analysis, showed a low and heterogeneous expression of the IL-2Rα and IL-2Rβ in almost all the cases studied [72,76]. Molecular studies further confirmed the weak and variable expression of IL-2Rα and IL-2Rβ, while the γ_c resulted highly detectable in all the cases studied. In agreement with these findings, low proliferative responses of CLL B cells could be elicited by IL-2 "in vitro" [72]. TNFα, however, was capable of upregulating IL-2Rα on CLL B cells and of increasing the binding and the mitogenic activity of IL-2 [76]. Moreover Trentin et al. also proposed that the IL-2R complex may have a role in the upregulation of colony-stimulating receptors (IL-3R, GM-CSFR, G-CFSR), further highlighting the complex network of cytokines each other interconnected [76]. In lymphoproliferative disease the leukemic B cell clone represents the expansion of B cells blocked at relatively late stage during differentiation and maturation. On this view a comparative study on the pattern of expression of cytokine receptors between normal B cells and a group of B cell leukemia, in addition to possibly represent a distinctive feature of a specific leukemia, could also be suggestive of mechanisms driving abnormal expansion of neoplastic cells. CLL B cells, for example, differ substantially from Hairy cells (HCL), with regards to the expression of the IL-2R chains [77,78]. HCL, in fact, have a strong constitutive IL-2Rα chain expression that is a feature of the classic form of HCL. Moreover the β chain of IL-2R is expressed at high levels in both classic as well as in the rare variant form of HCL, lacking IL-2Rα. Cells from both HCL forms show evident proliferation to exogenous IL-2, in agreement with considerable IL-2Rs expression and with the pivotal role played by the β chain in transducing IL-2 signal [77]. HCL and CLL may thus be considered distinct in terms of differentiative stage: the neoplastic B cells in HCL, morphologically and immunologically, resemble competent and activated B cells, unlike those of CLL that are in a relatively quiescent state. High IL-2Rs expression in HCL may support this opinion, suggesting that specific factors may have induced the activation of malignant cells, while weak IL-2Rs expression of CLL may reflect a "resting" state of the cells.

IL-15 is another cytokine of the IL-2 family, which shares the usage of the IL-2Rβ/γ_c chains with IL-2, and therefore has several biological properties in vitro similar to that of IL-2 [79-81]. However, the functional specificity of IL-15 is conferred by the existence of the IL-15Rα chain, which may interact with the IL-2Rβ/γ_c complex to form a high affinity receptor complex [82]. Differently from IL-2Rα, IL-15Rα is a high affinity receptor also when expressed as isolated chain. Several isoforms of IL-15Rα have been described [82] and IL15Rα expression has been found on many different cell types, including hematopoietic progenitor cells, non-hematopoietic cells [82] and some tumor cell lines [83]. Its role in IL-15 biology is extremely complex and unique among cytokine receptors, as it may serve to trans-present IL-15 molecule bound at the cell surface of an IL-15Rα+ presenting cell to apposing IL-2Rβ/γ_c+ cells [84]. IL-15Rα appeares crucial for IL-15 activity, since both IL-15- and IL-15Rα-deficient mice display similar phenotypic alterations, such as the defective development of NK cells and of intraepithelial T lymphocytes [85] Differently from IL-2, which is produced during the immune response and is tightly controlled at the transcriptional level, IL-15 gene is expressed by several cell types and IL-15 production is regulated at the post-transcriptional and post-translational levels [81]. A relevant source of IL-15, are stromal cells of lymphoid organs and of the bone marrow, where IL-15 plays an essential role in promoting differentiation of NK cells from their progenitors [81]. In addition, also activated monocyte/macrophages can produce IL-15, which is predominantly displayed as a membrane bound form [86]. IL-15 has been shown to promote proliferation of cells of different hematologic malignancies, such as acute T cell leukemias [87], cutaneous T cell lymphomas

[88], lymphoproliferative disorders of granular lymphocytes [89], the growth factor-dependent acute myleoid leukemia M07e [90], and hairy cell leukemia and CLL B cells [91]. In particular, recombinant IL-15 has been shown to support the proliferation of resting or activated CLL B cells, which express functional IL-2R β/γ_c complexes. The mitogenic effect of IL-15 on CLL B cells appeared similar to that of IL-2, and was blocked by monoclonal antibodies specific for the IL-2Rβ and the γ_c chains [91]. Moreover, the addition of graded doses of IL-15 to low-dose IL-2 produced a cumulative mitogenic effect on CLL B cells, suggesting that the two factors may act in concert in supporting CLL B cell proliferation. It is of note that, although the effects of IL-15 on CLL B cells seemed mainly mediated by the IL-2R β/γ_c complex, the possible role of the IL-15Rα has not yet been addressed in this type of leukemia. In this context, stimulation of IL-15Rα+ Burkitt lymphoma cells with IL-15 induced their proliferation and rescued them from C2-ceramide-induced apoptosis through recruitment of the tyrosine kinase Syk by IL-15Rα [92]. Thus, the study of IL-15/IL-15Ra pathway may certainly deserve further investigations in CLL B cells.

Recently we reported that IL-15 is capable of promoting proliferation of CLL B cells also after CD40 activation [93]. The anti-apoptotic effects of IL-15 on CD40-activated CLL-B cells are, however, counteracted by IL-21, the most recently identified member of the γ_c cytokine family [93]. IL-21 is produced by activated Th cells and mediates its biological activities through a specific IL-21R, which shares structural similarities with the IL-2Rβ chain [94,95]. IL-21 exerts regulatory effects on T, B and NK cells and on myeloid cells [96]. Indeed, IL-21 supports T cell proliferation and induces cytotoxic activity and IFNγ production by cytotoxic T cells and NK cells [94,96]. In view of these functions IL-21 has been studied as a potential anti-tumor agent in pre-clinical models of immuno or gene-therapy of solid tumors, where it showed a clearcut efficacy [96-98]. Regarding normal B cells, IL-21 may exert strikingly different effects, as it can induce either cell proliferation or growth arrest and terminal differentiation or apoptosis, depending on the differentiation and activation status of the B cells [99-101]. Although certain activities of IL-21 on B cells appear redundant with those of other cytokines, certainly IL-21 has unique functions, as revealed by the study of IL-21R-deficient mice, which display increased IgE production and reduced IgG1, and to a lesser extent IgG2 production [102]. In agreement with these data, IL-21 stimulated IgG1 production in vitro and prevented IgE class-switch [103]. IL-21 has also been shown to support the proliferation of acute T cell leukemia cells [104] and of multiple myeloma cells [105] in vitro. In view of these properties, we have explored the possible expression of IL-21R and the effects of IL-21 on CLL B cells [93]. Our data showed that IL-21R expression was rather heterogeneous on CLL B cells freshly isolated from different patients, as detected by flow cytometry, western blot and RT-PCR analyses. Low expression of surface IL-21R appeared more frequent in cases with CD38 and/or ZAP-70 positive and unmutated IgVH genes, suggesting a possible inverse correlation of IL-21R expression with these markers of CLL poor prognosis. Nonetheless, when CLL B cells were activated by CD40 triggering a clearcut increase of IL-21R mRNA and protein expression was observed in all the cases studied [93]. In addition, IL-21R was functional in CLL B cells and IL-21 triggered a cascade of signals involving JAK1 and JAK3 autophosphorylation, and phosphorylation of STAT-1, -3 and –5. Differently from IL-2 and IL-15, however, IL-21 did not induce proliferation of either resting or CD40-activated B cells but rather promoted a slowly proceeding apoptotic effect, which was more evident on CD40-activated than in resting CLL B cells [93]. Furthermore, it is noteworthy that IL-21 was capable of inhibiting proliferative and anti-apoptotic effects of IL-15 on activated CLL B cells, thus suggesting that not only IL-21 and IL-15 have opposite effects, but also that they counteract each other in the regulation of CLL B cell proliferation and survival. In addition, and in parallel with the apoptotic effect observed, we found that IL-21 induced CD23 down-regulation on the surface of CD40-activated CLL B cells (Figure 2). Down-modulation of CD23, which is characteristically highly expressed on CLL B cells, has been already demonstrated in CLL treated with proteosome inhibithors, which induced apoptosis [106]. As shown in figure 2 this effect was again opposite to that of IL-15, which, instead, clearly up-regulated the expression of CD23. CD23 down- or up-regulation thus appears strictly related to pro-or anti-apoptotic stimuli induced by cytokines in CLL. The pro-apoptotic effect of IL-21 was associated with activation of caspase-8 and caspase-3, as indicated by the use of specific fluorescent probes, and by the detection of cleaved forms of caspase-8 in IL-21 treated cells [93]. Caspase-8 activation was also confirmed by the finding of cleavage products of PARP, t-BID and p27KIP-1, which are all substrates of active caspase-8. In agreement with previous observations [25] showing a critical role of Bid in activated CLL B cells undergoing apoptosis, we also found that Bid, not expressed in resting cells, become detectable only after CD40 activation, and was cleaved to tBid after IL-21 treatment. The caspase-8 cleaved form t-Bid, a pro-apoptotic form of Bid capable to translocate to the mitochondrial membrane, appear then to act as a mediator of caspase-8 apoptotic effects [93]. This possibility was also reinforced by the evident drop in mitochondial membrane polarization in IL-21-treated activated CLL-B cells detected by DiOC-6 staining [93]. Sensitization of CLL B cells to IL-21 pro-apoptotic effects by the previous CD40-activation may be therefore related to the strong increase of IL-21R and also the "de novo" expression of Bid. One open question regards the mechanism linking IL-21 to the induction of a caspase-8-dependent apoptotic pathway, since the IL-21R has no death domain. Our data suggested that STAT-1 phosphorylation may be involved in IL-21 pro-apoptotic effects, since STAT-1 activation was less evident in resting than in CD40-activated CLL-B cells, which were more sensitive to the IL-21-mediated pro-apoptotic effects. In addition, IL-21 induced STAT-3 but no STAT-1 activation in human myeloma cells, which were stimulated to proliferate by IL-21 instead of undergoing apoptosis [105]. Moreover, the JAK-1/STAT-1 is involved in the induction of apoptosis in different cell types [107], and lymphocytes from STAT-1-deficient mice showed reduced apoptosis and increased proliferation [108]. Nonetheless the mechanism linking IL-21R to the induction of apoptosis in CLL B cells and the reasons for the opposite effects of IL-15, which shares signal transduction pathways with IL-21, may require further investigations.

In this context, another recent report confirmed that IL-21-treated CLL B cells tend to undergo apoptosis and also showed that they produce low levels of granzyme B. The combination of IL-21 with either CpG ODN activating toll-like recptor (TLR)-9 or anti-B cell receptor antibodies further increased the production of granzyme B by B-CLL cells and their apoptosis [109]. In addition, under these conditions, CLL B cells acquired the ability to induce apoptosis of untreated bystander CLL B cells. This effect could be inhibited by anti-granzyme B antibody. Collectively results from our study and from this report suggest that

IL-21, in combination with CPG-ODN or with CD40-triggering ligands, may require further evaluation for the development of new approaches to the therapy of B-CLL.

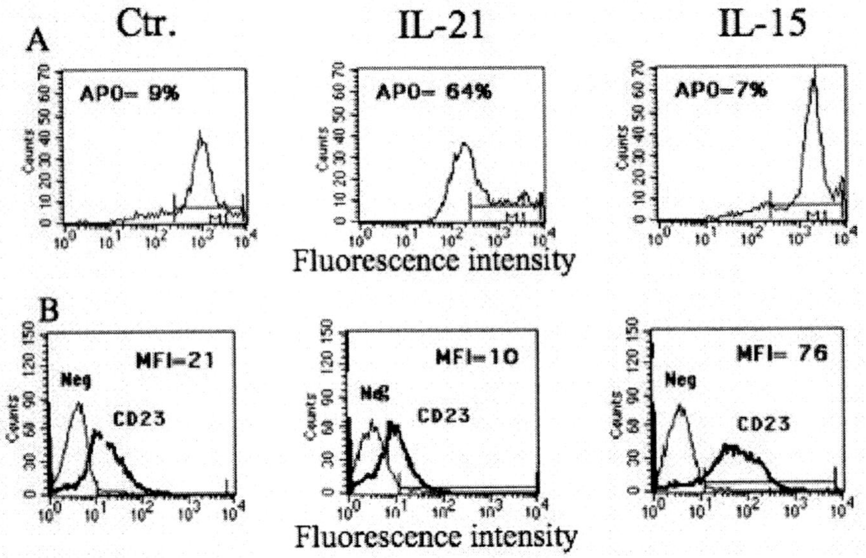

Figure 2. A- Opposite effects of IL-21 and IL-15 on CD40-activated CLL B cells: IL-21 induces apoptosis (APO) while IL-15 induces survival as evidenced by mitochondrial depolarization measured by DiOC6 staining. B-IL-21 and IL-15 differently modulate surface CD23 expression as detected by immunofluorescence and cytofluorimetric analysis.*MFI=mean fluorescence intensity

Not only IL-21 but also two unrelated cytokines, namely IL-5 and IL-10, have been shown in earlier studies to enhance the apoptotic rate of CLL B cells or to inhibit their proliferation. IL-5 is a Th2-cytokine, which stimulates eosinophil differentiation and function and acts through a heterodimeric receptor consisting of a cytokine-specific α chain and the common β chain that is shared with GM-CSF and IL-3 receptors. IL-5 increased spontaneous apoptosis of CLL B cells after 2 days in culture through a Bcl-2-independent mechanism, which was blocked by IL-4 pre-treatment [110,111].

Also IL-10 is a cytokine produced by Th2 cells and has been reported to influence CLL B cells survival. However, although there is a general agreement that IL-10 inhibits either spontaneous [112] or phorbol esther-induced [113,114] CLL-B cell proliferation, its role in regulating CLL B cell survival is controversial (Figure 3). Thus, an earlier report showed that IL-10 increased the apoptotic rate of CLL B cells, while subsequent studies proposed an opposite role of IL-10 [114,115] as it reduced the spontaneous or hydrocortisone-induced apoptotic cell rate. Jurlander et al. also studied the signaling properties of IL-10 on CLL B cells. IL-10R is a member of the interferon (IFN) receptor family. Upon ligand binding IL-10R mediated STAT-1 and –3 phosphorylation in CLL B cells, a signaling pattern similar to that of IFN-α (STAT-1 and –3) and IFN-γ (STAT-1). These authors also showed that IFNγ had biological effects similar to that of IL-10, protecting CLL B cells from hydrocortisone induced apoptosis. Furthermore detectable levels of IFNγ were demonstrated in malignant CLL B cells by in situ hybridization and flow cytometry, thus suggesting that leukemic cells are able to synthesize this cytokine. Moreover, IFNγ increased the survival of CLL B cells in vitro [116]. This observation together with findings of the presence of IFNγ in sera of CLL patients, suggests that this cytokine could be responsible for prolonging the life span of the malignant cells "in vivo" and that the anti-apoptotic effect could correlate with the severity of the disease [117]. Enhancement of cell viability "in vitro" and suppression of apoptosis were associated with a delay in Bcl-2 down-regulation [118]. An increased number of T cells (CD4+, CD8+) producing IFNγ have been further demonstrated in CLL [41,119,120]. In addition we reported that a unique subset of TCRα/β+/CD3+/CD8+/CD28- cells expressing CD30 and releasing IL-4 and IFNγ was expanded in CLL [41]. IL-4, as well as IFNγ, in CLL, appears to be anti-apoptotic (Figure 3), by mantaining Bcl-2 overexpression [121]. In agreement with our data Mu et al. described an increased cytoplasmic IL-4 content in resting and in activated CD8+ T cells in CLL, as compared to normal controls [122]. IL-4 receptors were further expressed on T and B cells and IL-4, when added to CLL B cells, induced apoptosis resistance, increased protein synthesis and rapid onset activation of STAT1, STAT5 and STAT6 [123-127]. From these observations it thus appears that T cell producing cytokines with anti-apoptotic effects are increased in CLL and this finding can be related to a pathogenetic role played by residual T cells in this disease. Our preliminar observation of an abnormal expansion of T cells highly expressing CD30 was then confirmed and extended to demonstrate that these cells inhibit CD40L-induced immunoglobulin class switching of CLL B cells through a CD30-CD30L dependent mechanism, again suggesting that T cells in CLL affect the normal process of B cell differentiation [128].

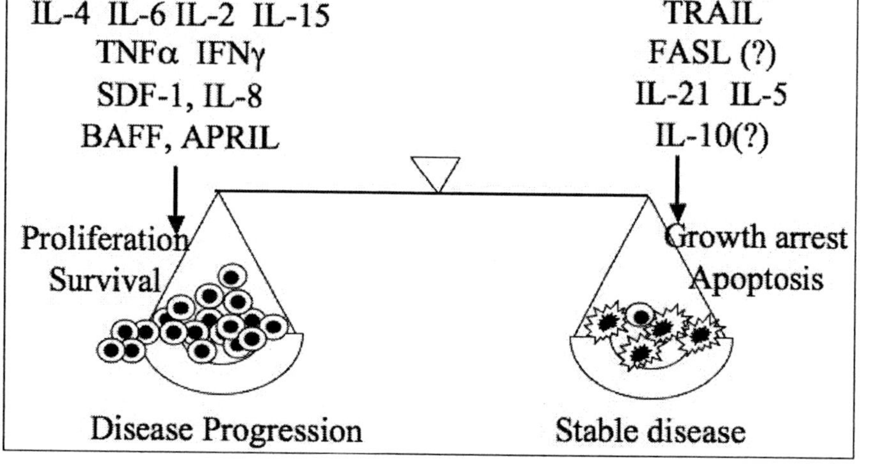

Figure 3. Accumulation of neoplastic B cells in CLL patients and disease progression may be influenced by environmental factors inducing survival and/or proliferation or, on the opposite, growth arrest and apoptosis.

Expression of receptors for cytokines such as IL-2 and IL-4 on CLL B cells may allow the development of targeted strategies of cytotoxic drug delivery. For this purpose, a recombinant fusion protein (denileukin diftitox or ONTAK), consisting of a fragment of diphteria toxin linked to IL-2, specifically binds to high or intermediate affinity IL-2R complexes. Upon binding to IL-2R-expressing cells, the fusion protein is internalized and may exert its selective cytocidal effect. Although the drug has been approved for the therapy of relapsing or refractory cutaneous T cell lymphoma, recent evidences indicate the possible usage of this drug for CLL therapy, in particular in combination with treatments augmenting the expression of high affinity receptors on these cells [129-131]. The IL-4-IL-4R pathway, which is functional in CLL, has been also recently targeted by the use of a recombinant IL-4-pseudomonas exotoxin fusion protein (IL-4PE) [132]: CLL B cells after IL-4PE “in vitro” treatment show increased apoptosis.

Among the cytokines that, over the last years, have been described as playing a regulatory role of CLL B cells growth there is also IL-8 (Figure 3). Malignant CLL B cells constitutively express IL-8 and IL-8 receptors (IL-8R) and the addition of exogenous IL-8 to CLL “in vitro” cultures prolongs their survival trhough Bcl-2 upregulation [133-136]. Moreover the detection of significant levels of IL-8 in serum of CLL patients has suggested that autocrine secretion of this factor may contribute to cell accumulation characteristic of the disease. Furthermore, plasmatic IL-8 in CLL correlates with RAI stages and serum beta2-microglobulin, and high IL-8 levels can be associated with increased risk of death [135]. The role of IL-8 “in vivo” is, however, still debated as a previous study based on cDNA microarray analysis identified a reduced IL-8 and IL-1β gene expression from CLL patients with low survival higher RAI stages [137].

Several studies further showed that also IL-6 may be present in plasma from CLL patients at elevated levels and in a stage-dependent manner [138,139]. Moreover levels of plasmatic IL-6 may represent a useful predictor for survival in CLL patients with advanced disease [140,141]. However, it is still unclear whether IL-6 is capable of promoting CLL growth, as demonstrated for some B-cell lymphoma and myeloma [142,143]. Data on IL-6 production “in vitro” by CLL B cells showed, in fact, high variability among the different cases and stages of the disease [144].

A significant correlation between IL-6 and the levels of a number of other cytokines in the plasma of CLL patients was further identified, including IL-1, IL-1 receptor antagonist (IL-1Ra) and tumor necrosis factor-α (TNF-α) [141]. This finding was not unexpected, as IL-1 and TNFα are potent inducers of IL-6. IL-1, for example, has been shown to have indirect hematopoietic activity by inducing gene expression of colony stimulating factor and of IL-6 [145]. However it is unclear whether IL-1 has a role “in vivo” and data on plasmatic levels appeared heterogeneous, probably depending on the disease state. Hulkonnen et al. reported that baseline plasma levels of IL-1β and IL-1Ra were lower in CLL plasma than in controls and varied between patients with different disease severity [139]. Plasma IL-1β was higher and IL-1Ra/IL-1β ratio lower in patients with non-progressive disease compared with those in progressive disease. However, since IL-1β is synthesized as inactive pro-IL-1β form converted to active form by interleukin-1 converting enzyme (ICE), also known as caspase-1, low levels of IL-1β could thus be due to low ICE activity. This hypotesis appear interesting as the ICE gene, located at chromosome region 11q22.2-22.3, is frequently affected in CLL [146-148] but a potential correlation between low ICE gene expression and the levels of plasmatic IL-1β deserve further investigations. We found that CLL B cells produce significant levels of IL-1β “in vitro” only after CD40 triggering (unpublished data). This finding may suggest that secretion of IL-1β might be related to an activated state of CLL B cells.

Conclusion

The elucidation of the mechanisms controlling resistance of CLL B cells to programmed cell death is of great interest in view of developing novel therapeutic approaches. Within the last 10 years it has been shown that many events required for apoptosis in normal B lymphocytes are altered in CLL, including both defects in death programme and an exacerbated response to survival signals. However, the finding that CD40 cross-linking on CLL B cells modulate the expression of death receptors (FAS, DR5), of pro-apoptotic cytokine receptors, or of specific molecules driving CLL to apoptosis (Bid, p73), making them more susceptible to cell death, may open new therapeutic possibilities. Further investigations are needed to determine whether activation via clinical-grade soluble CD40L or agonistic CD40 moAb, or adenoviral vector-mediated CD40L gene transfer approaches, could be effective as therapeutic strategy, in combination with particular cytokines, such as IL-21, which may trigger pro-apoptotic activity on CLL B cells and may also amplify immune responses against malignant B cells. Recent data suggest that also other stimuli such as BCR cross-linking or TLR agonists, such as oligo CpG, may potentiate the pro-apoptotic effects of IL-21, thus supplying more suitable reagents for combined therapies in CLL. A deeper knowledge of cytokines and of their receptors stimulating CLL B cell proliferation and survival, and the identification of their signal transduction pathways may provide new targets and tools for the development of molecular targeting approaches in this disease.

References

[1] Herr, I; Debatin, KM. Cellular stress response and apoptosis in cancer therapy. *Blood*, 2001, 98, 2603-2614.

[2] Wickremasinghe, RG; Hoffbrand, AV.Biochemical and genetic control of apoptosis: relevance to normal hematopoiesis and hematological malignancies. *Blood*, 1999, 93, 3587-3600 Review.

[3] Dancescu, M; Rubio-Trujillo, M; Biron, G; Delespesse, G; Sarfati, M. Interleukin-4 protects chronic lymphocytic leukemia B cells from death by apoptosis and upregulates bcl-2 expression. *J Exp Med* 1992, 176, 1319-1326.

[4] Panayioditis P, Ganeshaguru K, Jabbar SAB, Hoffbrand AV. Interleukin 4 inhibits apoptotic cell death and loss of bcl-2 protein in B-cell chronic lymphocytic leukaemia cells in vitro. *Br J Haematol*, 1993, 85, 39-45.

[5] Brass, U; Tretter, T; Schneller, F; Schuler, M; Huber, C; Peshel, C. IFN-alpha stimulates proliferation and cytokine secretion of CD40-stimulated B-cell chronic lymphocytic leukemia cells in vitro. *J Interferon Cytokine Res*, 1999, 19,335-343.

[6] Buschle, M; Campana, D; Carding, SR; Richard, C; Hoffbrand, V; Brenner, M. Interferon γ inhibits apoptotic cell death in B cell chronic lymphocytic leukemia. *J Exp Med*, 1993, 177, 213-218.

[7] Roue', G; Lancry, L; Duquesne, F; Salaun, V; Troussard, X; Sola, B. Upstream mediators of the Fas apoptotic transduction pathway are defective in B-chronic lymphocytic leukemia. *Leuk Res*, 2001, 25, 967-980.

[8] de Totero, D; Montera, M; Rosso, O; Clavio, M; Balleari, E; Foa, R, Gobbi, M. Resistance to CD95-mediated apoptosis of CD40-activated chronic lymphocytic leukemia B cells is not related to lack of DISC molecules expression. *The Hematology Journal*, 2004, 5, 152-160

[9] Romano, C; De Fanis, U; Sellitto, A; Chiurazzi, F; Guastafierro, S; Giunta, R; Tirelli, A; Rotoli, B; Lucifero, G. Induction of CD95 upregulation does not render chronic lymphocytic leukemia B-cells susceptible to CD95-mediated apoptosis. *Immunol Lett*, 2005, 97, 131-139.

[10] Eggermont, AM; ten Hagen, TL. Isolated limb perfusion for extremity soft-tissue sarcomas, in-transit metastases and other unresectable tumors: credits, debits, and future perspectives. *Curr Oncol Rep*, 2001, 3, 359-367. Review

[11] Digel, W; Stefanic, M; Schoniger, W; Buck, C; Raghavachar, A; Frickofen, N; Heimpel, H; Porzsol, F. Tumor necrosis factor induces proliferation of neoplastic B cells from chronic lymphocytic leukemia. *Blood*, 1989, 73, 1242-1246.

[12] Foa, R; Massaia, M; Cardona, S; Tos, AG; Bianchi, A; Attisano, C; Guarini, A; di Celle, PF; Fierro, MT. Production of tumor necrosis factor-alpha by B-cell chronic lymphocytic leukemia cells: a possible regulatory role of TNF in the progression of the disease. *Blood*, 1990, 76, 393-400.

[13] Jabbar, SAB; Hoffbrand, V; Wickresmashinghe, G. Regulation of transcription factors NFkB and AP-1 following tumor necrosis factor-α treatment of cells from B cell chronic lymphocytic leukemia patients. *Br J Haematol*, 1994, 86, 496-504.

[14] Gallego, A; Vargas, JA; Castejon, R; Citores, MJ; Romero, Y; Millan, I; Durantez, A. Production of intracellular IL-2, TNF-α, and IFN-γ by T cells in B-CLL. *Cytometry B Clin Cytom* 2003, 56, 23-29.

[15] Rossmann, ED; Lewin, N, Jeddi-Tehrani, M; Osterborg, A; Mellstedt H. Intracellular T cell cytokines in patients with B cell chronic lymphocytic leukaemia (B-CLL). *Eur J Haematol*, 2002, 68, 299-306.

[16] Rossi, D; Gaidano, G. Messengers of cell death: apoptotic signaling in health and disease. *Haematologica*, 2003, 88, 212-218

[17] de Totero, D; Tazzari, PL; Capaia, M; Montera, MP; Clavio, M; Balleari, E; Foa, R; Gobbi, M. CD40 triggering enhances fludarabine-induced apoptosis of chronic lymphocytic leukemia B-cells through autocrine release of tumor necrosis factor-α and interferon-γ and tumor necrosis factor receptor-I-II upregulation. *Haematologica*, 2003, 88,148-158.

[18] Trentin, L; Zambello, R; Agostini, C; Siviera, F; Adami, F; Marcolongo, R; Raimondi, R; Chisesi, S; Pizzolo, G; Semenzato, G. Expression and functional role of tumor necrosis factor receptors on leukemic cells from patients with type B chronic lymphoproliferative disorders. *Blood*, 1993, 81, 752-758.

[19] Bojarska-Junad, A; Rolinski, J; Wasik-Szczepaneko, E; Kaluzny, Z; Dmoszynska, A. Intracellular tumor necrosis factor production by T- and B- cells in B-cell chronic lymphocytic leukemia. *Haematologica*, 2002, 87, 490-499.

[20] Kitada, S; Zapata, JM; Andreeff, M; Reed, JC. Bryostatin and CD40-ligand enhance apoptosis resistance and induce expression of cell survival genes in B-cell chronic lymphocytic leukaemia. *Br J Haematol*, 1999, 106, 995-1004

[21] Himeji, D; Horiuchi, T; Tsukamoto, H; Hayashi, K; Watanabe, T; Harada, M. Characterization of caspase-8L: a novel isoform of caspase-8 that behaves as an inhibitor of the caspase cascade. *Blood*, 2002, 99, 4070-4078.

[22] Chu, P; Deforme, D; Pedersen, IM; Kim, Y; Kitada, S; Reed, JC; Kipps, TJ. Latent sensitivity of Fas-mediated apoptosis after CD40 ligation may explain activity of CD154 gene therapy in chronic lymphocytic leukemia. *Proc Natl Acad Sci USA* 2002, 99, 3854-3859.

[23] Pater, AP; Dicker, F; Mangiola, M; Welsh, K; Houghten, T; Ostresh, J; Nefzi, A; Redd, JC; Pinella, C; Kipps, TJ. Inhibitors of XIAP sensitize CD40-activated chronic lymphocytic leukemia cells to CD95-mediated apoptosis. *Blood*, 2005,106, 1742-1748.

[24] MacFarlane, M; Harper, N; Snowden, RT; Dyers, MJS; Barnett, GA; Pringle, JH; Cohen, GM. Mechanisms of resistance to TRAIL-induced apoptosis in primary B cell chronic lymphocytic leukaemia. *Oncogene*, 2002, 21, 6809-6818.

[25] Dicker, F; Kater, AP; Fukuda, T; Kipps, TJ. Fas ligand (CD178) and TRAIL synergistically induce apoptosis of CD40-activated chronic lymphocytic leukemia. *Blood*, 2005, 105, 3193-3198.

[26] Kater, AP; Evers, LM; Remmerswaal, EB; Jaspers, A; Oosterwijk, MF; van Lier, RA; van Oers, MH; Elderling, E. CD40 stimulation of B-cell chronic lymphocytic leukaemia cells enhances the anti-apoptotic profile, but also Bid expression and cells remain susceptible to autologous cytotoxic T-lymphocyte attack. *Br J Haematol*, 2004, 127, 404-415.

[27] Lagresle, C; Bella, C; Daniel, PT; Krammer, PH; Defrance, T. Regulation of germinal center B cell differentiation: role of the human APO-1/Fas-mediated apoptosis. 1995, *J Immunol,* 156, 2345-2348.

[28] Lederman, S; Yellin, MJ; Krichevsky, A; Belko, J; Lee, JJ; Chess, L. Identification of a novel surface protein on activated CD4+ T cells that induces contact-dependent B cells differentiation. *J Exp Med*, 1992, 175, 1091-1101.

[29] Bancherau, J; Bazan, F; Blanchard, D; Triere, F; Galizzi, JP; VanKooten, C; Rousset, F; Saeland, S. The CD40 antigen and its ligand. Annu Rev Immunol, 1994, 12, 881-922.

[30] Berberich, I; Shu, GL; Clark, EA. Cross-linking CD40 antigen on B cells rapidly activates nuclear factor-kB. *J Immunol*, 1994, 153, 4357-4366.

[31] Yellin, MJ; Sinning, J; Covey, LR; Sherman, W; Lee, JJ; Glickman-Nir, E; Sippel, KC; Rogers, J; Cleary, AM; Parker, Chess, L; Lederman, S. T lymphocyte T cell-B cell-activating molecule/CD40-Lmolecules induce normal B cell or chronic lymphocytic leukemia B cells to express CD80 (B7/BB-1) and enhance their costimulatory activity. *J Immunol*, 1994, 153: 666-674.

[32] Faris, M, Gaskin, F; Parsone, JT; Fu, SM. CD40 signaling pathway: anti-CD40 monoclonal antibody induces rapid dephosphorylation and phosphorylation of tyrosine-phosphorylated proteins including protein tyrosine kinases lyn, fyn, and syk and the appearence of a 28-kD tyrosine phosphorylated protein. *J Exp Med*, 1994, 179, 1923-1931.

[33] Lederman, S; Yellin, MJ; Cleary, AM; Pernis, A; Inghirami, G; Cohn, LE; Covey, LR; Lee, JJ; Rothman, P; Chess, L. T-BAM/CD40L on helper T-lymphocytes augments

lymphokine-induced B-cell Ig isotype switch recombination and rescue B-cells from programmed cell death. *J Immunol*, 1994; 152: 2163-2171.

[34] Splaswki, JB; Fu, SM; Lipsky, PE. Immunoregulatory role of CD40 in human B-cell differentiation. *J Immunol*, 1993, 150, 1276-1285.

[35] Grabstein, KH; Maliszewski, CR; Shanebeck, K; Sato, TA; Spriggs, MK; Fanslow, WC; Armitage; RJ. The regulation of T cell-dependent antibody formation in vitro by CD40 ligand and IL-2. *J Immunol*, 1993; 150, 3141-3147.

[36] Arruffo, A; Farrington, M; Hollenbaugh, D; Li, X; Milatovich, A; Nonoyama, S; Bajorath, J; Grosmaire, LS; Stenkamp,R;. Neubauer, M; et al. The CD40 ligand gp39 is defective in activated T cells from patients with X-linked hyper-IgM sindrome. *Cell*, 1993, 72, 291-300.

[37] Allen, RC,;Armitage, RJ; Conley, ME; Rosenblatt, H; Jenkins, NA; Copeland, NG; Bedell, MA; Edelhoff, S; Disteche, CM; Simoneaux, DK et al. CD40 ligand gene defects responsible for X-linked hyper-IgM sindrome. *Science*, 1993,259, 990-993.

[38] Di Santo, JP; Bonnefoy, JY; Gauchat, JF; Fisher, A; de Saint Basil, G. CD40 ligand mutations in X-linked immunodeficiency with hyper-IgM. *Nature*, 1993, 361, 541-543.

[39] Korthauer, U; Graf, D; Mages, HW; Triere, F; Padayachee, M; Malcom, S; Ugazio AG, Notarangelo,LD; Levinsky, RJ; Kroczeck, RA. Defective expression of T-cell CD40 ligand causes X-linked immunodeficiency with hyper-IgM. *Nature*, 1993, 3, 539-541.

[40] Cantwell, M; Hua, T; Pappas, J; Kipps, TJ. Acquired CD40-ligand deficiency in chronic lymphocytic leukemia. *Nat Med*, 1997, 3, 984-989.

[41] de Totero, D; Reato, G; Mauro, F; Cignetti, A; Ferrini, S; Guarini, A; Gobbi, M; Grossi, CE; Foa, R. IL4 production and increased CD30 expression by a unique CD8+ T cell subset in B-cell chronic lymphocytic leukemia. *Br J Haematol* 1999, 104, 589-599.

[42] Dazzi, F, D'Andrea, E; Biasi, G; De Silvestro, G; Gaidano, G; Schena, M; Tison, T; Vianello, F; Girolami, A; Caligaris-Cappio, F. Failure of B cells of chronic lymphocytic leukemia in presenting soluble and alloantigens. *Clin Immunol Immunopathol*, 1995, 75, 26-32.

[43] Van den Hove, LE; Van Gool, SW; Vandenberghe, P; Bakkus, M; Thielemans, K; Boogaerts, MA; Ceuppens, JL. CD40 triggering of chronic lymphocytic leukemia cells results in efficient alloantigen presentation and cytotoxic T lymphocyte induction by up-regulation of CD80 and CD86 costimulatory molecules. *Leukemia*, 1997, 11, 572-580.

[44] Kato, K; Cantwell, MJ; Sharma, S; Kipps, TJ. Gene transfer of CD40 ligand induces autologous immune recognition of chronic lymphocytic leukemia B cells. *J Clin Invest,* 1998, 101, 1133-1141.

[45] Wierda, WG; Cantwell, MJ; Woods, SJ; Rassenti, L; Prussak, CE; Kipps, TJ. CD40-ligand (CD154) gene therapy for chronic lymphocytic leukemia. *Blood* 2000, 96, 2917-2924.

[46] Battle, TE; Wierda, WG; Rassenti, LZ; Zahrieh, D; Neuberg, D; Kipps, TJ; Frank DA. In vivo activation of signal transducer and activator of transcription 1 after CD154 gene therapy for chronic lymphocytic leukemia is associated with clinical and immunologic response. *Clin Cancer Res* 2003, 9, 2166-2172.

[47] Haiat, S; Billard, C; Quiney, C; Ajchembaum-Cymbalista, F; Kolb, JP. Role of BAFF and APRIL in human B-cell chronic lymphocytic leukaemia. *Immunology*, 2006, 118, 281-292. Review.

[48] Mukopadhyay, A; Ni, J; Zhai, Y; Yu, GI; Aggarwal BB. Identification and characterization of a novel cytokine, THANK, a TNF homologue that activates apoptosis, nuclear factor-kappa B, and c-Jun NH2-terminal kinase. *J Biol Chem* 1999, 274, 15978-1981.

[49] Hatzoglou, A; Roussel, J; Bourgeade, MF; Rogier, E; Madry, C; Inoue, J; Devergne, O; Tapis A. TNF receptor family member BMCA (B cell maturation) associates with TNF-receptor-associated factor (TRAF 1), TRAF2, and TRAF3 and activated NF-kappa B, elk-1, c-Jun N-terminal kinase, and p38 mitogen-activated protein kinase. *J Immunol*, 2000, 165, 1322-1330.

[50] Thompson, JS; Bixler, SA; Qian, F; Vora, K; Scott, ML; Canchero, TG; Hession, C; Schneider, P; Sizing, ID; Mullen, C; Strauch, K;Safari, M; Benjamin, CD; Tshopp, J; Browning, JL; Ambrose , C.BAFF-R, a newly identified TNF receptor that specifically interacts with BAFF. *Science*, 2001, 293, 2108-2111.

[51] Briones, J; Timmermann, JM; Hilbert, DM; Levy, R. BLyS and BLyS receptor expression in non-Hodgkin's lymphoma. *Exp Hematol*, 2002, 30, 135-141.

[52] Novak, AJ; Bram, RJ; Kay, NE; Jelinek, DF. Aberrant expression of B-lymphocyte stimulator by B chronic lymphocytic leukemia cells: a mechanism for survival. *Blood*, 2002, 100, 2973-2979.

[53] Kern, C; Cornuel, JF; Billard, C;Tang, R;Rouillard, D; Stenou, V;Defrance, T; Aichembaum- Cymbalista, F; Simonin, PY; Feldblum, S, Kolb, JP. Involvement of BAFF and APRIL in the resistance to apoptosis of B-CLL through an autocrine pathway. *Blood*, 2004, 103, 679-688.

[54] Nishio, M; Endo, T; Tsukada ,N; Ohata, K; Kitada, S; Reed, JC; Zvaifler, NJ; Kipps TJ. Nurselike cells express BAFF and APRIL, which can promote survival of chronic lymphocytic leukemia cells via paracrine pathway distinct from that of SDF-1. *Blood*, 2005, 106, 1012-1020.

[55] Kitada, S; Andersen, J; Akar, S; Zapata, JM; Takayama, S; Krayeswski, S; Wang, HG; Zhang, X; Bullrich, F; Croce, CM; Rai, K; Hines, J; Reed, JC. Expression of apoptosis-regulating proteins in chronic lymphocytic leukemia: correlations with in vitro and in vivo chemoresponses. *Blood*, 1998, 91, 3379-3389

[56] Borner C. The Bcl-2 protein family: sensors and checkpoint for life-or-death decisions. *Mol Immunol*, 2003, 39, 615-647.

[57] Adachi M, Tefferi A, Greipp P, Kipps TJ, Tsujimoto Y. Preferential linkage of bcl-2 to immunoglobulin light chain gene in chronic lymphocytic leukemia. *J Exp Med*, 1990, 171, 559-564.

[58] Cimmino, A; Calin , GA; Fabbri, M; Iorio, MV; Ferracin, M; Shimizu, M; Wojcik, SE; Aqeilan, RI; Zupo, S; Dono, M; Rassenti, L; Alde, H; Volinia, S; Liu. CG; Kipps, TJ; Croce, CM. miR-15 and miR-16 induce apoptosis by targeting BCL2. *PNAS*, 2005, 102, 13944-13949.

[59] Calin, GA; Ferracin, M; Cimmino, A; Di Leva, G; Shimizu, M; Wojcik, SE; Iorio, MV; Visone, R; Sever, NI; Fabbri, M; Iuliano R; Palombo, T; Pichiorri, F; Roldo, C;

Garzon, R; Sevignani, C; Rassenti, L; Alder, H; Volinia, S; Liu, Cg; Kipps TJ; Negrini, M; Croce, CM. A microRNA signature associated with prognosis and progression in chronic lymphocytic leukemia. *New Engl J Med*, 2005, 353, 1793-1802.

[60] Grdisa, M; Tang, RP; Delmer, A; Levy, V; Gribben, JG; Ajchembaum-Cymbalista, F. Ligation of CD40 on B-cell chronic lymphocytic leukemia (B-CLL) cells promotes increased survival but does not impair fludarabine induced apoptosis. *Blood,* 2000, 96, 159a (abstract)

[61] Berrebi, A; Bassou, L; Shvidel, L; Vorst, E. Evaluation of in vitro apoptosis of B CLL cells: variations according to maturation stages and influence of dexamethasone, fludarabine and verapamil. *Blood*, 1998; 92, 162b (abstract).

[62] Romano, MF; Lamberti, A; Tassone, P; Alfinito, F; Costantini, S; Chiurazzi, F; Defrance, T; Bonelli, P; Tuccillo, F; Turco, MC; Venuta, S. Triggering of CD40 antigen inhibits fludarabine-induced apoptosis in B chronic lymphocytic leukemia cells. *Blood*, 1998, 92, 990-995.

[63] Dicker, F; Kater , AP; Prada, CE; Fukuda, T; Castro JE; Sun, G; Wang, JY; Kipps, TJ. CD154 induces p73 to overcome the resistance to apoptosis of chronic lymphocytic leukemia cells lacking functional p53. *Blood*, 2006, June 1 prepublished on line.

[64] Grell, M; Zimmermann, G; Gottfried, E; Chen, CM; Grunwald, U; Huang, DC; Wu, Lee YH; Durkop, H; Engelmann, H; Scheurich, P; Wajant, H; Strasser, A. Induction of cell death by tumor necrosis factor (TNF) receptor 2, CD40 and CD30: a role for TNF-R1 activation by endogenous membrane-anchored TNF. *Embo J*, 1999, 18, 3034-3043.

[65] Yang, YC; Hsu, TY; Chen, JY; Yang; CS; Lin, RH. Tumor necrosis factor-α-induced apoptosis in cord blood T lymphocytes: involvement of both tumour necrosis factor receptor types 1 and 2. *Br J Haematol*, 2001, 115, 435-441.

[66] Silvaram, S; Deshpande, CG; Ranganathan, R; Huang, X; Jaieh ,A; O' Brien, T; Huang, RW; Gregory, SA; Venugopal, P; Preisler HD. Tumor necrosis factor modulates CD20 expression on cells from chronic lymphocytic leukemia: a new role for TNF alpha? *Microsc Res Tech*, 2000, 50, 251-257.

[67] Kovanen, PE; Leonard, WJ. Cytokines and immunodeficiency diseases: critical roles of the gamma (c)-dependent interleukins 2,4,7,9, 15 and 21, and their signaling pathways. *Immunol Rev*, 2004, 202, 67-83

[68] Leonard WJ. Cytokines and immunodeficiency diseases. *Nature Rev Immunol,* 2001, 1, 200-208.

[69] Noguchi, M; Yi, H; Rosenblatt, HM; Filipovich, AH; Adelstein, S; Modi, WS; McBride, OW; Leonard, WJ.Interleukin-2 receptor γ chain mutation results in X-linked severe combined immunodeficiency in humans. *Cell*, 1993, 73, 147-157.

[70] Russell, SM; Tayebi, N; Nakajima, H; Riedy, MC; Roberts, JL; Aman, MJ; Magone, TS; Noguchi, M; Markert, ML; Buckley, RH; O'Shea, JJ; Leonard, WJ. Mutation of Jak3 in a patient with SCID: essential role of Jak3 in lymphoid development. *Science*, 1995, 270, 797-800

[71] Macchi, P; Villa, A; Giliani, S; Sacco, MG; Frattini, A; Porta, F; Ugazio, AG; Johnston, JA;Candotti, F; O'Shea, JJ et al. Mutations of Jak-3 gene in patients with autososmal severe combined immune deficiency (SCID). *Nature*, 1995, 377, 65-68.

[72] de Totero, D; di Celle, PF; Cignetti, A; Foa, R. The IL-2 receptor complex: expression and function on normal and leukemic B cells. *Leukemia*, 1995, 9, 1425-1431. Review.
[73] Malek, TR; Bayer, AL. Tolerance, not immunity, crucially depends on IL-2. *Nat Rev Immunol.* 2004, 9,665-674. Review.
[74] Nelson, BH; Willerford, DM. Biology of the interleukin-2 receptor. *Adv Immunol.* 1998, 70, 1-81. Review.
[75] Waldmann, T; Tagaya, Y; Bamford R. Interleukin-2, interleukin-15, and their receptors. *Int Rev Immunol*, 1998, 16, 205-226. Review.
[76] Trentin, L; Zambello, R; Agostini, C; Enthammer, C; Cerutti, A; Adami, F; Zamboni, S; Semenzato G. Expression and regulation of tumor necrosis factor, interleukin-2 and hematopoietic growth factor receptors in B-cell chronic lymphocytic leukemia. *Blood*, 1994, 12, 4249-4256.
[77] de Totero D, Tazzari, PL; Lauria, F; Raspadori, D; di Celle , PF; Carbone, A; Foa, R. Phenotypic analysis of hairy cell leukemia: " variant" cases express the interleukin-2 receptor beta chain, but not the alpha chain (CD25). *Blood*, 1993, 82, 528-535.
[78] de Totero, D; Carbone, A; Tazzari, PL; Raspadori, D; Ventura, A; Reato, G; Lauria, F; Foa, R. Expressionof the IL-2 receptor α,β and γ chain in hairy cell leukemia. *Leuk Lymphoma* 1994, 14 (suppl.1) 27-32. Review.
[79] Grabstein ,KH; Eisenman, J; Shanebeck, K; Rauch, C,; Srinivasan, S; Fung,V; Beers, C; Richardson, J; Schoenborn, MA; Ahdieh, M et al. Cloning of a T cell growth factor that interacts with the beta chain of the interleukin-2 receptor. *Science*, 1994, 264,965-968.
[80] Carson,WE; Giri, JG; Lindemann, MJ; Linett; ML; Ahdieh, M; Paxton, R; Anderson, D, Eisenmann, J; Grabstein, K; Caligiuri, MA. Interleukin (IL) 15 is a novel cytokine that activates human natural killer cells via components of the IL-2 receptor. *J Exp Med*, 1994, 180, 1395- 1403.
[81] Fehniger, TA; Caligiuri, MA. Interleukin 15: biology and relevance to human disease. *Blood*, 2001, 97, 14-32. Review
[82] Anderson, DM; Kumaki, S; Ahdieh, M; Bertles, J; Tometsko, M; Loomis, A; Giri, J; Copeland, NG; Gilbert, DJ; Jenkins, NA; et al. Functional characterization of the human interleukin-15 receptor alpha chain and close linkage of IL15RA and IL2RA genes. *J Biol Chem*, 1995, 15;270, 29862-29869.
[83] Pereno, R; Giron-Michel, J,; Gaggero, A; Cazes, E; Meazza, R; Monetti, M; Monaco, E; Mishal, Z; Jasmin, C; Indiveri ,F; Ferrini, S; Azzarone, B. IL-15/IL-15Ralpha intracellular trafficking in human melanoma cells and signal transduction through the IL-15Ralpha. *Oncogene*, 2000; 19, 5153-5162.
[84] Kobayashi, H; Dubois, S; Sato, N; Sabzevari, H; Sakai, Y, Waldmann, TA; Tagaya, Y. Role of trans-cellular IL-15 presentation in the activation of NK cell-mediated killing, which leads to enhanced tumor immunosurveillance. *Blood*, 2005,105, 721-727.
[85] Lodolce ,JP; Boone, DL; Chai, S; Swain,RE ;Dassopoulos, T; Trettin, S; Ma, A.IL-15 receptor maintains lymphoid homeostasis by supporting lymphocyte homing and proliferation. *Immunity*, 1998, 9, 669-676.
[86] Musso, T; Colosso, L; Zucca, M; Millesimo, M; Ravarino, D; Giovarelli, M; Malavasi, F; Ponzi, AN; Paus, R; Bulfone-Paus, S. Human monocytes constitutively express

membrane-bound, biologically active, and interferon-gamma-upregulated interleukin-15. *Blood*, 1999, 15, 93, 3531-3539.

[87] Bamford, RN; Battiata, AP; Burton, JD; Sharma, H; Waldmann, TA. Interleukin (IL) 15/IL-T production by the adult T-cell leukemia cell line HuT-102 is associated with a human T-cell lymphotrophic virus type I region /IL-15 fusion message that lacks many upstream AUGs that normally attenuates IL-15 mRNA translation. *Proc Natl Acad Sci U S A*, 1996, 93, 2897-2902

[88] Dobbeling, U; Dummer, R; Laine, E; Potoczna, N, Qin, JZ; Burg, G. Interleukin-15 is an autocrine/paracrine viability factor for cutaneous T-cell lymphoma cells. *Blood*, 1998, 92, 252-258.

[89] Zambello, R; Facco, M; Trentin, L; Sancetta, R; Tassinari, C; Perin, A; Milani, A; Pizzolo, G; Rodeghiero, F; Agostini, C; Meazza, R; Ferrini ,S; Semenzato G. Interleukin-15 triggers the proliferation and cytotoxicity of granular lymphocytes in patients with lymphoproliferative disease of granular lymphocytes. *Blood*, 1997, 89, 201-211.

[90] Meazza, R; Basso, S; Gaggero, A; Detotero, D;, Trentin, L; Pereno, R; Azzarone, B; Ferrini, S. Interleukin (IL)-15 induces survival and proliferation of the growth factor-dependent acute myeloid leukemia M-07e through the IL-2 receptor beta/gamma. *Int J Cancer*, 1998, 78, 189-195.

[91] Trentin, L; Cerutti, A; Zambello, R; Sancretta, R; Tassinari, C; Facco, M; Adami, F; Rodeghiero, F; Agostini, C; Semenzato, G. Interleukin-15 promotes the growth of leukemic cells of patients with B-cell chronic lymphoproliferative disorders. *Blood*, 1996, 87, 3327-3335.

[92] Bulanova, E; Budagian, V; Pohl, T; Krause, H; Durkop, H; Paus, R; Bulfone-Paus, S. The IL-15R alpha chain signals through association with Syk in human B cells. *J Immunol*, 2001, 167, 6292-6302.

[93] de Totero, D; Meazza, R; Zupo, S; Cutrona, G; Matis, S; Colombo, M; Balleari, E; Pierri, I; Fabbi, M; Capaia, M; Azzarone, B; Gobbi, M; Ferrarini, M; Ferrini, S. Interleukin-21 receptor (IL-21R) is up-regulated by CD40 triggering and mediates proapoptotic signals in chronic lymphocytic leukemia B cells. *Blood*, 2006,107, 3708-3715.

[94] Parrish-Novak, J; Dillon, SR; Nelson, A; Hammond, A; Sprecher, C; Gross, JA; Johnston, J; Madden, K; Xu, W; West, J; Schrader, S; Burkhead, S; Heipel, M; Brandt, C; Kuijper, JL; Kramer, J; Conklin, D; Presnell, SR; Berry, J; Shiota, F; Bort, S; Hambly, K; Mudri, S; Clegg, C; Moore, M Grant, FJ; Lofton-Day, C; Gilbert ,T; Rayond, F; Ching, A; Yao, L; Smith, D; Webster, P; Whitmore, T; Maurer, M; Kaushansky, K; Holly, RD; Foster, D. Interleukin 21 and its receptor are involved in NK cell expansion and regulation of lymphocyte function. *Nature*, 2000, 408, 57-63.

[95] Ozaki, K; Kikly, K; Michalovich, D; Ypung, PR; Leonard, WJ. Cloning of a type I cytokine receptor most related to the IL-2 receptor beta chain. *Procl Natl Acad Sci USA* 2000, 97, 11439-11444.

[96] Leonard, WJ; Spolski, R. Interleukin-21: a modulator of lymphoid proliferation, apoptosis and differentiation. *Nat Rev Immunol*. 2005, 5, 688-698. Review.

[97] Di Carlo, E; Comes, A; Orengo, AM; Rosso, O; Meazza, R; Musiani, P; Colombo, MP; Ferrini, S. IL-21 induces tumor rejection by specific CTL and IFN-gamma-dependent CXC chemokines in syngeneic mice. *J Immunol*, 2004, 172, 1540-1547.

[98] Comes, A; Rosso, O; Orengo, AM; Di Carlo, E; Sorrentino, C; Meazza, R; Piazza, T; Valzasina, B; Nanni ,P; Colombo, MP; Ferrini, S. CD25+ regulatory T cell depletion augments immunotherapy of micrometastases by an IL-21-secreting cellular vaccine. *J Immunol*, 2006, 176, 1750-1758.

[99] Ozaki,K, Spolski, R,Ettinger, R; Kim ,HP; Wang, G; Qi ,CF; Hwu, P; Shaffer, DJ; Akilesh, S;Roopenian ,DC; Morse, HC; Lipsky, PE; Leonard, WJ. Regulation of B cell differentiation and plasma cell generation by IL-21, a novel inducer of Blimp-1 and Bcl-6. *J Immunol*, 2004, 173, 5361-5371.

[100] Mehta, DS; Wurstel ,AL; Whitters, MJ; Young, DA; Collina, M; Grusby, MJ. IL-21 induces the apoptosis of resting and activated primary B cells. *J Immunol*, 2003, 170, 4111-4118.

[101] Jin, H; Carrio, R; Yu, A; Malek, TR,. Distinct activation signals determine whether IL-21 induces B cell costimulation, growth arrest or Bim-dependet apoptosis. *J Immunol*, 2004, 173, 657-665.

[102] Ozaki, K; Spolski, R; Feng, CG; Qi, CF; Cheng, J; Sher, A; Morse, HC; Liu ,C; Schwartzberg, PL; Leonard ,WJ. A critical role for IL-21 in regulating immunoglobulin production. *Science*, 2002 298, 1630-1634.

[103] Suto ,A; Nakajima, H; Hirose, K; Suzuki, K; Kagami, S; Seto, Y; Hoshimoto, A; Saito, Y; Foster, DC; Iwamoto, I. Interleukin 21 prevents antigen-induced IgE production by inhibiting germ line C(epsilon) transcription of IL-4-stimulated B cells. *Blood*, 2002, 100, 4565-4573.

[104] Ueda, M; Imada, K; Imura, A; Koga, H; Hishizawa, M; Uchiyama, T. Expression of functional interleukin-21 receptor on adult T-cell leukaemia cells. *Br J Haematol*. 2005, 128, 169-176.

[105] Brenne, AT; Baade, RO; Waage, A; Sundan, A; Borset, M; Hjort-Hansen. H. Interleukin-21 is a growth and survival factor for human myeloma cells. *Blood*, 2002, 99, 3756-3762.

[106] Duechler, M; Shehata, M; Schwarzmeier, JD; Hoelb, A; Hilgart, M; Hubmann R. Induction of apoptosis by proteasome inhibitors in B-CLL cells is associated with downregualtion of CD23 and inactivation of Notch2. *Leukemia*, 2005, 19, 260-267.

[107] Stephanou, A; Latchman, DS. STAT-1: a novel regulator of apoptosis. *Int J Exp Pathol*, 2003, 84, 239-244.

[108] Lee, CK; Smith; Gimeno, R; Gertner, R; Levy, DE. STAT1 affects lymphocyte survival and proliferation partially independent of its role downstream of IFN-gamma. *J Immunol*, 2000, 164, 1286-1292.

[109] Jahrsdorfer, B; Blackwell, SE; Wooldridge ,JE; Huang, J; Andreski, MW; Jacobus, LS; Taylor, CM; Weiner, GJ. B-Chronic Lymphocytic Leukemia Cells and other B cells can produce Granzyme B and gain cytotoxic potential after Interleukin-21-based activation. *Blood*, 2006 Jun 29; Epub ahead of print

[110] Mainou-Fowler, T; Craig, VA; Copplestone, JA; Hamon, MD; Prentice AG. Interleukin-5 (IL-5) increases spontaneous apoptosis of B-cell chronic lymphocytic

leukemia cells in vitro independently of bcl-2 expression and is inhibited by IL-4. *Blood*, 1994, 84, 2297-2304.

[111] Mainou-Fowler, T; Copplestone, JA; Prentice, AG. Effect of interleukins on the proliferation and survival of B cell chronic lymphocytic leukaemia cells. *J Clin Pathol*, 1995, 48; 482-487

[112] Fluckiger, AC; Garrone, P; Durand, I; Galizzi, JP; Banchereau J. Interleukin 10 (IL-10) upregulates functional high affinity IL-2 receptors on normal and leukemic B lymphocytes. *J Exp Med*, 1993, 178, 1473-1481

[113] Tangye, SG; Weston, KM; Raison, RL. Interleukin-10 inhibits the in vitro proliferation of human activated leukemic CD5+ B-cells. *Leuk Lymphoma*, 1998, 31, 121-130.

[114] Jurlander, J; Lai, CF; Tan, J; Chou, CC; Geisler,CH; Schriber, J; Blumenson, LE; Narula, SK; Baumann, H; Caligiuri, MA. Characterization of interleukin-10 receptor expression on B-cell chronic lymphocytic leukemia cells. *Blood*, 1997, 89, 4146-4152.

[115] Kitabayashi, A; Hirokawa, M; Miura, AB.The role of interleukin-10 (IL-10) in chronic B-lymphocytic leukemia: IL-10 prevents leukemic cells from apoptotic cell death. *Int J Hematol*, 1995, 62, 99-106.

[116] Buschle, M; Campana, D; Carding, S; Richard, C; Hoffbrand, AV; Brenner, MK. Interferon-γ inhibits apoptotic cell death in B cell chronic lymphocytic leukemia. *J Exp Med,* 1993, 177, 213-218.

[117] Rojas, R; Roman, J; Torres, A; Ramirez, R; Carracedo, J; Lopez, R; Garcia, JM; Martin, C; Pintado O. Inhibition of apoptotic cell death in B-CLL by interferon gamma correlates with clinical stage. *Leukemia*, 1996, 10, 1782-1788.

[118] Tange, SG; Raion, RL. Human cytokines suppress apoptosis of leukaemic CD5+ B cells and preserve expression of bcl-2. *Immunol Cell Biol*, 1997, 75,127-135.

[119] Zaki, M; Douglas, R; Patten, N; Bachinsky, M; Lamb, R; Nowell, P; Moore J. Disruption of the IFN-γ cytokine network in chronic lymphocytic leukemia contributes to resistance of leukemic B cells to apoptosis. *Leukemia Research*, 2000, 24, 611-621.

[120] De Totero, D; Gobbi M. High frequency of CD8+/CD28- cells expressing CD30, CD70 and KIR antigens in chronic lymphocytic leukemia may reflect expansion of chronically activated memory cells impaired in immune recognition. *Haematologica*, 2002, 87(suppl.II), 17-19.

[121] Dancescu, M; Rubio-Trujillo, M; Biron, G; Delespesse, G; Sarfati, R. Interleukin-4 protects chronic lymphocytic leukemia B cells from death by apoptosis and upregulates Bcl-2 expression. *J Exp Med*, 1992, 176, 1319-1326.

[122] Mu, X; Kay, NE; Gosland, MP; Jennings, CD. Analysis of blood T-cell cytokine expression in B-chronic lymphocytic leukemia: evidence for increased levels of cytoplasmic IL4 in resting and activated CD8 T cells. *Br J Haematol*, 1997, 96, 733-735.

[123] Kaminski, A; Demaine, A; Prentice, A. Cytoplasmic interleukin-4 (IL-4) and surface IL-4 receptor expression in patients with B- cell lymphocytic leukemia. *Blood*, 1998, 92, 2188-2189.

[124] Mainou-Fowler, T; Proctor, SJ; Miller, S; Dickinson, AM. Expression and production of interleukin 4 in B-cell chronic lymphocytic leukaemia. *Leuk Lymphoma*, 2001, 42, 689-698.

[125] Kay, NE; Han, L; Bone, N; Williams, G. Interleukin 4 content in chronic lymphocytic leukemia (CLL) B cells and blood CD8+ T cells from B-CLL patients: impact on clonal B-cell apoptosis. *Br J Haematol*, 2001, 112, 760-767.

[126] Rossmann, ED; Lewin, N; Jeddi-Tehrani, M; Osterborg, A; Mellstedt, H. Intracellular T cell cytokines in patients with B cell chronic lymphocytic leukemia (B-CLL). *Eur J Haematol*, 2002, 68, 299-306.

[127] Kay, NE; Pittner, BT. IL-4 biology: impact on normal and leukemic CLL B cells. *Leuk Lymphoma*, 2003, 44, 897-903.

[128] Cerruti, A; Kim, EC; Shah, S; Schattner, EJ; Zan, H; Schaffer, A; Casali P. Dysregulation of CD30+ T cells by leukemia impairs isotype switching in normal B cells. *Nat Immunol*, 2001, 2, 150-156.

[129] Gorgun, G; Foss, F. Immunomodulatory effects of RXR rexinoids: modulation of high-affinity IL-2R expression enhances susceptibility to denileukin diftitox. *Blood,* 2002, 100, 1399-403.

[130] Frankel ,AE; Fleming, DR; Hall, PD; Powell, BL; Black, JH; Leftwich, C; Gartenhaus, R. A phase II study of DT fusion protein denileukin diftitox in patients with fludarabine-refractory chronic lymphocytic leukemia. *Clin Cancer Res*, 2003, 9, 3555-3561.

[131] Frankel, AE; Surendranathan, A; Black, JH; White, A; Ganjoo, K; Cripe LD. Phase II clinical studies of denileukin diftitox diphtheria toxin fusion protein in patients with previously treated chronic lymphocytic leukemia. *Cancer*, 2006, 106: 2158-2164.

[132] Kay, NE; Bone, ND; Lee, YK; Jelinek, DF; Leland, P; Battle, TE; Frank, DA; Puri, RK. A recombinant IL-4 pseudomonas exotoxin inhibits protein synthesis and overcomes apoptosis resistance in human CLL B cells. *Leuk Res*, 2005, 29, 1009-1018.

[133] Di Celle, PF; Carbone, A; Marchis, D, Zhou, D; Sozzani, S; Zupo, S; Pini, M; Mantovani, A; Foa R. Cytokine gene expression in B-cell chronic lymphocytic leukemia: evidence of constitutive interleukin-8 (IL-8) mRNA expression and secretion of biologically active IL-8 protein. *Blood*, 1994, 84, 220-228.

[134] Francia di Celle, P; Mariani, S; Riera, L; Stacchino, A; Reato, G; Foa, R. Interleukin-8 induces the accumulation of B-cell chronic lymphocytic leukemia cells by prolonging survival in an autocrine fashion. *Blood*, 1996, 87, 4382-4389.

[135] Molica, S; Vitelli, G; Levato, D; Levato, L; Dattilo, A; Gandolfo, GM. Clinico-biological implications of increased serum levels of interleukin-8 in B-cell chronic lymphocytic leukemia. *Haematologica*, 1999, 84,208-211.

[136] Wierda, WG; Johnson, MM; Do, KA; Manshouri , T; Dey, A; O'Brien, S; Giles, FJ, Kantarajian, H; Thomas, D; Faderl, S; Lerner, S; Keating ,M; Albitar, M. Plasma interleukin 8 level predicts for survival in chronic lymphocytic leukaemia. *Br J Haematol*, 2003, 120, 452-456.

[137] Stratowa, C; Loffler, G; Lichter, P; Stilgenbauer, S; Haberl, P, Schweifer, N; Dohner, H; Wilgenbus, KK. CDNA microarray gene expression analysis of B-cell chronic lymphocytic leukemia proposes potential new prognostic markers involved in lymphocyte trafficking. *Int J Cancer*, 2001, 91, 474-480.

[138] Reittie, JE; Yong, KL; Panayioditis, P; Hoffbrand, AV. Interleukin-6 inhibits apoptosis and tumor necrosis factor induced proliferation of B-chronic lymphocytic leukemia. *Leuk Lymphoma*, 1996, 22, 83-90.

[139] Hulkkonen, J; Vilpo, J; Vilpo, I; Hurme, M. Interleukin-1β, interleukin-1 receptor antagonist and interleukin-6 plasma levels and cytokine gene polymorphisms in chronic lymphocytic leukemia: correlation with prognostic parameters. *Haematologica,* 2000, 85, 600-606.

[140] Fayad, L; Keating, MJ; Reuben, JM; O'Brien, S; Bang-Ning, L; Lerner, S; Kurzrock, R. Interleukin-6 and interleukin-10 levels in chronic lymphocytic leukemia: correlation with phenotypic characteristics and outcome. *Blood*, 2001, 97, 256-263.

[141] Lai, R; O'Brien, S; Maushouri, T; Rogers, A; Kantarajaian, H; Keating, M; Albitar, M. Prognostic value of plasma interleukin-6 levels in patients with chronic lymphocytic leukemia. *Cancer*, 2002, 95, 1071-1075.

[142] Barut, B; Cochran, M; O'Hara, C; Anderson, K. Response pattern of hairy cells leukemia to B-cell mitogens and growth factors. *Blood*, 1990, 76, 2091-2097.

[143] Burger, R; Wendler, J; Antoni, K; Helm, G; Kalden, J; Gramatzki, M. Interleukin-6 production in B-cell neoplasias and Castelman's disease: evidence for an additional paracrine loop. *Ann Hematol*, 1994, 69:25-31.

[144] Hulkonnen, J; Vilpo, J; Vilpo, L; Hurme M. Diminished production of interleukin-6 in chronic lymphocytic leukemia (B-CLL) cells from patients at advanced stages of disease. Tampere CLL group. *Br J Haematol* 1998, 100, 478-483.

[145] Bagby GC, Jr. Interleukin-1 and hematopoiesis. *Blood Rev*, 1989, 3, 152-161.

[146] Dohner, H; Stilgenbauer, S; James, MR; Bennere, A Weilguni, T; Bentz, M; Fisher, K; Hunstein, W;Lichter, P. 11q deletions identify a new subset of B-cell chronic lymphocytic leukemia characterized by extensive nodal involvement and inferior prognosis. *Blood*, 1997, 89, 2516-2522.

[147] Karhu, R; Knuutila, S; Kallioniemi, OP; Siltonen, S; Aine, R; Vilpo, L; Vilpo, G Frequent loss of the 11q14-24 region in chronic lymphocytic leukemia: a study by comparative genomic hybridization. Tampere CLL Group. *Genes Chromosomes Cancer*, 1997, 19, 286.290

[148] Zhu,Y; Monni, O; El-Rifai, W; Siitonen, SM; Vilpo, J; ,Knutila, S. Discontinuous deletions 11q23 in B cell chronic lymphocytic leukemia. *Leukemia*, 1999, 13, 708-712.

In: Focus on Cell Apoptosis Research
Editor: Daniel W. Cho, pp. 203-231

ISBN: 1-60021-433-9

Chapter VIII

CELL DEATH MECHANISMS IN PANCREATITIS

Madhav Bhatia**, *Sharmila Adhikari,
Tamizhselvi Ramasamy and Yang Cao
Department of Pharmacology, National University of Singapore, Singapore

ABSTRACT

Acute pancreatitis has numerous causes, culminating in a common pathway of changes within the pancreatic acinar cell. The pancreatic acinar cell being the functional unit of the exocrine pancreas, it synthesizes, stores, and secretes digestive enzymes. Evidence has accumulated of the significance of apoptotic cell death in the systemic manifestation associated with acute pancreatitis and studies suggest that apoptotic cell death may play a considerable role in affecting mortality and morbidity in severe acute pancreatitis. Apoptosis in pancreatitis is mediated by caspase activation and mitochondrial dysfunction in acinar cells. It is suggested that caspases play an important role in apoptosis. The expression of caspases is a frequent event in chronic pancreatitis and its distribution pattern may reflect the functions of this protease in its role in the apoptotic pathway in atrophic acinar. Also acute pancreatitis modifies mRNA level of both pro and anti-apoptotic genes simultaneously. Numerous studies suggest that pancreatic bcl2, bax and p53 mRNA levels increased extensively in the mild acute pancreatitis and the levels were unchanged during severe acute pancreatitis. This article presents a brief overview of the apoptotic machinery in pancreatitis, including specific and indirect death signals and the underlying molecular events in cell.

Keywords: pancreatic acinar cell, acute pancreatitis, caspases, apoptotic genes

---

* Correspondence concerning this article should be addressed to Madhav Bhatia, Ph.D. Cardiovascular Biology Research Programme, Department of Pharmacology, Centre for Life Sciences, National University of Singapore, Yong Loo Lin School of Medicine, 28 Medical Drive, #03-02, Singapore 117456. Tel: (65)-6516-8256; Fax: (65)-6775-7674; e-mail. mbhatia@nus.edu.sg.

Introduction

Acute pancreatitis is a severe disease with a significant morbidity and mortality [1]. The overall mortality rate of acute pancreatitis is around 10%. However, among severe cases in which the functions of multiple organs are compromised, the mortality rate can reach 20-30% [2]. Clinically, the severity of acute pancreatitis varies considerably. Edematous pancreatitis is the mildest form of the disease, characterized by interstitial edema and in this form the acinar cell structure remains intact. On the contrary, severe acute pancreatitis is characterized by massive destruction of pancreatic parenchyma and infiltration of leukocytes [2]. It has been hypothesized that milder forms of experimental pancreatitis are associated with more apoptosis and the relatively severe forms, with more necrosis [3,4]. Most of the reports support the idea that pancreatic acinar cell apoptosis acts as a self-defense mechanism against the development of pancreatic necrosis, extensively reviewed elsewhere [5,6].

In pancreatitis, acinar cell death occurs by both apoptosis and necrosis [7-10]. Apoptosis plays a key role in a variety of biological events, including cell differentiation, organogenesis and various disease conditions [11,12]. "Apoptosis" is one of the most frequently used words in biology and medicine today. It was introduced by Kerr, and colleagues in 1972 to describe a form of cell death distinct from necrosis [13]. Apoptosis was initially recognized as "programmed cell death," a type of cell death that serves to eliminate excessive or unwanted cells, in the course of organ development. Apoptotic cells are phagocytosed by neighbouring cells, while necrotic cells cause local and general inflammatory reactions. Several reports demonstrate that a chemical at a low concentration may cause apoptosis, while at a higher (toxic) concentration it may cause necrosis [14-18]. Since apoptotic processes require energy, depletion of intracellular ATP results in the switch from apoptosis to necrosis [19-21]. It is well established that reactive oxygen species (ROS) are intimately related to apoptosis and a burst in the intracellular level of ROS triggers the switch from apoptosis to necrosis [16,22,23]. Necrosis has been considered a passive degenerative phenomenon induced by direct toxic or physical injuries, which most often occurs accidentally. Nuclear DNA is randomly cleaved as a consequence of cellular degeneration. Leakage of the cytoplasm through plasma membrane disruption induces cellular inflammatory responses [24,25].

There are two major signaling pathways of apoptosis, the death receptor-mediated extrinsic pathway and the mitochondrial intrinsic pathway. In the death receptor signaling pathway, signal is provided by the interaction between the ligand and death receptor. Initiated by caspase 8 or 10 followed by the activation of effector caspases such as caspase 3 [26-28]. In the mitochondrial pathway, permeabilization of the mitochondrial outer membrane, results in the release of apoptotic factors such as cytochrome c followed by the activation of caspase cascade. The regulatory role of Bcl-2 family proteins in mitochondrial pathway of apoptosis is well established [29,30]. Although the two pathways appear to be distinct, death receptor signaling may mediate apoptosis via mitochondrial pathway via tBid [31]. This review outlines the mechanisms of apoptosis and necrosis and its involvement in pancreatitis.

Mechanism of Cell Death

Two Alternative Forms of Cell Death

The concept of apoptosis or programmed cell death (PCD), with unique and dynamic morphological and biochemical features is distinguishable from necrosis (Table 1). Apoptosis is a type of cell death that does not involve an inflammatory response and occurs in a tightly controlled manner. In contrast to necrosis, apoptosis involves activation of catabolic enzymes prior to cytolysis. Many cellular changes that take place during this form of death have an important role in recognition and elimination of those cells by neighboring phagocyte cells [32,33].

Table 1. Morphological and biochemical features

Apoptosis	Necrosis
• Shrinkage of the cell and its organelles and condensation of chromatin	• Cell swelling and rupture of the plasma membrane
• No leakage of cellular contents	• Leakage of cellular constituents
• Apoptotic bodies engulfed by phagocytes	• Cellular and nuclear lysis
• Activation of specific cysteine proteases, the caspases, and internucleosomal DNA fragmentation	• Absent in necrotic cells
• Regulated signaling mechanisms	• Poorly regulated
• No inflammatory response	• Acute inflammatory response

The Phases of Death

The death programme can be divided into four different phases. The 'initiation' phase, during which cells receive signals that result in the activation of the death programme. The 'commitment' phase, the point after which death signals become irreversible. The 'amplification' phase, where multiple caspases are recruited to cooperate in the destruction of the cell, and finally, the 'demolition' phase, where a panoply of active caspases dismantle cellular structures, either directly, or via activation of other enzymes such as CAD/DFF45 [34].

Caspase Activation Mechanisms

Caspases

A group of cysteine proteases - caspases [35,36], are essential for programmed cell death in a variety of species [37]. Cysteine protease activity can be detected in all cells undergoing apoptosis, regardless of their origin or the death stimulus. Caspases, the interleukin-1β-converting enzyme family proteases, are highly homologous to *Caenorhabditis elegans* cell death gene *CED-3*. Fourteen caspases have been identified so far, all of which share some common properties: they are all aspartate-specific cysteine proteases; they all have a conservative pentapeptide active site 'QACXG' (X can be R, Q or D); their precursors are all zymogens known as procaspases. The N-terminal of the prodomain in procaspases contains a highly diverse structure required for caspase activation; and they are all capable of autoactivating as well as activating other caspases, to produce a heterodimer with a big and a small subunit, and two heterodimers form an enzymatic active heterotetramer [38]. Based on their homology in amino acid sequences, caspases are divided into three subfamilies:

Subfamily I: Apoptosis activator/initiator- Caspase-2, 8, 9 and 10
Subfamily II: Apoptosis executioner/effector- Caspase-3, 6 and 7
Subfamily III: Inflammatory mediator- Caspase- 1, 4, 5, 11, 12, 13, and 14

Generally, there are two pathways through which the caspase family proteases can be activated: one is the death receptor-mediated pathway; the other is the mitochondrion-mediated pathway.

Death Receptor-Mediated Apoptotic Pathway

Cell death signals, such as Fas ligand (FasL) and tumor necrosis factor (TNF)-2, can be specifically recognized by their corresponding death receptors, such as Fas or TNF receptor (TNFR)-1, in the plasma membrane. Their binding will in turn activate the death receptors. Fas can bind to the Fas-associated death domain (FADD) (or TNFR-associated death domain, TRADD) and cause FADD aggregation and the emergence of death effector domains (DEDs) [39]. The death receptors (CD95/Fas/APO-1, TNFR1, DR3/WSL-1/ TRAMP, DR4/TRAIL-R1, DR5/TRAIL-R2, DR6) are a subset of the TNF/NGF receptor family of cell surface molecules that possess a common motif within their cytoplasmic tails, called the death domain [28,40]. The death domains of these receptors are responsible for recruiting adaptor molecules that, in turn, recruit caspases to the receptor complex. This simple strategy greatly increases the probability of caspase activation by facilitating cross processing of caspases within the cluster. This proximity-induced processing is thought to drive apical caspase activation events that are initiated by all of the death receptors. Thus, death receptors can drive caspase activation in a very direct manner [41,42]. The exposed DEDs interact with the DEDs in the prodomain of procaspase-8, which will induce the oligomerization of procaspase-8 localized on the cytosolic side of the plasma membrane. Then a massive molecule complex known as the death-inducing signal complex (DISC) is formed. In DISC, two linear subunits of procaspase-8 compact to each other followed by procaspase-8 autoactivation to caspase-8 [43]. The activation of the downstream pathways of caspase-8 varies with different cell types. In Type I cells (cells of some lymphoid cell lines), caspase-8 is vigorously activated and can directly activate the downstream procaspases (e.g. procaspase-3). In Type II cells (other than Type I cells), caspase-8 is only mildly activated and unable to activate procaspase-3 directly. However, it can activate the mitochondrion-mediated pathway by truncating Bid (a pro-apoptotic Bcl-2 family member), a kind of proapoptotic protein in the cytosol, into its active form, tBid. tBid will trigger the activation of the mitochondrion pathway: cytochrome c, apoptosis-inducing factor (AIF) and other molecules are released from mitochondria, leading to the initiation of apoptosis [44] (Fig. 1).

Caspase-10 can function independently of caspase-8 in initiating Fas- and TNF-related apoptosis in lymphoid cells [45].

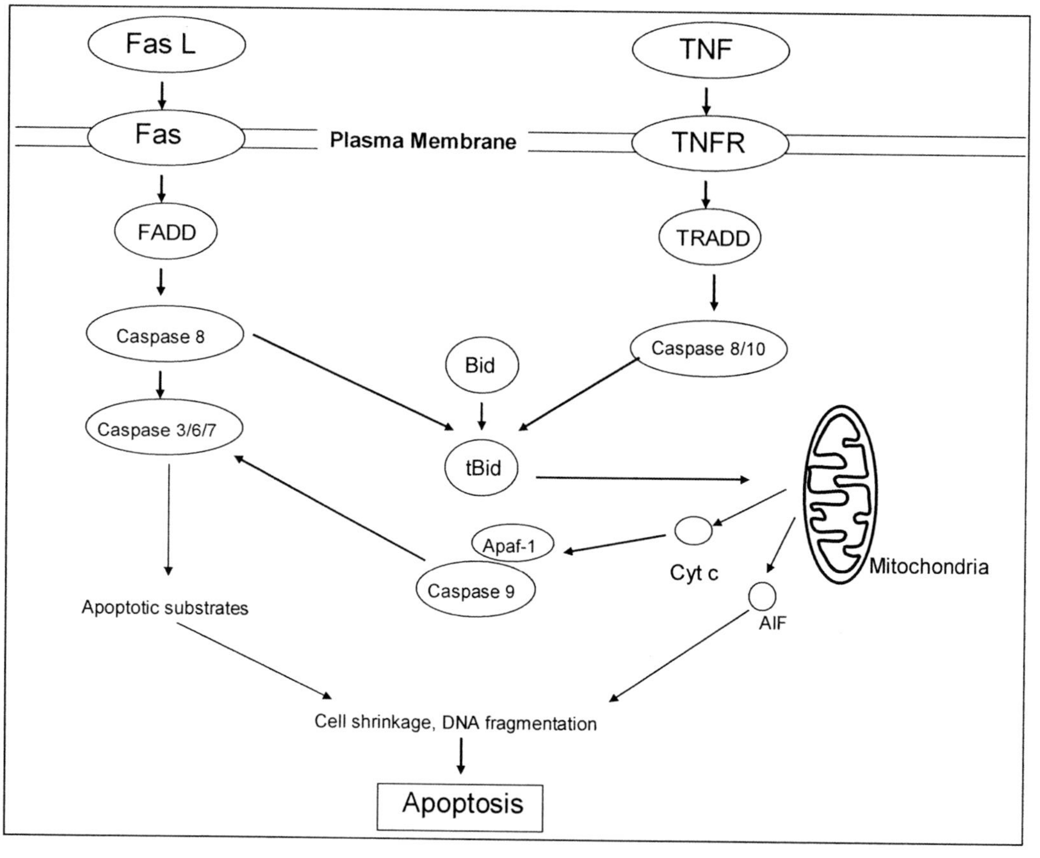

Fig. 1. Death receptor-dependent procaspase-activation pathway. AIF, apoptosis-inducing factor; Apaf-1, apoptotic protease activation factor-1; Cyto c, cytochrome c; FADD, Fas-associated death domain; TNF, tumor necrosis factor; TNFR, TNF receptor; TRADD, TNFR-associated death domain.

Mitochondrion-Mediated Apoptotic Pathway

The mitochondrial pathway of apoptosis is dependent on the process of mitochondrial outer membrane permeabilization, which leads to the release of proteins from the mitochondrial intermembrane space into the cytosol. Mitochondria play a major role in apoptosis triggered by many stimuli. They integrate death signals through Bcl-2 family members and coordinate caspase activation through the release of cytochrome c as a result of the outer mitochondrial membrane becoming permeable [46]. In general terms, antiapoptotic members of the Bcl-2 family tend to inhibit the release of mitochondrial proteins, whereas proapoptotic members (such as BID, BAX and BAK) favor release [47]. Proteins of the intermembrane space are released to the cytosol [48]. One is holocytochrome c, which binds to cytosolic, monomeric apoptotic protease activating factor–1 (APAF-1). The interaction with cytochrome c permits a 2′-deoxyadenosine 5′-triphosphate (dATP)–mediated conformational change in APAF-1, promoting APAF-1 oligomerization to form an "apoptosome". The apoptosome then binds the proform of a protease, caspase-9. The oligomerization of caspase-9 on the apoptosome activates the protease; initiator caspases such as caspase-9 can only be activated by dimerization [49].

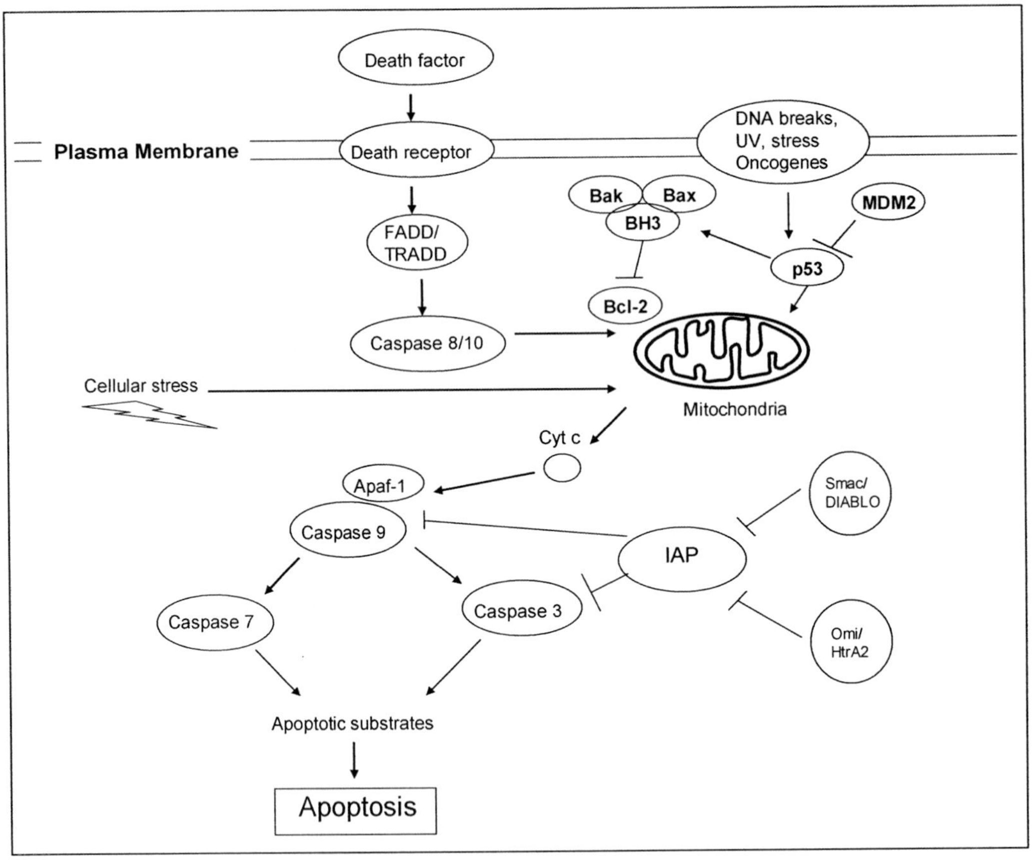

Fig. 2. Mitochondrion-mediated caspase-dependent pathway. Apaf-1, apoptotic protease activation factor-1; Cyto c, cytochrome c; FADD,Fas-associated death domain; TRADD, tumour necrosis factor receptor-associated death domain. IAP, inhibitor of apoptosis proteins.

The active caspase-9 cleaves two other caspases, caspase-3 and caspase-7. These are "executioner" caspases, and, unlike initiator caspases, these can only be activated by cleavage at specific sites, which are targeted by initiator caspases. Once activated, the executioner caspases orchestrate apoptosis through the cleavage of key substrates within the cell, resulting in demolition and packaging of the dying cell for removal by phagocytic cells (Fig 2). The process of apoptosis in mitochondrial outer membrane permeabilization is regulated at the apoptosome and each of the caspases. The latter are controlled by inhibitor of apoptosis proteins (IAPs) that bind and ubiquitinate the caspases for proteasomal degradation. The function of the IAPs, in turn, is blocked by IAP inhibitors. These include Smac (also known as DIABLO) and Omi (also known as HtrA2), which compete with caspases for IAP binding. Like cytochrome c, Smac and Omi are sequestered in the mitochondrial intermembrane space and are only available to regulate the cytosolic IAPs upon mitochondrial outer membrane permeabilization (Bcl-2, IAP will be discussed in detail below) [48].

Upon activation by DNA damage-induced or oncogene-induced signaling pathways, p53 promotes the expression of a number of genes that are involved in apoptosis, including those encoding death receptors [50,51] and proapoptotic members of the Bcl-2 family [52-54]. In most cases, p53-induced apoptosis proceeds through mitochondrial release of cytochrome c, which leads to caspase activation [55]. Furthermore, p53 modifications can influence the ability of p53 to bind its negative regulator Mdm2 (a p53-inducible E3 ubiquitin ligase capable of targeting p53 for degradation) [56].

Caspase-Independent Apoptosis

Characterization of *C. elegans* ced-3 mutants revealed that PCD underwent even in the absence of *ced-3* [57] which indicated that cell death can occur in the absence of caspase activation. Subsequent findings were consistent with the existence of *ced-3*-independent PCD [58]. Some studies have suggested caspase-independent PCD in mammals. Genetic studies on PCD induced by BH3-domain only proteins, such as tBID, BIM and BAD, showed that these proteins, which have been shown to promote caspase activation and apoptosis, can also kill cells independently of Apaf-1 and downstream caspases. Specifically, Apaf-1 $^{-/-}$ mouse embryonic fibroblasts (MEFs) could still die in response to the overexpression of BH3-domain-only proteins [59]. Caspase activation was not detected in the dying cells when assayed with fluorogenic substrates for caspase-2, -3, -6 or -7, nor could cell death be blocked by the pan-caspase inhibitor zVAD.fmk [59].

Also other mammalian mediators of caspase-independent PCD have been identified as apoptosis-inducing factor (AIF), a conserved mitochondrial oxidoreductase [60] and endonuclease G [61]. In the early mouse embryo, PCD occurs in the ectoderm to create the pro-amniotic cavity. An in vitro model of this event uses an aggregate of embryonic stem (ES) cells in which the inner ectodermal cells undergo cell death [60]. Both have been suggested to be capable of inducing apoptotic changes in the nucleus (i.e., DNA fragmentation), but neither have been demonstrated to be responsible for other stereotypical changes that occur in apoptosis (e.g., exposure of phosphatidylserine and proteolysis of vital substrates [60,61].

Caspases are proteases, it is possible that caspase-independent cell death might require other proteases. Indeed, there is weak, yet suggestive, evidence that might support roles for cathepsins, calpains and granzyme B in caspase-independent PCD. Cathepsins are lysosomal proteases, and mice lacking cystatin B, an inhibitor of cathepsins B, H, L and S, show cerebellar cell death. This suggests that cathepsins might mediate cell death, although it is also possible that the phenotype was caused by a cathepsin-independent effect of cystatin B [62]. Calpains are a family of calcium-dependent proteases that have also been described as possible facilitators of cell death. In breast cancer cells, vitamin-D-induced cell death appeared to be caspase-independent [63] and promoted by calpains [64].

Necrotic Signaling Pathway

The signaling mechanisms mediating necrotic cell death are, in general, poorly understood. Necrosis-like programmed cell death is mediated by a coordinated series of signaling events that can be triggered via death receptors [65,66]. Fas kills activated primary T cells efficiently in the absence of active caspases, which results in necrotic morphological changes and late mitochondrial damage but no cytochrome c release. This Fas ligand–induced caspase-independent death is absent in T cells that are deficient in either Fas-associated death domain (FADD) or receptor-interacting protein (RIP). RIP is also required for necrotic death induced by tumor necrosis factor (TNF) and TNF-related apoptosis-inducing ligand (TRAIL). Thus, Fas, TRAIL and TNF receptors can initiate cell death by two alternative pathways, one relying on caspase-8 and the other dependent on the kinase RIP [67-69]. Evidence suggests that Fas may also cause necrosis. It was found that, when overexpressed in a highly sensitive fibrosarcoma cell line, Fas triggers cell death in a caspase-independent necrotic manner [70]. In addition, artificial aggregation of FADD initiates cell death with necrotic morphology in caspase-8–deficient Jurkat cells [71]. Using genetically deficient Jurkat T cells, the kinase receptor-interacting protein (RIP) was identified as a crucial component not only of Fas-mediated, but also of TNF and TNF-related apoptosis-inducing ligand (TRAIL)-triggered caspase-independent cell death.

Apart from caspase-8 and caspase-10, FLIP is currently the only protein known to bind to the DED domain of FADD. FLIP was shown to enhance RIP recruitment to Fas [72]. Overexpression of FLIP is toxic to cells [73] and thus its implication in the necrotic pathway cannot be excluded. FADD has an opposite observed effect in the TNF-triggered necrotic pathway where a FADD-interacting protein has an anti-necrotic effect. A likely candidate is caspase-8, the inhibition of which can sensitize cells to TNF-induced necrosis [66]. Necrosis occurs as a consequence of acute ATP depletion. Several ATP-dependent ion channels become ineffective leading to disruption of actin cytoskeleton, cell swelling, membrane blebbing and eventual collapse of the cell [74,75]. Also Na^{+} overloading Ca^{2+} accumulation and changes in mitochondrial permeability, may also participate in necrotic process [76-78]. Opening of high conductance permeability transition (PT) pores in the mitochondrial inner membrane initiates the mitochondrial permeability transition (MPT). Ca^{2+}, inorganic phosphate, alkaline pH, and reactive oxygen species (ROS) are a few of the many agents that promote the MPT, whereas the immunosuppressive drug, cyclosporin A (CsA), Mg^{2+}, acidic pH, and phospholipase inhibitors, including trifluoperazine, dibucaine, and quinacrine, block opening of the PT pores [79,80] (fig. 3). Interestingly, cathepsin and calpain proteases have been described as key regulators of necrotic cell death in C. elegans [81].

Regulating Factors

The activation and inactivation of caspases are regulated by various proteins, ions and other factors, such as IAP, Bcl-2 family proteins, calpain, Ca^{2+}, FLICE-Inhibitory Protein (FLIP), Gran B, cytokine response modifier A (Crm A), p35 and NF-κB.

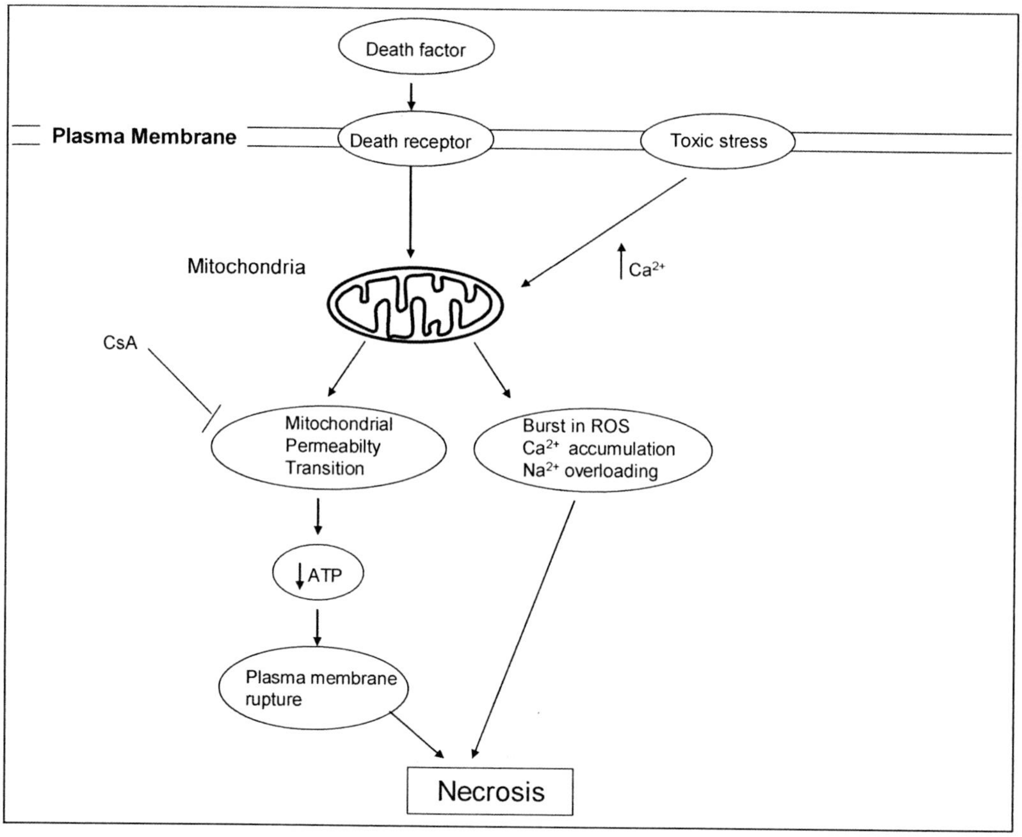

Fig. 3. Cell death by necrosis. CsA, Cyclosporin A; ROS, Reactive oxygen species.

IAP

In the mid 1990s, a screen for viral-encoded suppressers of apoptosis led to the identification of two death-suppressing genes from *Cydai pomonella granulovirus* (CpGV) and *Orgyia pseudotsugata nucleopolyhedrovirus* (OpNPV), each with a high degree of similarity to the other [82,83]. These viral proteins, Cp-IAP and Op-IAP, defined the first members of a family of inhibitors of apoptosis proteins (IAP) containing one or more baculoviral IAP repeats (BIR). This BIR domain consists of approximately 70 amino acids with a conserved spacing of cysteine and histidine residues, which suggests that this structure represents a novel zinc binding fold [84]. Homology to this BIR domain has been used to isolate a number of cellular proteins that inhibit apoptosis. Other members of the IAP family include XIAP (X-linked mammalian inhibitor of apoptosis protein) c-IAP1, c-IAP2, NAIP (neuronal apoptosis inhibitory protein), livin and Survivin in mammals; DIAP1, DIAP2, and Deterin in Drosophila, and CeBIR1 and CeBIR2 in *C. elegans*; TnIAP and SfIAP in Lepidopteran; and SpIAP and ScIAP in yeast [84,85]. Various members of the family of inhibitory proteins contain one to three N- terminal baculovirus BIR domains and one conservative C-terminal RING (really interesting new gene) domain. The IAP family (X-IAP, c-IAP-1, c-IAP-2, N-AIP) appear to negatively regulate caspase activities by acting as pseudosubstrates for the caspases, binding to active caspases and neutralizing their activity [86,87]. The activity of mammalian IAP can be inhibited by Smac/DIABLO released from mitochondria. As the four residue sequence (Ala1-Val2-Pro3-Ile4) in the newly formed N-terminus of Smac/DIABLO can recognize and bind to the caspase-9-binding site of XIAP, so that XIAP will be inactivated, its inhibiting effect on caspase-9 will in turn be relieved [88]. IAP family proteins may also have other functions besides caspase inhibition. For example, IAP family members in yeast can neither unite caspases nor induce apoptosis [85].

Bcl-2 Family Proteins

The mitochondria-dependent pathway for apoptosis is governed by Bcl-2 family proteins. Mitochondria play a major role in apoptosis triggered by many stimuli. They integrate death signals through Bcl-2 family members and coordinate caspase activation through the release of cytochrome c as a result of the outer mitochondrial membrane becoming permeable [46,89]. Bcl-2 belongs to a growing family of proteins that regulate apoptosis [89,90].

According to functional and structural criteria, the members can be divided into two groups. Group I proteins are all anti-apoptotic proteins, including A1/Bfl1, Bcl-2, Bcl-w, Bcl-xL, Boo/Diva, Mcl-1, NR-13 and Nrf3 in mammals, BHRF-1, E1B19K, Ks-Bcl-2, LMW5-HL and ORF16 in bacteria, and Ced-9 in *C. elegans* [91-93]. They all have four short Bcl-2 homology (BH) domains: BH1, BH2, BH3 and BH4. Group II proteins are all pro-apoptotic proteins, including Bad, Bak, Bax, Bcl-rambo, Bcl-xS, Bid, Bik, Bim, Blk, BNIP3, Bok/Mtd, Hrk and Nip3 in mammals, and Egl-1 in *C. elegans* [92].

Bcl-2 family proteins contain at least one of the four conserved regions called Bcl-2 homology domains (BH1–BH4). These motifs are formed by helices and enable the different members of the family to form either homo- or heterodimers and to regulate each other [94,95]. The Bcl-2-related proteins display either anti-apoptotic or pro-apoptotic function. The balance between Bcl-2 family members defines whether a cell will live or die. The members that inhibit apoptosis, such as Bcl-2 and Bcl-X_L, harbor at least three BH domains. Over-expression of Bcl-2 and Bcl-X_L inhibits apoptosis induced by a variety of stimuli, in particular those that utilize the mitochondria-dependent pathway [93,96]. This is often associated with protection against the release of cytochrome *c* as well as collapse of membrane potential [97,98]. Moreover, co-expression of Bcl-2, Bcl-X_L or other anti-apoptotic Bcl-2 family members can rescue yeast from the cytotoxic effects of Bax. Studies of the effects of Bax on mitochondria in intact yeast have provided evidence of hyper-polarization (based on cationic dyes) and matrix alkalinization (based on pH-GFP), with concomitant cytosol acidification [47,99-101]. Furthermore, Bcl-2 or Bcl-X_L prevents these effects of Bax on mitochondria and pH regulation. Moreover, channel formation by pore forming Bcl-2 family proteins in synthetic membranes (including anti-apoptotic Bcl-2, Bcl-X_L and pro-apoptotic Bax and Bid) is also markedly enhanced by low pH again with optima in the pH 4-5 range [102]. Among the death promoters, some proteins (e.g. Bax, Bak) contain BH1, BH2 and BH3, and closely resemble Bcl-2. Increased Bax expression is often correlated with apoptotic cell death. Bax and Bak, originally localized in the cytoplasm, can translocate to the mitochondrial outer membrane after an apoptotic program starts. Following the translocation, they will undergo conformation changes, oligomerization and insertion into the mitochondrial outer membrane to elevate the mitochondrion permeability transition pores [39]. The Bcl-2 mediated protection of mitochondria is most likely due to the prevention of pore opening caused by Bax- or Bak- dependent mechanisms [103]. Up-regulation and/or translocation of Bax to the mitochondria is a second major pathway for mitochondrial perturbation preceding the onset of apoptosis. At the mitochondria, Bax invokes cytochrome *c* release and loss of membrane potential [104], possibly via direct pore formation [105] or by association with the voltage-dependent anion channel [106].

Other Bcl-2 family proteins (e.g. Bid, Bad) possess only the BH3 domain, which is essential for both their interaction with other family members and their death-promoting activity [107] these are called 'BH3-only proteins'. The three-dimensional structures of Bid [60] and Bcl-X_L [108] are similar, suggesting that Bid may also form ion channels. Bid requires the cooperation of relatives such as Bax for its effect [109]. Unlike most other members of the Bcl-2 family, BID is normally localized to the cytoplasm. However, caspase-8 can catalyze the cleavage of BID, the C-terminal portion of which then translocates to the mitochondria where it integrates into the outer membrane and provokes the release of cytochrome c - an important accessory molecule for downstream caspase activation events [34]. This caspase-8 mediated activation of Bid represents an important mechanism accounting for cross-talk between the death receptor (extrinsic) and mitochondrial (intrinsic) pathways [110].

Most Bcl-2 family proteins also contain a C-terminal ~20-residue hydrophobic domain that targets them to intracellular membranes, predominantly to the outer membrane of mitochondria [111]. In contrast to Bcl-2, Bcl-X_L, and many of the Bcl-2 family proteins, the pro-apoptotic proteins Bid, Bim, and BAD lack C-terminal transmembrane domains. These proteins are normally found in the cytosol but can be induced to target mitochondria. The principal mechanism by which Bcl-2 family proteins regulate apoptosis is probably by controlling cytochrome-*c* release [112]. In addition to cytochrome c, Bcl-2 family proteins have been reported to control the release of other proteins from mitochondria. These proteins include: caspases [113-115], apoptosis inducing factor [116] and Smac/Diablo, the inhibitor of IAP family proteins [117,118].

Calpain and Ca2+

Calpain is a kind of Ca^{2+}-dependent cysteine protease of the papainase family. It is generally believed that calpain is activated in both necrosis and apoptosis. Calpain and caspase-3 share many common substrates, including fodrin, Ca^{2+}-dependent protein kinase and ADP ribosyltransferase/PARP [119]. In apoptosis induced by ER stress, calpain's functions are particularly salient because of the perturbed Ca^{2+} homeostasis. In the brain cells of rats suffering from unilateral hypoxia-ischemia, m-calpain first cleaved procaspase-3 into 29 kDa fragments to facilitate its further cleavage and activation [120]. Cisplatin, a kind of anticancer agent, can cause ER stress and apoptosis. During this process, the activation of procaspase-12 by cisplatin is dependent on Ca^{2+} and calpain [121]. In addition, calpain can also cleave Bcl-xL in its loop region, which will convert Bcl-xL to a proapoptotic molecule from an anti-apoptotic one [122]. Bcl-2 family proteins emerge as major regulators of cellular calcium handling. In addition, apoptotic factors released from the mitochondria and activated caspase enzymes may also control Ca^{2+} transport proteins [123].

FLICE-Inhibitory Protein (FLIP)

A subgroup of membrane receptors of the TNF superfamily, collectively known as the death receptors, induces apoptosis upon ligand binding [124]. CD95 (Fas) is the most extensively studied member of the death receptor family. Upon ligation of the Fas receptor with its ligand (FasL), receptor clustering at the cell membrane results in conformational changes, enabling recruitment of an adaptor molecule FADD, and procaspase-8 (also known as FLICE) to the membrane-bound receptor in a complex termed the death inducing signaling complex (DISC). Recruitment of procaspase-8 to this complex permits its activation, initiating a caspase cascade that ultimately leads to cell death. Therefore, a level of control can also occur at the level of this receptor complex. A family of FLICE-inhibitory proteins (FLIPs) has recently been identified, with both viral and mammalian members [73,125-128]. The viral proteins (v-FLIPs) contain two tandem death effector domains (DED), while the human homolog (c-FLIP$_L$) contains two DED regions along with an inactive caspase-8–like catalytic domain at the C-terminus. The short form of this protein (c-FLIP$_S$) is similar in structure to the viral FLIPs. FLIPs are thought to be recruited to the DISC through their DED domains, thus preventing procaspase-8 recruitment and activation. Therefore, receptor-mediated cell death can be regulated at the level of the most proximal caspase, caspase-8. Inhibition of apoptosis by FLIPs through the inactivation of caspase-8 occurs in several apoptotic model systems, including germinal center B-cells, human peripheral blood T-lymphocytes, and mast cells [129-131].

Gran B and Crm A

Gran B is a kind of serine protease with an important role in apoptosis in cytotoxic T cells. Gran B can activate various procaspases, such as procaspase-3, procaspase-7, procaspase-8, procaspase-9 and procaspase-10, to initiate apoptosis [131,132]. Also it was found that Gran B could cleave Bid to initiate the mitochondrion-mediated activation pathway [132].

CrmA, a 38-kD cytokine response modifier protein of the cowpox virus, is a protease inhibitor of the serpin family initially shown to inhibit interleukin-1 beta converting enzyme (ICE) and cytokine-induced apoptosis [133-135]. CrmA is a unique member of the serpin family in that it can inhibit both cysteine proteases (caspases) and serine proteases, such as granzyme B [136]. CrmA is a potent inhibitor of diverse apoptotic model systems such as Fas- and TNF-induced apoptosis [135], serum withdrawal [137], and hypoxia-mediated apoptosis [138]. While crmA is most effective at inhibiting caspase-1 (ICE) and caspase-8, it can also inhibit with decreasing effectiveness caspase-10, -6, -3, and -7 [139]. In general, crmA rescues cells primarily from apoptotic signals involving death receptors, or other mechanisms in which caspase-8 (or caspase-10) acts as the apical protease [140]. In contrast, crmA is ineffective at preventing cell death owing to stress or genotoxic damage, which suggests that apoptotic stimuli that act directly at the level of mitochondria are relatively resistant to crmA inhibition. Thus, apoptotic stimuli activating caspases, such as caspase-9, as the initial protease are essentially crmA-insensitive.

The activity of Gran B can be inhibited by Crm A, a kind of serpin from the vaccinia virus. Crm A, a strong inhibitor of caspase-1 and caspase-8, and a weak inhibitor of caspase-3 and caspase-6, can prevent the cross-link of Fas and inactivate Gran B [132].

p35

Another caspase inhibitor identified from a novel gene product of the baculovirus *Autographa californica* nuclear polyhedrosis virus, designated p35, prevents cell death in various apoptotic model systems from insects to mammals [141-143]. p35, also blocks cell death induced with a wide variety of apoptotic induction signals including ionizing radiation, Fas, and TNF [144,145]. p35, with the ability of binding to caspases to cleave and inactivate them, is an effective inhibitor of caspases from caspase-1 to caspase-8 [146]. The crystal structure of the caspase-8/p35 complex was recently solved and revealed that caspase inhibition occurs owing to a covalent thioester linkage between p35 and the active site of caspase-8 [147]. It is interesting to note, p35 inhibits the activity of a number of different caspases; however, it was recently reported that p35 does not prevent caspase-9 activity in a cell-free system of mammalian caspase activation [148].

Proteins such as IAPs, FLIPs, crmA, and p35 prevent apoptosis at the level of caspases to inhibit aberrant cell death. This variety of different regulating proteins used to inhibit specific caspase activities depends on the particular stimuli employed to initiate apoptosis. One advantage of this diverse capacity of various caspase inhibitors to prevent apoptosis is the ability to use these proteins to define a specific caspase cascade that a particular stimulus engages to commit the cell to death. Additionally, structural studies examining the interactions between caspase inhibitors and specific initiator or effector caspases offers new insights into the design of synthetic proteins with increased selectivity for their use in therapeutic approaches for controlling apoptosis.

NF-κB

The NF-κB transcription factors are critical players in the control of the apoptotic response to a variety of stimuli. They are best known for their pivotal roles in immune, inflammatory and acute phase responses, but they also play important role in apoptosis [149-151]. Whereas NF-κB is most commonly involved in suppressing apoptosis by transactivating the expression of anti-apoptotic genes, it can promote programmed cell death in response to certain death-inducing signals and in certain cell types. In this context, in mature T cells NF-κB triggered the Fas-death cascade by directly activating the expression of Fas ligand (FasL) [152-154]. TNFα is a key cytokine that regulates immune responses, cell differentiation and apoptosis. Activation of TNF receptors TNFR1 or TNFR2 by TNFα represents the essential pathway balancing the induction of cell death with its inhibition through activation of NF-κB [155]. Also, endogenous caspase-8 mediates TNF-α-induced activation of NF-κB via FLASH (FLICE-associated huge protein) [156]. Bcl-2 and Bcl-xL, which block apoptosis, have also been shown to suppress NF-κB activation induced by TNF in endothelial cells [157].

Pancreatic Acinar Cell Death in Acute Pancreatitis

Progressive pancreatic acinar cell death is a hallmark in both clinical and experimental acute pancreatitis, the mechanisms of which remain poorly understood. [6]. Necrosis has been classically considered the major form of cell death in acute pancreatitis, and the degree of pancreatic necrosis is strongly associated with the prognosis of this disease [158,159]. On the other hand, apoptosis was suggested to mediate the destruction of acinar cells and induce atrophy of the organ [160-162]. However, careful biochemical and morphological examination of experimental models of acute pancreatitis has shown that severe acute pancreatitis (e.g. that induced by pancreatic duct ligation in the opossum, by choline-deficient and ethionine supplemented diet in the mouse, and by caerulein-hyperstimulation in the mouse) is associated primarily with necrosis but little apoptosis, whereas mild acute pancreatitis (e.g. that induced by pancreatic duct ligation and by caerulein-hyperstimulation in the rat) is associated primarily with apoptotic cell death and little necrosis [3,10]. In other words, the severity of acute pancreatitis is inversely related to the extent of acinar cell apoptosis. Moreover, induction of apoptosis reduces the severity of experimental pancreatitis, while inhibition of it worsens the disease [4,163-166]. These suggested that apoptosis is a teleologically beneficial form of cell death in acute pancreatitis. Hence, elucidating the molecular mechanisms of pancreatic cell apoptosis as well as how it benefits against acute pancreatitis would provide valuable insights into clinical pancreatic disorders and offer potential pharmacological interventions.

Mechanism of Apoptosis in Acute Pancreatitis

Caspase Activation

Caspase activation has been observed in isolated pancreatic acinar cells *in vitro*, stimulated with apoptotic inducers (e.g. cholecystokinin, crambene, hydrogen sulphide, and menadione), as well as in experimental pancreatitis induced by caerulein *in vivo* [167-171]. Stimulation of isolated rat pancreatic acini with cholecystokinin (CCK), which serves as a model for human acute edematous pancreatitis *in vitro*, stimulates death signaling pathways in rat pancreatic acinar cells, including caspase activation, cytochrome c release, and mitochondrial depolarization, leading to apoptosis. The mitochondrial dysfunction is mediated by upstream caspase(s). CCK causes mitochondrial alterations through both PTP (permeability transition pore)-dependent (cytochrome c release) and PTP-independent (mitochondrial depolarization) mechanisms [171]. Some pathogenic factors, such as oxidative stress, a major pathogenic factor in acute pancreatitis, induce apoptosis of acinar cells involving activation of caspase-3, the executioner caspase, which degrades the DNA repair protein Ku70 and Ku80 [172].

In addition to mediate apoptosis, caspases also regulate other processes in the pancreatic acinar cell that play key roles in pancreatitis. The initiator caspase, caspase 8 has been suggested to mediate degradation of plectin, which represents a critical biochemical event during CCK-induced secretory blockade and cell injury in acute pancreatitis [173]. Caspases also down regulate the intra-acinar cell activation of trypsin, which initiates autodigestion and acute pancreatitis [171]. Particularly, caspases negatively regulate necrosis through degradation of receptor-interacting protein (RIP), a key mediator of "programmed" necrosis and thus protect from necrosis in pancreatitis [171,172,174]. This may explain the inverse correlation between the extent of apoptosis on the one hand and necrosis along with the severity of the disease on the other hand observed in experimental models of pancreatitis.

Apoptotic Associated Genes

In addition to caspase, the Bcl-2 family of apoptotic regulators is another one of the functional components comprised in the apoptosis pathway. Acute pancreatitits modifies mRNA level of both pro and anti-apoptotic genes simultaneously. Pancreatic bclXL, bax, and p53 mRNA levels increased significantly in a temporal fashion during induction of acute pancreatitis [175].

The increase in expression of p53 and Bax, and a decrease in Bcl-2 expression are also found in glucose/glucose oxidase (G/GO), an oxidative stress induced apoptosis in pancreatic acinar AR42J cells (20). However, in the enhanced severe acute pancreatitis model, pancreatic expression of bax and Bcl-2 remains unchanged, which may due to the decreases of the sensitivity of acinar cells to apoptotic stimuli during the deleterious course of acute pancreatitis [166]. Besides Bcl-2 family, SIP gene has recently been identified, whose expression localized in acinar cell, and demonstrated that as a stress-inducible gene, it is overexpressed during pancreatitis and promoted cell apoptosis [176].

Neutrophils

Neutrophils activation is an important determinant of severity of acute pancreatitis. Neutrophil apoptosis is one of resolutions for such inflammatory responses. On the contrary, delays in such apoptotic response are associated with persistent inflammation and subsequent tissue damage [177,178]. However, the infiltration of neutrophils may itself contribute to regulate pancreatic acinar cell death in acute pancreatitis. The depletion of neutrophils by administration of antineutrophil serum or using ICAM-1 neutralizing antibody, results in a significant increase in acinar cells undergoing apoptosis on the one hand, and significantly reduced the extent of necrosis on the other hand [179]. Yet in the experimental model of taurocholate-induced severe acute pancreatitis, neutrophils, via a TNF-α-dependent mechanism, is involved in the development of both apoptotic and necrotic forms of acinar cell death [180].

Cytokines

In addition to their role in inducing inflammation, cytokines, such as TNF-α and interleukin-1β, which are released by neutrophils (and to a lesser extent, by acinar cells) have also been shown to induce apoptosis in pancreas. Pancreatic acinar cells have been reported to produce and release TNF-α, which is not only involved in the development of pancreatitis, but also mediated apoptosis in acinar cell suspension *in vitro* as well as *in vivo* in the caerulein model of acute pancreatitis [9,181]. Other cytokines, such as interleukin 6 plays a role in initiating the inflammatory as well as antiapoptotic responses in severe acute pancreatitis. Blocking IL-6 suppresses STAT-3 (a transcription factor which down regulates cell apoptosis) activation in the pancreas, and consequently attenuates the severity of severe acute pancreatitis by promotion of pancreatic acinar cell apoptosis [164].

Agents which Induce Pancreatic Acinar Cell Apoptosis but do not Cause Acute Pancreatitis

CCK as well as hydrogen sulfide is an inducer of pancreatic acinar cell apoptosis. They can, however, also induce or promote inflammation in acute pancreatitis. In addition to CCK and hydrogen sulfide, other inducers of pancreatic acinar cell apoptosis have been used to investigate cell death in relation to acute pancreatitis. Examples of these compounds are menadione and crambene (1-cyano-2-hydroxy-3-butene-CHB). Menadione is a quinone that is metabolized by flavoprotein reductase to semiquinone, which can be oxidized back to quinone in the presence of molecular oxygen. In this redox cycle the superoxide anion radical, hydrogen peroxide and other reactive oxygen species are generated [182]. Menadione can cause elevations in the cytosolic Ca^{2+} concentration contributing to cell death [182]. In a recent study, menadione has been shown to evoke repetitive cytosolic Ca^{2+} spikes, partial mitochondrial depolarization, cytochrome c release and apoptosis of pancreatic acinar cells *in vitro* [170].

Crambene is a stable plant nitrile found in many cruciferous vegetables. Crambene is a potent inducer of "phase II" detoxification enzymes, including certain glutathione Stransferases and quinone reductase, which are important enzymes associated with conjugation and elimination of reactive chemical intermediates and carcinogens [183,184]. Also, it has been shown to cause cell cycle arrest of the Hep G2 cells or colonic adenocarcinoma cells in G2/M phase [170]. The intrinsic pathway of apoptosis mediated by loss of mitochondrial membrane potential, release of mitochondrial cytochrome C as well as activation of caspase 9 is suggested to be the primary pathway involved in crambene induced pancreatic acinar cell apoptosis. [168]. Upon intravenous administration to rat as well as mouse, crambene induces apoptosis of the pancreatic acinar cells [185,186]. Moreover, induction of pancreatic acinar cell apoptosis with crambene protects mice against acute pancreatitis induced by caerulein [4]. This is despite the fact that crambene does not alter the interaction of CCK to its receptor on pancreatic acinar cells [4]. The maximal protective effect of crambene against acute pancreatitis is observed when the pancreatic acinar cells have become committed to apoptosis [4]. A possible role of the gap junction communication (GJC) and connexin gene Cx32 has been found in crambene-induced apoptosis [166]. Mice deficient in Cx32 gene were found to be resistant to apoptosis induced by crambene and were more susceptible to acute pancreatitis than their wild-type counterparts [166], which indicate that defective GJC in Cx32-deficient mice prevented acinar cell apoptosis at a step upstream of caspase-3 activation. Besides crambene, an extract of Artemisia asiatica (DA-9601) has also been shown its beneficial effect in caerulein-induced pancreatitis in rats by induction of pancreatic acinar cells apoptosis [163].

Conclusion

Cell death plays an essential role in normal development and functioning of multicellular organisms and that abnormalities in this process can cause disease. Research studies on the molecular mechanisms of apoptosis and its regulation has led to the deeper understanding of the pathophysiology of many diseases. Several strategies have emerged for therapeutic intervention in a wide range of ailments, including cancer, autoimmune disorders, immunodeficiency, inflammation, ischemic heart diseases, stroke and neurodegenerative diseases [187]. Numerous novel approaches are currently being followed employing gene therapy and antisense strategies in order to target specific apoptotic regulators [188]. These approaches should facilitate efforts to rationally manipulate the apoptotic machinery for therapeutic gain.

Both apoptotic and necrotic forms of cell death are seen in clinical, as well as experimental acute pancreatitis [5,6]. The mechanisms of pancreatic acinar cell death - apoptosis and necrosis - in acute pancreatitis are just beginning to be identified. The vast array of molecules involved in both the apoptotic and necrotic cell death pathways offers several potential targets for therapeutic intervention. The extent of pancreatic acinar cell apoptosis has been shown to be inversely related to the severity of the disease, suggesting that apoptosis is a teleologically beneficial form of cell death in acute pancreatitis. Indeed, in animal models, induction of apoptosis has been found to have a protective action against acute pancreatitis. It is, therefore, reasonable to speculate that selective induction of pancreatic acinar cells apoptosis may be of value in clinical acute pancreatitis. As a first step, the effect prophylactic induction of apoptosis in individuals with high risk of pancreatitis (such as those undergoing ERCP procedure) should be investigated. More basic research into the mechanisms of apoptosis of pancreatic acinar cells and how it benefits against acute pancreatitis is, however, needed in order for these studies to be taken into the clinic. Active research in this direction will facilitate a transition of our knowledge of pancreatic acinar cell death in the form of necrosis *versus* apoptosis from bench to bedside.

REFERENCES

[1] Banks PA. Medical management of acute pancreatitis and complications. In: Go VW, DiMagno EP, Gardner JD, Lebenthal E, Reber HA, Scheele GA, Eds. *The Pancreas: Biology, Pathobiology, and Disease*. New York: Raven Press.1993 pp593–611.

[2] Leung, P. S. and Ip, S. P. (2006) Pancreatic acinar cell: Its role in acute pancreatitis. *Int J Biochem Cell Biol 38*, 1024-1030.

[3] Gukovskaya, A. S., Perkins, P., Zaninovic, V., Sandoval, D., Rutherford, R., Fitzsimmons, T., Pandol, S. J. and Poucell-Hatton, S. (1996) Mechanisms of cell death after pancreatic duct obstruction in the opossum and the rat. *Gastroenterology 110*, 875-884.

[4] Bhatia, M., Wallig, M. A., Hofbauer, B., Lee, H. S., Frossard, J. L., Steer, M. L. and Saluja, A. K. (1998) Induction of apoptosis in pancreatic acinar cells reduces the severity of acute pancreatitis. *Biochem Biophys Res Commun 246*, 476-483.

[5] Bhatia, M. (2004) Apoptosis of pancreatic acinar cells in acute pancreatitis: is it good or bad? *J Cell Mol Med 8*, 402-409.

[6] Bhatia, M. (2004) Apoptosis versus necrosis in acute pancreatitis. *Am J Physiol Gastrointest Liver Physiol 286*, G189-196.

[7] Kaiser, A. M., Saluja, A. K., Lu, L., Yamanaka, K., Yamaguchi, Y. and Steer, M. L. (1996) Effects of cycloheximide on pancreatic endonuclease activity, apoptosis, and severity of acute pancreatitis. *Am J Physiol 271*, C982-993.

[8] Walker, N. I., Winterford, C. M., Williamson, R. M. and Kerr, J. F. (1993) Ethionine-induced atrophy of rat pancreas involves apoptosis of acinar cells. *Pancreas 8*, 443-449.

[9] Gukovskaya, A. S., Gukovsky, I., Zaninovic, V., Song, M., Sandoval, D., Gukovsky, S. and Pandol, S. J. (1997) Pancreatic acinar cells produce, release, and respond to tumor necrosis factor-alpha. Role in regulating cell death and pancreatitis. *J Clin Invest 100*, 1853-1862.

[10] Kaiser, A. M., Saluja, A. K., Sengupta, A., Saluja, M. and Steer, M. L. (1995) Relationship between severity, necrosis, and apoptosis in five models of experimental acute pancreatitis. *Am J Physiol 269*, C1295-1304.

[11] Steller, H. (1995) Mechanisms and genes of cellular suicide. *Science 267*, 1445-1449.

[12] Thompson, C. B. (1995) Apoptosis in the pathogenesis and treatment of disease. *Science 267*, 1456-1462.

[13] Kerr, J. F., Wyllie, A. H. and Currie, A. R. (1972) Apoptosis: a basic biological phenomenon with wide-ranging implications in tissue kinetics. *Br J Cancer 26*, 239-257.

[14] Dive, C. and Hickman, J. A. (1991) Drug-target interactions: only the first step in the commitment to a programmed cell death? *Br J Cancer 64*, 192-196.

[15] Eastman, A. (1993) Apoptosis: a product of programmed and unprogrammed cell death. *Toxicol Appl Pharmacol 121*, 160-164.

[16] Bonfoco, E., Krainc, D., Ankarcrona, M., Nicotera, P. and Lipton, S. A. (1995) Apoptosis and necrosis: two distinct events induced, respectively, by mild and intense insults with N-methyl-D-aspartate or nitric oxide/superoxide in cortical cell cultures. *Proc Natl Acad Sci U S A 92*, 7162-7166.

[17] Lieberthal, W., Triaca, V. and Levine, J. (1996) Mechanisms of death induced by cisplatin in proximal tubular epithelial cells: apoptosis vs. necrosis. *Am J Physiol 270*, F700-708.

[18] Benchoua, A., Guegan, C., Couriaud, C., Hosseini, H., Sampaio, N., Morin, D. and Onteniente, B. (2001) Specific caspase pathways are activated in the two stages of cerebral infarction. *J Neurosci 21*, 7127-7134.

[19] Eguchi, Y., Shimizu, S. and Tsujimoto, Y. (1997) Intracellular ATP levels determine cell death fate by apoptosis or necrosis. *Cancer Res 57*, 1835-1840.

[20] Leist, M., Single, B., Castoldi, A. F., Kuhnle, S. and Nicotera, P. (1997) Intracellular adenosine triphosphate (ATP) concentration: a switch in the decision between apoptosis and necrosis. *J Exp Med 185*, 1481-1486.

[21] Ha, H. C. and Snyder, S. H. (1999) Poly(ADP-ribose) polymerase is a mediator of necrotic cell death by ATP depletion. *Proc Natl Acad Sci U S A 96*, 13978-13982.

[22] Nobel, C. S., Burgess, D. H., Zhivotovsky, B., Burkitt, M. J., Orrenius, S. and Slater, A. F. (1997) Mechanism of dithiocarbamate inhibition of apoptosis: thiol oxidation by dithiocarbamate disulfides directly inhibits processing of the caspase-3 proenzyme. *Chem Res Toxicol 10*, 636-643.

[23] Hampton, M. B. and Orrenius, S. (1997) Dual regulation of caspase activity by hydrogen peroxide: implications for apoptosis. *FEBS Lett 414*, 552-556.

[24] Hawkins, H. K., Ericsson, J. L., Biberfeld, P. and Trump, B. F. (1972) Lysosome and phagosome stability in lethal cell injury. Morphologic tracer studies in cell injury due to inhibition of energy metabolism, immune cytolysis and photosensitization. *Am J Pathol 68*, 255-258.

[25] Alison, M. R. and Sarraf, C. E. (1994) Liver cell death: patterns and mechanisms. *Gut 35*, 577-581.
[26] Gupta, S. (2000) Molecular steps of cell suicide: an insight into immune senescence. *J Clin Immunol 20*, 229-239.
[27] Krammer, P. H. (2000) CD95's deadly mission in the immune system. *Nature 407*, 789-795.
[28] Ashkenazi, A. and Dixit, V. M. (1998) Death receptors: signaling and modulation. *Science 281*, 1305-1308.
[29] Kroemer, G. and Reed, J. C. (2000) Mitochondrial control of cell death. *Nat Med 6*, 513-519.
[30] Martinou, J. C. and Green, D. R. (2001) Breaking the mitochondrial barrier. *Nat Rev Mol Cell Biol 2*, 63-67.
[31] Li, H., Zhu, H., Xu, C. J. and Yuan, J. (1998) Cleavage of BID by caspase 8 mediates the mitochondrial damage in the Fas pathway of apoptosis. *Cell 94*, 491-501.
[32] Devitt, A., Moffatt, O. D., Raykundalia, C., Capra, J. D., Simmons, D. L. and Gregory, C. D. (1998) Human CD14 mediates recognition and phagocytosis of apoptotic cells. *Nature 392*, 505-509.
[33] Fadok, V. A., Bratton, D. L., Rose, D. M., Pearson, A., Ezekewitz, R. A. and Henson, P. M. (2000) A receptor for phosphatidylserine-specific clearance of apoptotic cells. *Nature 405*, 85-90.
[34] Slee, E. A., Adrain, C. and Martin, S. J. (1999) Serial killers: ordering caspase activation events in apoptosis. *Cell Death Differ 6*, 1067-1074.
[35] Thornberry, N. A. and Lazebnik, Y. (1998) Caspases: enemies within. *Science 281*, 1312-1316.
[36] Alnemri, E. S., Livingston, D. J., Nicholson, D. W., Salvesen, G., Thornberry, N. A., Wong, W. W. and Yuan, J. (1996) Human ICE/CED-3 protease nomenclature. *Cell 87*, 171.
[37] Nicholson, D. W. and Thornberry, N. A. (1997) Caspases: killer proteases. *Trends Biochem Sci 22*, 299-306.
[38] Launay, S., Hermine, O., Fontenay, M., Kroemer, G., Solary, E. and Garrido, C. (2005) Vital functions for lethal caspases. *Oncogene 24*, 5137-5148.
[39] Fan, T. J., Han, L. H., Cong, R. S. and Liang, J. (2005) Caspase family proteases and apoptosis. *Acta Biochim Biophys Sin (Shanghai) 37*, 719-727.
[40] Pan, G., Bauer, J. H., Haridas, V., Wang, S., Liu, D., Yu, G., Vincenz, C., Aggarwal, B. B., Ni, J. and Dixit, V. M. (1998) Identification and functional characterization of DR6, a novel death domain-containing TNF receptor. *FEBS Lett 431*, 351-356.
[41] Yang, X., Chang, H. Y. and Baltimore, D. (1998) Autoproteolytic activation of pro-caspases by oligomerization. *Mol Cell 1*, 319-325.
[42] Muzio, M., Stockwell, B. R., Stennicke, H. R., Salvesen, G. S. and Dixit, V. M. (1998) An induced proximity model for caspase-8 activation. *J Biol Chem 273*, 2926-2930.
[43] Wang, Z. B., Liu, Y. Q. and Cui, Y. F. (2005) Pathways to caspase activation. *Cell Biol Int 29*, 489-496.

[44] Scaffidi, C., Fulda, S., Srinivasan, A., Friesen, C., Li, F., Tomaselli, K. J., Debatin, K. M., Krammer, P. H. and Peter, M. E. (1998) Two CD95 (APO-1/Fas) signaling pathways. *Embo J 17*, 1675-1687.

[45] Wang, J., Chun, H. J., Wong, W., Spencer, D. M. and Lenardo, M. J. (2001) Caspase-10 is an initiator caspase in death receptor signaling. *Proc Natl Acad Sci U S A 98*, 13884-13888.

[46] Hengartner, M. O. (2000) The biochemistry of apoptosis. *Nature 407*, 770-776.

[47] Gross, A., Pilcher, K., Blachly-Dyson, E., Basso, E., Jockel, J., Bassik, M. C., Korsmeyer, S. J. and Forte, M. (2000) Biochemical and genetic analysis of the mitochondrial response of yeast to BAX and BCL-X(L). *Mol Cell Biol 20*, 3125-3136.

[48] Danial, N. N. and Korsmeyer, S. J. (2004) Cell death: critical control points. Cell 116, 205-219.

[49] Boatright, K. M. and Salvesen, G. S. (2003) Mechanisms of caspase activation. *Curr Opin Cell Biol 15*, 725-731.

[50] Owen-Schaub, L. B., Zhang, W., Cusack, J. C., Angelo, L. S., Santee, S. M., Fujiwara, T., Roth, J. A., Deisseroth, A. B., Zhang, W. W., Kruzel, E. and et al. (1995) Wild-type human p53 and a temperature-sensitive mutant induce Fas/APO-1 expression. *Mol Cell Biol 15*, 3032-3040.

[51] Muller, M., Wilder, S., Bannasch, D., Israeli, D., Lehlbach, K., Li-Weber, M., Friedman, S. L., Galle, P. R., Stremmel, W., Oren, M. and Krammer, P. H. (1998) p53 activates the CD95 (APO-1/Fas) gene in response to DNA damage by anticancer drugs. *J Exp Med 188*, 2033-2045.

[52] Miyashita, T. and Reed, J. C. (1995) Tumor suppressor p53 is a direct transcriptional activator of the human bax gene. *Cell 80*, 293-299.

[53] Oda, E., Ohki, R., Murasawa, H., Nemoto, J., Shibue, T., Yamashita, T., Tokino, T., Taniguchi, T. and Tanaka, N. (2000) Noxa, a BH3-only member of the Bcl-2 family and candidate mediator of p53-induced apoptosis. *Science 288*, 1053-1058.

[54] Chipuk, J. E., Kuwana, T., Bouchier-Hayes, L., Droin, N. M., Newmeyer, D. D., Schuler, M. and Green, D. R. (2004) Direct activation of Bax by p53 mediates mitochondrial membrane permeabilization and apoptosis. *Science 303*, 1010-1014.

[55] Schuler, M., Bossy-Wetzel, E., Goldstein, J. C., Fitzgerald, P. and Green, D. R. (2000) p53 induces apoptosis by caspase activation through mitochondrial cytochrome c release. *J Biol Chem 275*, 7337-7342.

[56] Shieh, S. Y., Ikeda, M., Taya, Y. and Prives, C. (1997) DNA damage-induced phosphorylation of p53 alleviates inhibition by MDM2. *Cell 91*, 325-334.

[57] Ellis, H. M. and Horvitz, H. R. (1986) Genetic control of programmed cell death in the nematode C. elegans. *Cell 44*, 817-829.

[58] Shaham, S., Reddien, P. W., Davies, B. and Horvitz, H. R. (1999) Mutational analysis of the Caenorhabditis elegans cell-death gene ced-3. *Genetics 153*, 1655-1671.

[59] Cheng, E. H., Wei, M. C., Weiler, S., Flavell, R. A., Mak, T. W., Lindsten, T. and Korsmeyer, S. J. (2001) BCL-2, BCL-X(L) sequester BH3 domain-only molecules preventing BAX- and BAK-mediated mitochondrial apoptosis. *Mol Cell 8*, 705-711.

[60] Chou, J. J., Li, H., Salvesen, G. S., Yuan, J. and Wagner, G. (1999) Solution structure of BID, an intracellular amplifier of apoptotic signaling. *Cell 96*, 615-624.

[61] Li, L. Y., Luo, X. and Wang, X. (2001) Endonuclease G is an apoptotic DNase when released from mitochondria. *Nature 412*, 95-99.

[62] Pennacchio, L. A., Bouley, D. M., Higgins, K. M., Scott, M. P., Noebels, J. L. and Myers, R. M. (1998) Progressive ataxia, myoclonic epilepsy and cerebellar apoptosis in cystatin B-deficient mice. *Nat Genet 20*, 251-258.

[63] Mathiasen, I. S., Lademann, U. and Jaattela, M. (1999) Apoptosis induced by vitamin D compounds in breast cancer cells is inhibited by Bcl-2 but does not involve known caspases or p53. *Cancer Res 59*, 4848-4856.

[64] Mathiasen, I. S., Sergeev, I. N., Bastholm, L., Elling, F., Norman, A. W. and Jaattela, M. (2002) Calcium and calpain as key mediators of apoptosis-like death induced by vitamin D compounds in breast cancer cells. *J Biol Chem 277*, 30738-30745.

[65] Kuida, K., Zheng, T. S., Na, S., Kuan, C., Yang, D., Karasuyama, H., Rakic, P. and Flavell, R. A. (1996) Decreased apoptosis in the brain and premature lethality in CPP32-deficient mice. *Nature 384*, 368-372.

[66] Vercammen, D., Beyaert, R., Denecker, G., Goossens, V., Van Loo, G., Declercq, W., Grooten, J., Fiers, W. and Vandenabeele, P. (1998) Inhibition of caspases increases the sensitivity of L929 cells to necrosis mediated by tumor necrosis factor. *J Exp Med 187*, 1477-1485.

[67] Chan, F. K., Shisler, J., Bixby, J. G., Felices, M., Zheng, L., Appel, M., Orenstein, J., Moss, B. and Lenardo, M. J. (2003) A role for tumor necrosis factor receptor-2 and receptor-interacting protein in programmed necrosis and antiviral responses. *J Biol Chem 278*, 51613-51621.

[68] Holler, N., Zaru, R., Micheau, O., Thome, M., Attinger, A., Valitutti, S., Bodmer, J. L., Schneider, P., Seed, B. and Tschopp, J. (2000) Fas triggers an alternative, caspase-8-independent cell death pathway using the kinase RIP as effector molecule. *Nat Immunol 1*, 489-495.

[69] Lin, Y., Choksi, S., Shen, H. M., Yang, Q. F., Hur, G. M., Kim, Y. S., Tran, J. H., Nedospasov, S. A. and Liu, Z. G. (2004) Tumor necrosis factor-induced nonapoptotic cell death requires receptor-interacting protein-mediated cellular reactive oxygen species accumulation. *J Biol Chem 279*, 10822-10828.

[70] Vercammen, D., Brouckaert, G., Denecker, G., Van de Craen, M., Declercq, W., Fiers, W. and Vandenabeele, P. (1998) Dual signaling of the Fas receptor: initiation of both apoptotic and necrotic cell death pathways. *J Exp Med 188*, 919-930.

[71] Kawahara, A., Ohsawa, Y., Matsumura, H., Uchiyama, Y. and Nagata, S. (1998) Caspase-independent cell killing by Fas-associated protein with death domain. *J Cell Biol 143*, 1353-1360.

[72] Kataoka, T., Budd, R. C., Holler, N., Thome, M., Martinon, F., Irmler, M., Burns, K., Hahne, M., Kennedy, N., Kovacsovics, M. and Tschopp, J. (2000) The caspase-8 inhibitor FLIP promotes activation of NF-kappaB and Erk signaling pathways. *Curr Biol 10*, 640-648.

[73] Irisarri, M., Plumas, J., Bonnefoix, T., Jacob, M. C., Roucard, C., Pasquier, M. A., Sotto, J. J. and Lajmanovich, A. (2000) Resistance to CD95-mediated apoptosis through constitutive c-FLIP expression in a non-Hodgkin's lymphoma B cell line. *Leukemia 14*, 2149-2158.

[74] Carini, R., Autelli, R., Bellomo, G., Dianzani, M. U. and Albano, E. (1995) Sodium-mediated cell swelling is associated with irreversible damage in isolated hepatocytes exposed to hypoxia or mitochondrial toxins. *Biochem Biophys Res Commun 206*, 180-185.

[75] Okada, Y., Maeno, E., Shimizu, T., Dezaki, K., Wang, J. and Morishima, S. (2001) Receptor-mediated control of regulatory volume decrease (RVD) and apoptotic volume decrease (AVD). *J Physiol 532*, 3-16.

[76] Barros, L. F., Stutzin, A., Calixto, A., Catalan, M., Castro, J., Hetz, C. and Hermosilla, T. (2001) Nonselective cation channels as effectors of free radical-induced rat liver cell necrosis. *Hepatology 33*, 114-122.

[77] Lemasters, J. J. (1999) V. Necrapoptosis and the mitochondrial permeability transition: shared pathways to necrosis and apoptosis. *Am J Physiol 276*, G1-6.

[78] Levitsky, J., Gurell, D. and Frishman, W. H. (1998) Sodium ion/hydrogen ion exchange inhibition: a new pharmacologic approach to myocardial ischemia and reperfusion injury. *J Clin Pharmacol 38,* 887-897.

[79] Gunter, T. E. and Pfeiffer, D. R. (1990) Mechanisms by which mitochondria transport calcium. *Am J Physiol 258*, C755-786.

[80] Kim, J. S., He, L. and Lemasters, J. J. (2003) Mitochondrial permeability transition: a common pathway to necrosis and apoptosis. *Biochem Biophys Res Commun 304*, 463-470.

[81] Syntichaki, P., Xu, K., Driscoll, M. and Tavernarakis, N. (2002) Specific aspartyl and calpain proteases are required for neurodegeneration in C. elegans. *Nature 419*, 939-944.

[82] Crook, N. E., Clem, R. J. and Miller, L. K. (1993) An apoptosis-inhibiting baculovirus gene with a zinc finger-like motif. *J Virol 67*, 2168-2174.

[83] Birnbaum, M. J., Clem, R. J. and Miller, L. K. (1994) An apoptosis-inhibiting gene from a nuclear polyhedrosis virus encoding a polypeptide with Cys/His sequence motifs. *J Virol 68*, 2521-2528.

[84] Deveraux, Q. L. and Reed, J. C. (1999) IAP family proteins--suppressors of apoptosis. *Genes Dev 13*, 239-252.

[85] Uren, A. G., Coulson, E. J. and Vaux, D. L. (1998) Conservation of baculovirus inhibitor of apoptosis repeat proteins (BIRPs) in viruses, nematodes, vertebrates and yeasts. *Trends Biochem Sci 23*, 159-162.

[86] Deveraux, Q. L., Takahashi, R., Salvesen, G. S. and Reed, J. C. (1997) X-linked IAP is a direct inhibitor of cell-death proteases. *Nature 388*, 300-304.

[87] Deveraux, Q. L., Roy, N., Stennicke, H. R., Van Arsdale, T., Zhou, Q., Srinivasula, S. M., Alnemri, E. S., Salvesen, G. S. and Reed, J. C. (1998) IAPs block apoptotic events induced by caspase-8 and cytochrome c by direct inhibition of distinct caspases. *Embo J 17*, 2215-2223.

[88] Costantini, P., Bruey, J. M., Castedo, M., Metivier, D., Loeffler, M., Susin, S. A., Ravagnan, L., Zamzami, N., Garrido, C. and Kroemer, G. (2002) Pre-processed caspase-9 contained in mitochondria participates in apoptosis. *Cell Death Differ 9*, 82-88.

[89] Adams, J. M. and Cory, S. (1998) The Bcl-2 protein family: arbiters of cell survival. *Science 281*, 1322-1326.

[90] Chao, D. T. and Korsmeyer, S. J. (1998) BCL-2 family: regulators of cell death. *Annu Rev Immunol 16*, 395-419.

[91] Fu, Y. F. and Fan, T. J. (2002) Bcl-2 family proteins and apoptosis. *Sheng Wu Hua Xue Yu Sheng Wu Wu Li Xue Bao (Shanghai) 34*, 389-394.

[92] Milosevic, J., Hoffarth, S., Huber, C. and Schuler, M. (2003) The DNA damage-induced decrease of Bcl-2 is secondary to the activation of apoptotic effector caspases. *Oncogene 22*, 6852-6856.

[93] Gross, A., McDonnell, J. M. and Korsmeyer, S. J. (1999) BCL-2 family members and the mitochondria in apoptosis. *Genes Dev 13*, 1899-1911.

[94] Kelekar, A. and Thompson, C. B. (1998) Bcl-2-family proteins: the role of the BH3 domain in apoptosis. *Trends Cell Biol 8*, 324-330.

[95] Oltvai, Z. N., Milliman, C. L. and Korsmeyer, S. J. (1993) Bcl-2 heterodimerizes in vivo with a conserved homolog, Bax, that accelerates programmed cell death. *Cell 74*, 609-619.

[96] Green, D. R. and Reed, J. C. (1998) Mitochondria and apoptosis. *Science 281*, 1309-1312.

[97] Kluck, R. M., Bossy-Wetzel, E., Green, D. R. and Newmeyer, D. D. (1997) The release of cytochrome c from mitochondria: a primary site for Bcl-2 regulation of apoptosis. *Science 275*, 1132-1136.

[98] Rosse, T., Olivier, R., Monney, L., Rager, M., Conus, S., Fellay, I., Jansen, B. and Borner, C. (1998) Bcl-2 prolongs cell survival after Bax-induced release of cytochrome c. *Nature 391*, 496-499.

[99] Matsuyama, S., Llopis, J., Deveraux, Q. L., Tsien, R. Y. and Reed, J. C. (2000) Changes in intramitochondrial and cytosolic pH: early events that modulate caspase activation during apoptosis. *Nat Cell Biol 2*, 318-325.

[100] Minn, A. J., Kettlun, C. S., Liang, H., Kelekar, A., Vander Heiden, M. G., Chang, B. S., Fesik, S. W., Fill, M. and Thompson, C. B. (1999) Bcl-xL regulates apoptosis by heterodimerization-dependent and -independent mechanisms. *Embo J 18*, 632-643.

[101] Harris, M. H., Vander Heiden, M. G., Kron, S. J. and Thompson, C. B. (2000) Role of oxidative phosphorylation in Bax toxicity. *Mol Cell Biol 20*, 3590-3596.

[102] Schendel, S. L., Azimov, R., Pawlowski, K., Godzik, A., Kagan, B. L. and Reed, J. C. (1999) Ion channel activity of the BH3 only Bcl-2 family member, BID. *J Biol Chem 274*, 21932-21936.

[103] Huigsloot, M., Tijdens, I. B., Mulder, G. J. and van de Water, B. (2002) Differential regulation of doxorubicin-induced mitochondrial dysfunction and apoptosis by Bcl-2 in mammary adenocarcinoma (MTLn3) cells. *J Biol Chem 277*, 35869-35879.

[104] Gross, A., Jockel, J., Wei, M. C. and Korsmeyer, S. J. (1998) Enforced dimerization of BAX results in its translocation, mitochondrial dysfunction and apoptosis. *Embo J 17*, 3878-3885.

[105] Schlesinger, P. H., Gross, A., Yin, X. M., Yamamoto, K., Saito, M., Waksman, G. and Korsmeyer, S. J. (1997) Comparison of the ion channel characteristics of proapoptotic BAX and antiapoptotic BCL-2. *Proc Natl Acad Sci U S A 94*, 11357-11362.

[106] Tsujimoto, Y. and Shimizu, S. (2000) VDAC regulation by the Bcl-2 family of proteins. *Cell Death Differ 7*, 1174-1181.
[107] Tsujimoto, Y. (2003) Cell death regulation by the Bcl-2 protein family in the mitochondria. *J Cell Physiol 195*, 158-167.
[108] Sattler, M., Liang, H., Nettesheim, D., Meadows, R. P., Harlan, J. E., Eberstadt, M., Yoon, H. S., Shuker, S. B., Chang, B. S., Minn, A. J., Thompson, C. B. and Fesik, S. W. (1997) Structure of Bcl-xL-Bak peptide complex: recognition between regulators of apoptosis. *Science 275*, 983-986.
[109] Desagher, S., Osen-Sand, A., Nichols, A., Eskes, R., Montessuit, S., Lauper, S., Maundrell, K., Antonsson, B. and Martinou, J. C. (1999) Bid-induced conformational change of Bax is responsible for mitochondrial cytochrome c release during apoptosis. *J Cell Biol 144*, 891-901.
[110] Yin, X. M., Wang, K., Gross, A., Zhao, Y., Zinkel, S., Klocke, B., Roth, K. A. and Korsmeyer, S. J. (1999) Bid-deficient mice are resistant to Fas-induced hepatocellular apoptosis. *Nature 400*, 886-891.
[111] Goping, I. S., Gross, A., Lavoie, J. N., Nguyen, M., Jemmerson, R., Roth, K., Korsmeyer, S. J. and Shore, G. C. (1998) Regulated targeting of BAX to mitochondria. *J Cell Biol 143*, 207-215.
[112] Desagher, S. and Martinou, J. C. (2000) Mitochondria as the central control point of apoptosis. *Trends Cell Biol 10*, 369-377.
[113] Mancini, M., Nicholson, D. W., Roy, S., Thornberry, N. A., Peterson, E. P., Casciola-Rosen, L. A. and Rosen, A. (1998) The caspase-3 precursor has a cytosolic and mitochondrial distribution: implications for apoptotic signaling. *J Cell Biol 140*, 1485-1495.
[114] Krajewski, S., Krajewska, M., Ellerby, L. M., Welsh, K., Xie, Z., Deveraux, Q. L., Salvesen, G. S., Bredesen, D. E., Rosenthal, R. E., Fiskum, G. and Reed, J. C. (1999) Release of caspase-9 from mitochondria during neuronal apoptosis and cerebral ischemia. *Proc Natl Acad Sci U S A 96*, 5752-5757.
[115] Susin, S. A., Lorenzo, H. K., Zamzami, N., Marzo, I., Brenner, C., Larochette, N., Prevost, M. C., Alzari, P. M. and Kroemer, G. (1999) Mitochondrial release of caspase-2 and -9 during the apoptotic process. *J Exp Med 189*, 381-394.
[116] Susin, S. A., Lorenzo, H. K., Zamzami, N., Marzo, I., Snow, B. E., Brothers, G. M., Mangion, J., Jacotot, E., Costantini, P., Loeffler, M., Larochette, N., Goodlett, D. R., Aebersold, R., Siderovski, D. P., Penninger, J. M. and Kroemer, G. (1999) Molecular characterization of mitochondrial apoptosis-inducing factor. *Nature 397*, 441-446.
[117] Du, C., Fang, M., Li, Y., Li, L. and Wang, X. (2000) Smac, a mitochondrial protein that promotes cytochrome c-dependent caspase activation by eliminating IAP inhibition. *Cell 102*, 33-42.
[118] Verhagen, A. M., Ekert, P. G., Pakusch, M., Silke, J., Connolly, L. M., Reid, G. E., Moritz, R. L., Simpson, R. J. and Vaux, D. L. (2000) Identification of DIABLO, a mammalian protein that promotes apoptosis by binding to and antagonizing IAP proteins. *Cell 102*, 43-53.
[119] Wang, K. K. (2000) Calpain and caspase: can you tell the difference? *Trends Neurosci 23*, 20-26.

[120] Blomgren, K., Zhu, C., Wang, X., Karlsson, J. O., Leverin, A. L., Bahr, B. A., Mallard, C. and Hagberg, H. (2001) Synergistic activation of caspase-3 by m-calpain after neonatal hypoxia-ischemia: a mechanism of "pathological apoptosis"? *J Biol Chem 276*, 10191-10198.

[121] Mandic, A., Hansson, J., Linder, S. and Shoshan, M. C. (2003) Cisplatin induces endoplasmic reticulum stress and nucleus-independent apoptotic signaling. *J Biol Chem 278*, 9100-9106.

[122] Nakagawa, T. and Yuan, J. (2000) Cross-talk between two cysteine protease families. Activation of caspase-12 by calpain in apoptosis. *J Cell Biol 150*, 887-894.

[123] Hajnoczky, G., Davies, E. and Madesh, M. (2003) Calcium signaling and apoptosis. *Biochem Biophys Res Commun 304*, 445-454.

[124] Schmitz, I., Kirchhoff, S. and Krammer, P. H. (2000) Regulation of death receptor-mediated apoptosis pathways. *Int J Biochem Cell Biol 32*, 1123-1136.

[125] Thome, M., Schneider, P., Hofmann, K., Fickenscher, H., Meinl, E., Neipel, F., Mattmann, C., Burns, K., Bodmer, J. L., Schroter, M., Scaffidi, C., Krammer, P. H., Peter, M. E. and Tschopp, J. (1997) Viral FLICE-inhibitory proteins (FLIPs) prevent apoptosis induced by death receptors. *Nature 386*, 517-521.

[126] Hu, S., Vincenz, C., Buller, M. and Dixit, V. M. (1997) A novel family of viral death effector domain-containing molecules that inhibit both CD-95- and tumor necrosis factor receptor-1-induced apoptosis. *J Biol Chem 272*, 9621-9624.

[127] Irmler, M., Thome, M., Hahne, M., Schneider, P., Hofmann, K., Steiner, V., Bodmer, J. L., Schroter, M., Burns, K., Mattmann, C., Rimoldi, D., French, L. E. and Tschopp, J. (1997) Inhibition of death receptor signals by cellular FLIP. *Nature 388*, 190-195.

[128] Tschopp, J., Irmler, M. and Thome, M. (1998) Inhibition of fas death signals by FLIPs. *Curr Opin Immunol 10*, 552-558.

[129] Kirchhoff, S., Muller, W. W., Krueger, A., Schmitz, I. and Krammer, P. H. (2000) TCR-mediated up-regulation of c-FLIPshort correlates with resistance toward CD95-mediated apoptosis by blocking death-inducing signaling complex activity. *J Immunol 165*, 6293-6300.

[130] Yoshikawa, H., Nakajima, Y. and Tasaka, K. (2000) Enhanced expression of Fas-associated death domain-like IL-1-converting enzyme (FLICE)-inhibitory protein induces resistance to Fas-mediated apoptosis in activated mast cells. *J Immunol 165*, 6262-6269.

[131] Adrain, C., Murphy, B. M. and Martin, S. J. (2005) Molecular ordering of the caspase activation cascade initiated by the cytotoxic T lymphocyte/natural killer (CTL/NK) protease granzyme B. *J Biol Chem 280*, 4663-4673.

[132] Barry, M., Heibein, J. A., Pinkoski, M. J., Lee, S. F., Moyer, R. W., Green, D. R. and Bleackley, R. C. (2000) Granzyme B short-circuits the need for caspase 8 activity during granule-mediated cytotoxic T-lymphocyte killing by directly cleaving Bid. *Mol Cell Biol 20*, 3781-3794.

[133] Ray, C. A., Black, R. A., Kronheim, S. R., Greenstreet, T. A., Sleath, P. R., Salvesen, G. S. and Pickup, D. J. (1992) Viral inhibition of inflammation: cowpox virus encodes an inhibitor of the interleukin-1 beta converting enzyme. *Cell 69*, 597-604.

[134] Komiyama, T., Ray, C. A., Pickup, D. J., Howard, A. D., Thornberry, N. A., Peterson, E. P. and Salvesen, G. (1994) Inhibition of interleukin-1 beta converting enzyme by the cowpox virus serpin CrmA. An example of cross-class inhibition. *J Biol Chem 269*, 19331-19337.

[135] Tewari, M. and Dixit, V. M. (1995) Fas- and tumor necrosis factor-induced apoptosis is inhibited by the poxvirus crmA gene product. *J Biol Chem 270*, 3255-3260.

[136] Komiyama, T., Quan, L. T. and Salvesen, G. S. (1996) Inhibition of cysteine and serine proteinases by the cowpox virus serpin CRMA. *Adv Exp Med Biol 389*, 173-176.

[137] Miura, M., Zhu, H., Rotello, R., Hartwieg, E. A. and Yuan, J. (1993) Induction of apoptosis in fibroblasts by IL-1 beta-converting enzyme, a mammalian homolog of the C. elegans cell death gene ced-3. *Cell 75*, 653-660.

[138] Gurevich, R. M., Regula, K. M. and Kirshenbaum, L. A. (2001) Serpin protein CrmA suppresses hypoxia-mediated apoptosis of ventricular myocytes. *Circulation 103*, 1984-1991.

[139] Zhou, Q. and Salvesen, G. S. (2000) Viral caspase inhibitors CrmA and p35. *Methods Enzymol 322*, 143-154.

[140] Bertin, J., Armstrong, R. C., Ottilie, S., Martin, D. A., Wang, Y., Banks, S., Wang, G. H., Senkevich, T. G., Alnemri, E. S., Moss, B., Lenardo, M. J., Tomaselli, K. J. and Cohen, J. I. (1997) Death effector domain-containing herpesvirus and poxvirus proteins inhibit both Fas- and TNFR1-induced apoptosis. *Proc Natl Acad Sci U S A 94*, 1172-1176.

[141] Bump, N. J., Hackett, M., Hugunin, M., Seshagiri, S., Brady, K., Chen, P., Ferenz, C., Franklin, S., Ghayur, T., Li, P. and et al. (1995) Inhibition of ICE family proteases by baculovirus antiapoptotic protein p35. *Science 269*, 1885-1888.

[142] Xue, D. and Horvitz, H. R. (1995) Inhibition of the Caenorhabditis elegans cell-death protease CED-3 by a CED-3 cleavage site in baculovirus p35 protein. *Nature 377*, 248-251.

[143] Miller, L. K. (1997) Baculovirus interaction with host apoptotic pathways. *J Cell Physiol 173*, 178-182.

[144] Datta, R., Kojima, H., Banach, D., Bump, N. J., Talanian, R. V., Alnemri, E. S., Weichselbaum, R. R., Wong, W. W. and Kufe, D. W. (1997) Activation of a CrmA-insensitive, p35-sensitive pathway in ionizing radiation-induced apoptosis. *J Biol Chem 272*, 1965-1969.

[145] Beidler, D. R., Tewari, M., Friesen, P. D., Poirier, G. and Dixit, V. M. (1995) The baculovirus p35 protein inhibits Fas- and tumor necrosis factor-induced apoptosis. *J Biol Chem 270*, 16526-16528.

[146] Stennicke, H. R., Ryan, C. A. and Salvesen, G. S. (2002) Reprieval from execution: the molecular basis of caspase inhibition. *Trends Biochem Sci 27*, 94-101.

[147] Xu, G., Cirilli, M., Huang, Y., Rich, R. L., Myszka, D. G. and Wu, H. (2001) Covalent inhibition revealed by the crystal structure of the caspase-8/p35 complex. *Nature 410*, 494-497.

[148] Vier, J., Furmann, C. and Hacker, G. (2000) Baculovirus P35 protein does not inhibit caspase-9 in a cell-free system of apoptosis. *Biochem Biophys Res Commun 276*, 855-861.

[149] Silverman, N. and Maniatis, T. (2001) NF-kappaB signaling pathways in mammalian and insect innate immunity. *Genes Dev 15*, 2321-2342.

[150] Karin, M. and Lin, A. (2002) NF-kappaB at the crossroads of life and death. *Nat Immunol 3*, 221-227.

[151] Li, Q. and Verma, I. M. (2002) NF-kappaB regulation in the immune system. *Nat Rev Immunol 2*, 725-734.

[152] Kasibhatla, S., Brunner, T., Genestier, L., Echeverri, F., Mahboubi, A. and Green, D. R. (1998) DNA damaging agents induce expression of Fas ligand and subsequent apoptosis in T lymphocytes via the activation of NF-kappa B and AP-1. *Mol Cell 1*, 543-551.

[153] Lin, B., Williams-Skipp, C., Tao, Y., Schleicher, M. S., Cano, L. L., Duke, R. C. and Scheinman, R. I. (1999) NF-kappaB functions as both a proapoptotic and antiapoptotic regulatory factor within a single cell type. *Cell Death Differ 6,* 570-582.

[154] Zheng, Y., Ouaaz, F., Bruzzo, P., Singh, V., Gerondakis, S. and Beg, A. A. (2001) NF-kappa B RelA (p65) is essential for TNF-alpha-induced fas expression but dispensable for both TCR-induced expression and activation-induced cell death. *J Immunol 166*, 4949-4957.

[155] Kucharczak, J., Simmons, M. J., Fan, Y. and Gelinas, C. (2003) To be, or not to be: NF-kappaB is the answer--role of Rel/NF-kappaB in the regulation of apoptosis. *Oncogene 22*, 8961-8982.

[156] Jun, J. I., Chung, C. W., Lee, H. J., Pyo, J. O., Lee, K. N., Kim, N. S., Kim, Y. S., Yoo, H. S., Lee, T. H., Kim, E. and Jung, Y. K. (2005) Role of FLASH in caspase-8-mediated activation of NF-kappaB: dominant-negative function of FLASH mutant in NF-kappaB signaling pathway. *Oncogene 24*, 688-696.

[157] Badrichani, A. Z., Stroka, D. M., Bilbao, G., Curiel, D. T., Bach, F. H. and Ferran, C. (1999) Bcl-2 and Bcl-XL serve an anti-inflammatory function in endothelial cells through inhibition of NF-kappaB. *J Clin Invest 103*, 543-553.

[158] Kloppel, G. and Maillet, B. (1993) Pathology of acute and chronic pancreatitis. *Pancreas 8*, 659-670.

[159] Nevalainen, T. J. and Aho, H. J. (1992) Standards of morphological evaluation and histological grading in experimental acute pancreatitis. *Eur Surg Res 24 Suppl 1*, 14-23.

[160] Walker, N. I. (1987) Ultrastructure of the rat pancreas after experimental duct ligation. I. The role of apoptosis and intraepithelial macrophages in acinar cell deletion. *Am J Pathol 126*, 439-451.

[161] Walker, N. I., Winterford, C. M. and Kerr, J. F. (1992) Ultrastructure of the rat pancreas after experimental duct ligation. II. Duct and stromal cell proliferation, differentiation, and deletion. *Pancreas 7*, 420-434.

[162] Jimi, A., Kojiro, M., Miyasaka, K., Kono, A. and Funakoshi, A. (1997) Apoptosis in the pancreas of genetically diabetic rats with a disrupted cholecystokinin (CCK-A) receptor gene. *Pancreas 14*, 109-112.

[163] Hahm, K. B., Kim, J. H., You, B. M., Kim, Y. S., Cho, S. W., Yim, H., Ahn, B. O. and Kim, W. B. (1998) Induction of apoptosis with an extract of Artemisia asiatica attenuates the severity of cerulein-induced pancreatitis in rats. *Pancreas 17*, 153-157.

[164] Chao, K. C., Chao, K. F., Chuang, C. C. and Liu, S. H. (2006) Blockade of interleukin 6 accelerates acinar cell apoptosis and attenuates experimental acute pancreatitis in vivo. *Br J Surg 93*, 332-338.

[165] Lupia, E., Goffi, A., De Giuli, P., Azzolino, O., Bosco, O., Patrucco, E., Vivaldo, M. C., Ricca, M., Wymann, M. P., Hirsch, E., Montrucchio, G. and Emanuelli, G. (2004) Ablation of phosphoinositide 3-kinase-gamma reduces the severity of acute pancreatitis. *Am J Pathol 165*, 2003-2011.

[166] Frossard, J. L., Rubbia-Brandt, L., Wallig, M. A., Benathan, M., Ott, T., Morel, P., Hadengue, A., Suter, S., Willecke, K. and Chanson, M. (2003) Severe acute pancreatitis and reduced acinar cell apoptosis in the exocrine pancreas of mice deficient for the Cx32 gene. *Gastroenterology 124*, 481-493.

[167] Gukovsky, I., Reyes, C. N., Vaquero, E. C., Gukovskaya, A. S. and Pandol, S. J. (2003) Curcumin ameliorates ethanol and nonethanol experimental pancreatitis. *Am J Physiol Gastrointest Liver Physiol 284*, G85-95.

[168] Cao, Y., Adhikari, S., Ang, A. D., Clement, M. V., Wallig, M. and Bhatia, M. (2006) Crambene induces pancreatic acinar cell apoptosis via the activation of mitochondrial pathway. *Am J Physiol Gastrointest Liver Physiol*

[169] Cao, Y., Adhikari, S., Ang, A. D., Moore, P. K. and Bhatia, M. (2006) Mechanism of Induction of Pancreatic Acinar Cell Apoptosis by Hydrogen Sulfide. *Am J Physiol Cell Physiol*

[170] Gerasimenko, J. V., Gerasimenko, O. V., Palejwala, A., Tepikin, A. V., Petersen, O. H. and Watson, A. J. (2002) Menadione-induced apoptosis: roles of cytosolic Ca(2+) elevations and the mitochondrial permeability transition pore. *J Cell Sci 115*, 485-497.

[171] Gukovskaya, A. S., Gukovsky, I., Jung, Y., Mouria, M. and Pandol, S. J. (2002) Cholecystokinin induces caspase activation and mitochondrial dysfunction in pancreatic acinar cells. Roles in cell injury processes of pancreatitis. *J Biol Chem 277*, 22595-22604.

[172] Song, J. Y., Lim, J. W., Kim, H., Morio, T. and Kim, K. H. (2003) Oxidative stress induces nuclear loss of DNA repair proteins Ku70 and Ku80 and apoptosis in pancreatic acinar AR42J cells. *J Biol Chem 278*, 36676-36687.

[173] Beil, M., Leser, J., Lutz, M. P., Gukovskaya, A., Seufferlein, T., Lynch, G., Pandol, S. J. and Adler, G. (2002) Caspase 8-mediated cleavage of plectin precedes F-actin breakdown in acinar cells during pancreatitis. *Am J Physiol Gastrointest Liver Physiol 282*, G450-460.

[174] Mareninova, O. A., Sung, K. F., Hong, P., Lugea, A., Pandol, S. J., Gukovsky, I. and Gukovskaya, A. S. (2006) Cell death in pancreatitis: caspases protect from necrotizing pancreatitis. *J Biol Chem 281*, 3370-3381.

[175] Gomez, G., Lee, H. M., He, Q., Englander, E. W., Uchida, T. and Greeley, G. H., Jr. (2001) Acute pancreatitis signals activation of apoptosis-associated and survival genes in mice. *Exp Biol Med (Maywood) 226*, 692-700.

[176] Tomasini, R., Samir, A. A., Vaccaro, M. I., Pebusque, M. J., Dagorn, J. C., Iovanna, J. L. and Dusetti, N. J. (2001) Molecular and functional characterization of the stress-induced protein (SIP) gene and its two transcripts generated by alternative splicing. SIP induced by stress and promotes cell death. *J Biol Chem 276*, 44185-44192.

[177] O'Neill, S., O'Neill, A. J., Conroy, E., Brady, H. R., Fitzpatrick, J. M. and Watson, R. W. (2000) Altered caspase expression results in delayed neutrophil apoptosis in acute pancreatitis. *J Leukoc Biol 68*, 15-20.

[178] Salomone, T., Tosi, P., Raiti, C., Guariento, A., Tomassetti, P., Migliori, M., Saieva, C., Romboli, M. and Gullo, L. (2002) Apoptosis in the peripheral blood mononuclear cells as a self-limitation process in human acute pancreatitis. *Pancreatology 2*, 204-210.

[179] Fujimoto, K., Hosotani, R., Doi, R., Wada, M., Lee, J. U., Koshiba, T., Miyamoto, Y. and Imamura, M. (1997) Role of neutrophils in caerulein-induced pancreatitis in rats: possible involvement of apoptosis. *Digestion 58*, 421-430.

[180] Rau, B., Paszkowski, A., Esber, S., Gansauge, F., Poch, B., Beger, H. G. and Moller, P. (2001) Anti-ICAM-1 antibody modulates late onset of acinar cell apoptosis and early necrosis in taurocholate-induced experimental acute pancreatitis. *Pancreas 23*, 80-88.

[181] Satoh, A., Gukovskaya, A. S., Edderkaoui, M., Daghighian, M. S., Reeve, J. R., Jr., Shimosegawa, T. and Pandol, S. J. (2005) Tumor necrosis factor-alpha mediates pancreatitis responses in acinar cells via protein kinase C and proline-rich tyrosine kinase 2. *Gastroenterology 129*, 639-651.

[182] Monks, T. J., Hanzlik, R. P., Cohen, G. M., Ross, D. and Graham, D. G. (1992) Quinone chemistry and toxicity. *Toxicol Appl Pharmacol 112*, 2-16.

[183] Niedoborski, T. E., Klein, B. P. and Wallig, M. A. (2001) Rapid isolation and purification of 1-cyano-2-hydroxy-3-butene (crambene) from Crambe abyssinica seed meal using immiscible solvent extraction and high-performance liquid chromatography. *J Agric Food Chem 49*, 3594-3599.

[184] Wallig, M. A., Kingston, S., Staack, R. and Jefferey, E. H. (1998) Induction of rat pancreatic glutathione S-transferase and quinone reductase activities by a mixture of glucosinolate breakdown derivatives found in Brussels sprouts. *Food Chem Toxicol 36*, 365-373.

[185] Bhatia, M., Brady, M., Kang, Y. K., Costello, E., Newton, D. J., Christmas, S. E., Neoptolemos, J. P. and Slavin, J. (2002) MCP-1 but not CINC synthesis is increased in rat pancreatic acini in response to cerulein hyperstimulation. *Am J Physiol Gastrointest Liver Physiol 282*, G77-85.

[186] Bhatia, M., Slavin, J., Cao, Y., Basbaum, A. I. and Neoptolemos, J. P. (2003) Preprotachykinin-A gene deletion protects mice against acute pancreatitis and associated lung injury. *Am J Physiol Gastrointest Liver Physiol 284*, G830-836.

[187] Nicholson, D. W. (2000) From bench to clinic with apoptosis-based therapeutic agents. *Nature 407*, 810-816.

[188] Fischer, U. and Schulze-Osthoff, K. (2005) Apoptosis-based therapies and drug targets. *Cell Death Differ 12 Suppl 1*, 942-961.

In: Focus on Cell Apoptosis Research
Editor: Daniel W. Cho, pp. 233-249

ISBN: 1-60021-433-9

Chapter IX

THE ROLE OF APOPTOSIS IN PSORIASIS AND LICHEN PLANUS

Marija Kaštelan and Larisa Prpić Massari
Department of Dermatology, Clinical Hospital Centre and Medical Faculty
University of Rijeka, Croatia

ABSTRACT

Apoptosis is an active physiological process of genetically programmed cell death. It is a unique process maintaining the homeostasis of variety of physiological systems in the body including the skin. In skin, cell death by apoptosis is an essential way of regulating keratinocyte proliferation and epidermal growth. Terminal differentiation of keratinocyte represents simply a special form of apoptotic death. The process by which keratinocytes undergo apoptosis is a multistep program that includes several mechanisms. One is induced by binding of specific death ligands to death receptors on target cells such as binding of FasL expressed on activated cytotoxic T cells to Fas expressed on keratinocytes. Another is mediated by the release of effectors cell granules such as cytolytic molecule perforin and serine proteases named granzymes. The apoptotic process is not only involved in maintaining skin homeostasis but also has a role in the pathogenesis of a variety of skin diseases. Moreover, failure of cells to undergo apoptosis might be the main cause of some dermatosis. In skin, different apoptotic stimuli may simultaneously trigger multiple apoptotic pathways. Apoptotic keratinocytes along with dysfunctional apoptotic process occur in some skin diseases such as psoriasis and lichen planus. The evidence is accumulating about psoriasis as a T cell-mediated disease. In psoriasis, conflicting results are obtained regarding the role of apoptosis in development of psoriatic lesions. There is a decreased apoptosis of lesional keratinocytes with decreased level of p53 activation. Moreover, it seems that psoriatic keratinocytes has increased resistance to apoptosis due to as yet not well defined mechanisms. Multiple studies revealed increased level of antiapototic Bcl-x_L protein in psoriasis. However, several authors demonstrated enhanced up regulation of perforin apoptotic pathway, both, in peripheral blood of psoriatic patients and in psoriatic lesions. A hallmark of lichen planus is the formation of colloid bodies representing apoptotic keratinocytes.

Infiltrating T cells seems to be responsible for damage of basal keratinocytes in lichen planus. Experimental data provided evidence for a role of perforin in the apoptosis of lesional keratinocytes. In addition, enhanced Fas-mediated death of epidermal keratinocytes in lichen lesions is also actively involved in the apoptosis of adjacent basal keratinocytes. Here, we review the complex role of several apoptotic mechanisms in the etiopathogenesis of papulosquamous diseases, psoriasis and lichen planus.

Introduction

Apoptosis is a physiological process of programmed cell death responsible for the homeostasis of a variety of physiological systems in the body including skin. The term apoptosis is coined by Kerr et al. in 1972. to describe an active process with characteristic morphological features [1]. Apoptosis is morphologically characterized by cell shrinkage, nuclear condensation, cellular fragmentation and finally phagocytosis by neighbouring cells [2]. In skin, cell death by apoptosis is a unique way of regulating keratinocyte proliferation, differentiation and epidermal growth as well [3,4]. Balance between cell death and cell proliferation maintains homeostasis of epidermal compartment. It seems that complex programme of apoptotic cell death, by maintaining the gradients of antiapoptotic and proapoptotic factors, regulates the timing of apoptosis in the epidermis and also epidermal differentiation [5]. In addition, terminal differentiation of keratinocytes might simply represent a special form of apoptotic death [6,7]. However, there are some data that keratinocyte terminal differentiation and apoptosis are distinct cellular events triggered by different stimuli [8-10].

The process by which keratinocytes undergo apoptosis is a multistep program that include several mechanisms. Briefly, the «extrinsic» apoptotic pathway is mediated by binding of Fas ligand (FasL) or TNF to specific death receptors thus leading to the activation of different caspases in an ordered sequence [11,12]. The «intrinsic» pathway involves mitochondrial release of multiple proapoptotic factors that ends in final activation of the caspase-3 [13]. Both, apoptosis induced through death receptors and mitochondria, ends in the activation of the effector caspase-3, responsible for the cleavage of a variety of intracellular substrates and for the activation of caspase-activated deoxyribonuclease that initiates DNA cleavage [14]. Finally, the process ends in the formation of apoptotic bodies and its phagocytosis by neighbouring macrophages and dendritic cells.

The process of apoptosis is controlled by proteins of the Bcl-2 family, i.e. several proapoptotic (Bax, Bak, Bad) and antiapoptotic (Bcl-2, $Bclx_L$) proteins [13]. Keratinocyte apoptosis can be blocked by upregulation of apoptotic inhibitors and the members of the inhibitor of apoptosis protein (IAP) family [13-15]. So far, it seems that keratinocyte apoptotic responses are generally dependent on p53 whose levels are potentially suppressed in normal keratinocytes by a number of regulators [16]. The p53 can activate the transcription of proapoptotic regulators of the mitochondrial (Bax, Noxa, Apaf-1) and death receptor pathways (Fas, death receptors 4 and 5) or on the other hand repress that of antiapoptotic (Bcl-2, $Bclx_L$) proteins [17].

Therefore, it seems that dysfunctional apoptosis has an important role in the pathogenesis of several skin diseases [6,18]. Some of them, such as alopecia areata and lichen planus are characterized by increased keratinocyte apoptosis, while the others such as psoriasis and skin cancers are associated with decreased keratinocyte apoptosis [6]. Here, we shall review the role of apoptosis in the pathogenesis of two papulosquamous diseases, psoriasis and lichen planus, one with decreased and the other one with increased keratinocyte apoptosis.

Apoptosis in Psoriasis

Psoriasis is a chronic hypeproliferative inflammatory skin disease characterized by abnormal keratinocyte hyperproliferation and differentiation, intraepidermal accumulation of neutrophilic granulocytes and inflammatory vascular changes [19]. Psoriasis is now generally recognized as a T cell mediated inflammatory disease [20,21]. Immunologically, psoriasis is characterized by a dense mononuclear inflammatory infiltrate particularly in the dermis that mostly consists of T cells [22]. Infiltrating T cells create a type 1 cytokine microenvironment within the psoriatic lesions with type 1 cytokines IFNγ, TNFα and IL-2, but not IL-4 or IL-5 [23-25]. Such particular cytokine pattern is sufficient in inducing typical psoriatic changes, mainly keratinocyte hyperproliferation (IL-6, IFNγ) and accumulation of neutrophil granulocytes (IL-8) in the epidermis [24,26]. Activated keratinocytes also produce a wide range of cytokines that enhance activation and chemotaxis of T cells in the skin (IL-6, IL-8) or even induce the increased expression of particular adhesion molecules (IL-1) important in T cell trafficking [20,22]. Therefore, T cells and keratinocytes are equal partners in the persistent dialogue during the creation of psoriatic lesions.

Although, the precise mechanism of how activated T cells trigger psoriasis is as yet unknown, it seems that the specific immune reaction to a putative antigen mediated by T cells leads to the creation of psoriatic lesions [27]. The real nature of putative psoriatic antigen is as yet unknown, but it is now recognized that bacterial endotoxins acting as superantigens or a particular epidermal self-antigens are capable of triggering psoriasis [28,29].

In the last decade, evidence suggesting psoriasis as a T cell mediated disease was accumulating. The important role of T cells in psoriasis pathogenesis was first suggested after observed effectiveness of cyclosporin in the treatment of psoriasis [30]. Cyclosporin was found to selectively inhibit IL-2 and IFNγ release from CD4+ T cells. Further evidence suggesting also an important role for T lymphocytes in eliciting psoriasis includes the successful use of diphtheria fusion toxin (DAB_{389}IL-2) in psoriasis treatment [31]. DAB_{389}IL-2 was found to be a potent immunotoxin only to T cells expressing interleukin-2 receptor (IL-2R). Finally, experiments on a severe combined immunodeficient (SCID) mouse confirmed the key role of activated T cells in the initiation and persistence of psoriatic process [32]. In a SCID mouse, only injection of autologous CD4+ T cells from the psoriatic patient could induce transformation of engrafted symptomless skin into a typical psoriatic lesion.

Recent findings suggest, that in lesions, proinflammatory cytokines released by activated CD4+ and CD8+ T cells stimulate keratinocyte proliferation and abnormal differentiation [33]. It is proposed that intraepidermal infiltration of CD4+ T cells is the primary cellular event that precedes the epidermal hyperplasia in the acute psoriatic lesion [34]. It is then followed by the selective activation and proliferation of specific subsets of CD8+ T cells in the epidermis of psoriatic plaque. Thus, it seems that the initiation, perpetuation and maintenance of psoriatic lesions are mediated by activated T cells [34].

Perforin Apoptotic Pathway in Psoriasis

The role of cell cytotoxicity mediated by cytotoxic CD8+ T lymphocytes (CTLs) and natural killer (NK) cells in the pathogenesis of psoriasis are still not completely understood. Apoptosis, induced by CTL, is mediated either via the release of effector cell granules, i.e. perforin and granzymes, or by binding ligands to different death receptors on target cells [35]. Perforin and granzyme B are stored within lysosomes of CTLs and NK cells. Upon activation, CTL and NK cells release perforin and granzyme B that afterwards cooperatively induce target cell death. Perforin is a pore forming protein that leads to osmotic lysis of target cell thus enabling granzymes to enter the cell and trigger apoptosis [36]. Binding of ligands (FasL, TNF) expressed on CTL/NK cells to death receptors on the target cells, induces recruitment of several specific adapter molecules and via activation of caspase-8 trigger cell death along the common apoptotic pathway [2,35]. The expression of perforin is up regulated in activated CD8+ T cells, but in NK cells the expression is constitutively very high and stable [36]. The perforin expression could be also stimulated in some activated CD4+ T cells. CD8+ T cells and NK cells use both, the perforin/granzyme and Fas/FasL pathway while CD4+ T cells preferentially use Fas/FasL pathway [37]. Currently, there are very few data regarding the role of perforin-mediated cytotoxicity in psoriasis.

However, the role of perforin-mediated cytotoxicity has been demonstrated in various autoimmune diseases [38,39]. *In vitro* and *in vivo* studies suggested that CTLs cytotoxicity might be mediated by cytotoxic granules in certain inflammatory diseases in humans [40]. In addition, it seems that T-cell cytotoxicity against keratinocytes is mediated by perforin in some inflammatory skin diseases [41]. In our previous work we found the systemic upregulation of perforin molecule in the exacerbation phase of psoriasis [42]. This increase was mostly related to CD8+P+ cells, a subpopulation with cytotoxic potential and rapid killing cells bearing MHC class I molecules and appropriate antigen. [43,44]. These cells could participate directly in the switching off of the inflammatory process via HLA class I-mediated recognition of the psoriasis epitope(s) [45]. In skin lesions, we observed increased perforin expression in the epidermis of psoriatic lesions compared with non-lesional and healthy skin [46]. Moreover, the perforin expression in lesions was almost three times higher in epidermis than in dermis suggesting significant accumulation of perforin positive cells in the lesional epidermis. Austin et al. reported an up-regulated expression of another cytolytic protein, T-cell-restricted intracellular antigen (TIA-1), in the epidermal CD8+ T cells in psoriatic lesions speculating about a key role of CD8+ CTLs in mediating psoriatic phenotype [47]. The well-known HLA-Cw*0602 association with psoriasis is in line with the proposed role of CD8+ T cells. On the other hand, Yawalkar et al. also recently showed an increase of perforin and granzyme-B expression in psoriatic lesions as compared with normal and non-lesional skin [48]. Immunoreactivity for these cytotoxic proteins was found in both, CD4+ and CD8+ cells, but not in CD56+ cells.

In our study, perforin-expressing cells were found suprabasaly in close contact with damaged keratinocytes suggesting that T lymphocytes induce damage to keratinocytes by releasing cytolytic molecules [46]. A spongiform degeneration of basement membrane was seen at the site where dermal perforin-positive cells migrate into the epidermis disrupting focally the epidermal basement membrane. Based on our data, it seems likely that apoptotic keratinocytes might trigger an injury response program causing regenerative hyperplasia of keratinocytes, a hallmark of psoriasis [49]. More damaged, apoptotic keratinocytes might constantly stimulate wound repair process thus closing the vicious circle that drive keratinocyte hyperproliferation. Another possibility is that the failure to eliminate these abnormal keratinocytes could result in the persistence of chronic inflammatory conditions, constantly recruiting specific T cells [49].

Fas/FasL Apoptotic Pathway in Psoriasis

As mentioned before, Fas ligand induces apoptosis through binding of Fas (CD95) on the target cells [35]. Fas is marginally expressed in normal basal keratinocytes, but its secretion can be upregulated by TH1 type cytokine, IFN-γ. Nickoloff et al. previously showed that IFN-γ has a central role in psoriatic process [34]. Moreover, its expression is significantly elevated in psoriatic epidermis. Several reports observed higher Fas expression in the psoriatic lesional epidermis, but not in the nonlesional one [50,51]. Because the expression of Fas is induced by IFN-γ in keratinocytes, this might explain the upregulation of Fas in the psoriatic epidermis, which is characterized by TH1 cytokine pattern.

While Fas is upregulated in the psoriatic epidermis, the data on FasL expression in the epidermis are controversial. Some studies showed that FasL is low or absent in both, normal and psoriatic epidermis, while the others, using the same antibodies, found out that FasL was expressed in almost all cell layers of normal epidermis [52-54]. Takahashi et al. observed only slight expression of FasL mRNA in both, normal and psoriatic epidermis [51]. Thus, it seems that despite marked expression of Fas in psoriatic epidermis, psoriatic keratinocytes lack FasL, thereby no activation of Fas occurs. However, Fas could be activated by UVB irradiation without FasL suggesting the potential of Fas of psoriatic involved epidermis to induce apoptosis [55-57].

Psoriatic keratinocytes have been shown to have the ability to produce granzyme B and express FasL, thereby being extremely resistant to killing by Fas-bearing cells [58]. If epidermal keratinocytes have the ability to produce cytolytic molecules, perforin and granzyme and/or express FasL molecule on the surface, they might also have the potential to protect the epidermis from T cell mediated damage by inducing Fas mediated death of neighbouring Fas positive T cells [37,58]. Hence, increased epidermal thickness as a hallmark of psoriatic epidermis might also depend on the relative balance between T cells bearing FasL and keratinocytes. These data suggest that the susceptibility of the psoriatic epidermis to undergo immune-mediated damage may be dependent on its ability to express FasL and the density of its Fas membrane receptor.

Bcl-2 Family Proteins in Psoriasis

Psoriasis can be viewed as a disease of epidermal hyperproliferation, increased epidermal turnover with abnormal terminal differentiation and impaired barrier function of the epidermis [59]. Psoriatic keratinocytes are characterized by both, hyperplasia and incomplete differentiation. Moreover, psoriatic keratinocytes are now recognized to display abnormal resistance to apoptosis what has been implicated as one of the key pathogenetic mechanisms in psoriasis [60].

As mentioned before, the Bcl-2 family of proteins control cell death either preventing (Bcl-2, Bcl-x_L) or promoting cell death (Bax, Bad, Bak) [61-63]. The balance between these proteins determines the susceptibility of cells to undergo apoptotic cell death. So far, the expression of many regulatory molecules of proliferation has been well clarified in psoriatic epidermis [50,64-66]. However, data on the expression of Bcl-2 family proteins in psoriatic lesions are controversial. Bianchi et al. showed that Bcl-2 is expressed in the basal keratinocytes of normal but not psoriatic epidermis, while Wrone-Smith et al. reported that Bcl-2 was not expressed in normal or psoriatic epidermis [50,57]. Takahashi et al. observed overexpression of Bcl-2 in basal to spinous cell layers, with its higher level in nonlesional than in lesional epidermis [51]. They also reported the increased expression of Bcl-x_L and Bax molecules in the psoriatic epidermis. However, the significance of this finding remains to be determined, because these molecules are antagonizing each other. Tomkova et al. also demonstrated an increased expression of pro-apoptotic Bax protein in the suprabasal cell layers of psoriatic and normal epidermis, while Fukuya et al. observed strong diffuse immunostaining of Bax throughout the epidermis of psoriatic lesions [67,68]. Thus, it is difficult to determine whether the psoriatic epidermis is in pro-apopototic or anti-apoptotic condition in terms of Bcl-2 related proteins [50]. Further studies are necessary to clarify the role of Bcl-2 and Bax protein in psoriasis.

Several studies reported the overexpression of the cell survival protein Bcl-x_L in all layers of psoriatic epidermis [51,60,69]. Because Bcl-x_L blocks apoptosis, the increased epidermal thickness characteristic of psoriasis might be related to a defect in the apoptotic pathway. It seems that upregulation of Bcl-x_L might contribute to the hyperproliferation, albeit not to the differentiation of psoriatic keratinocytes [69]. Keratinocytes derived from psoriatic plaques resist induction of apoptosis having a particular phenotype that suggests prolonged survival compared with keratinocytes from normal skin [60]. Thus, it is likely that epidermal acanthosis may result from not only an increase in frequency of cells undergoing proliferation but also from a decrease in cells undergoing cell death [50,51]. However, the underlying mechanisms of this dysfunctional control of psoriatic keratinocyte apoptosis are as yet unknown. How Bcl-x_L is upregulated in the psoriatic epidermis and whether the overexpression of the Bcl-x_L in psoriatic keratinocytes is responsible for their enhanced survival is as yet not clear. It is also unclear whether overexpression of the Bcl-x_L in psoriasis is a primary pathological event or is it induced by upregulation of particular psoriatic cytokines. So far, it seems that overexpression of Bcl-x_L in psoriatic lesions suggests that the apoptosis of psoriatic keratinocytes is prevented, thus leading to epidermal acanthosis.

Recent data indicate that tumor necrosis factor-α (TNF-α) induces mostly the synthesis of anti-apoptotic Bcl-x and Bcl-2 proteins and to a lesser extent the pro-apoptotic Bax protein in psoriatic keratinocytes [50,70]. Increased concentrations of TNF-α have been found in psoriatic lesions compared with nonlesional and healthy skin [71]. It seems that this single cytokine, overproduced by innate (dendritic cells, NK cells) and adaptive (T lymphocytes) immune cells, has a potential in inducing psoriatic inflammatory cascade. Moreover, the effectiveness of new biologicals, targeting TNF-α in psoriasis treatment is to date very well established [72-74]. Hence, it is likely that the neutralization of TNF-α changes the intracellular Bax/Bcl-2 ratio in psoriatic epidermis, thereby making the psoriatic keratinocytes susceptible to apoptotic signals [71].

Another cytokine known to upregulates the expression of Bcl-2 molecule is interferon-γ (IFN-γ) which level is over 10-fold greater in the psoriatic epidermis than in normal epidermis [75,76]. IFN-γ induces proliferation of psoriatic keratinocytes and growth arrest in normal keratinocytes [77-79]. Moreover, IFN-γ triggers apoptosis in terminally differentiated epidermal keratinocytes of normal skin, while contrariwise, psoriatic keratinocytes fail to undergo apoptosis, although an overexpression of IFN-γ is present in psoriasis [50,80]. It seems that IFN-γ, by upregulation of Bcl-2, which blocks apoptosis, might prevent the apoptosis of psoriatic keratinocytes [75]. Thus, this abnormal response of psoriatic keratinocytes to IFN-γ contributes significantly to epidermal hyperproliferation in psoriasis.

It has been recently reported that delayed apoptosis of psoriatic keratinocytes may be mediated at least in part by keratinocyte-derived IL-15 [81]. The observed overexpression of IL-15 cytokine and IL-15R in psoriatic epidermis suggests that this cytokine may inhibit keratinocyte cell death in the upper layers of the epidermis thus leading to epidermal hyperplasia in psoriasis. Moreover, it is interesting that IL-15 is a potent chemoattractant and inhibitor of lymphocyte apoptosis as well as a T cell and NK cell growth factor [82-85]. Considering psoriasis as a T-cell mediated immune disease, keratinocyte-derived IL-15 may also account for T cell accumulation and survival in psoriatic lesions [81].

Tomkova et al. demonstrated that the Bcl-2 homologous protein, pro-apoptotic Bad, is constitutively expressed in the lower suprabasal compartment of normal human epidermis [86,87]. Psoriatic epidermis showed an apparently weak and predominantly diffuse Bad staining pattern in the mid and upper compartment of epidermis with significantly lower expresssion in the upper than the lower epidermal layers [86]. It has been suggested that Bad expression may contribute to the longevity of psoriatic keratinocytes, because its assumed inactivation in the upper epidermis interferes with Bcl-x_L overexpression, which has been well documented in psoriasis [67,86].

The terminal deoxynucleotidyl transferase mediated nick-end labeling (TUNEL) method has been widely used for the detection of apoptotic cells [88]. Despite the known antiapoptotic features of psoriatic epidermis, several reports documented that most keratinocytes in all layers of psoriatic epidermis are TUNEL-positive [50,57,89]. On the other hand, some studies showed a discrepancy between TUNEL-positive cells and chromatin-condensed apoptotic cells in the epidermis suggesting that TUNEL might not be a reliable marker of apoptosis in epidermis [10, 60,90]. Moreover, Kawashima et al. recently showed that the increase in numbers of double-strand DNA breaks in psoriatic lesions was not the result of apoptotic processes but, rather, caused by active DNA replication [69]. In addition, Laporte et al. reported that apoptotic cells are decreased by one-third in the established psoriatic epidermis compared with the normal skin [89]. Since TUNEL-positive cells in psoriatic lesions lack morphological evidence of apoptosis, it has been suggested that they are proliferating and not apoptotic keratinocytes [60,69,90].

Taken all together, it is clear that apoptotic process plays a complex role in controlling epidermal proliferation in psoriasis. However, we still have to elucidate the details of how apoptosis is involved in psoriasis pathogenesis. Future challenges also include the clinical implications of these data directed toward development of new apoptosis-based therapeutics in psoriasis treatment.

Apoptosis in Lichen Planus

Lichen planus is a chronic inflammatory mucocutaneous disorder, characterized histologicaly by epidermal basal cell destruction and a band-like inflammatory cell infiltrate in papillary dermis [91,92]. Research in this field is at present dominated by the hypothesis that lichen planus is T-cell-mediated autoimmune disorder with a lymphocytotoxic process against the epidermal basal layer that seems to be the primary site of the immunologic injury [93]. However, immunological changes, specific for lichen planus, occur both systemically and locally, in the peripheral blood as well as in the skin and mucous lesions [93]. The skin and mucous lesions are strongly infiltrated with T lymphocytes.

Conflicting results were obtained regarding the dominance of CD4+ and CD8+ T lymphocytes, with some authors finding predominance of CD4+ and others observing an accumulation of CD8+ cells in the inflammatory infiltrate [92,94,95]. However, it seems that majority of subepidermal and intraepidermal mononuclear cells are of CD8+ phenotype while CD4+ cells mainly infiltrate the deep lamina propria [96]. According to our published data, the incidence of CD8+ cells was much more higher in lesional epidermis indicating a significant accumulation of these cells in the epidermal compartment [97]. There are also some interesting observations of Zhou XY et al. showing significantly higher number of intraepithelial CD8+ T cells in region of basal membrane disruption than in regions of basal membrane continuity [98]. Consequently, it seems that CD4+ cells that infiltrates predominately dermis activate cytotoxic CD8+ cells, causing their migration towards epidermis and local destruction of the basement membrane [99,100]. Based on these results, CD8+ cells seem to be the major effector's cell population in developing lichen planus lesions.

The formation of colloid bodies, also called hyaline, cytoid or Civatte bodies [101], in the lower epidermis and upper dermis is a characteristic histological feature of lichen planus.

Although the mechanism of colloid body formation is unclear, electron microscopic studies of colloid bodies in lichen planus lesions showed condensed chromatin and fragmented nuclei of basal and suprabasal cells [102,103,104]. By in situ nick end labelling, pyknotic nuclei of colloid bodies were positively stained [105]. These findings suggest that DNA fragmentation as well as apoptotic process occur in basal and suprabasal cells, and that colloid bodies are subsequently formed in the process of intraepidermal cell apoptosis [105].

The process of programmed cell death or apoptosis is mediated by different «extrinsic» and «intrinsic» signals. They both cause activation of caspase-3 responsible for the cleavage of a variety of intracellular substrates and finally DNA [2]. This process ends in the formation of apoptotic bodies and its phagocytosis by neighbouring cells, macrophages and dendritic cells [13,14].

As mentioned before, a characteristic histological feature of lichen planus is the formation of colloid bodies that represent apoptotic keratinocytes and suggest an active role of different apoptotic mechanisms in the evolution of lichen planus [37]. It seems that apoptotic process in lichen planus is mainly mediated by CD8+ cytotoxic T lymphocytes (CTLs) through two distinct pathways, perforin/granzyme and Fas/FasL pathway [97]. Perforin, a pore forming protein induces target cell lysis subsequently enabling granzymes to enter the cell and activate specific cellular proteases, named caspases that finally fragment the DNA [44,106]. Generally, perforin expression is a marker of functionally activated CTLs *in situ* [44,106]. Currently, there are very few data in literature suggesting the importance of perforin-mediated cytotoxicity in pathogenesis of lichen planus [97,105].

Analysing the lichen planus lesions, we found a significant up-regulation of perforin in the epidermis [97]. Almost two times higher percentage of perforin-positive cells was observed in the lesional epidermis than in lesional dermis suggesting their accumulation in the epidermal compartment [97]. Perforin-positive cells were found in close proximity to apoptotic keratinocytes indicating that these cells may be involved in inducing damage to keratinocytes [97]. Shimizu et al. also observed higher incidence of perforin-positive and granzyme B-positive CD8+ cells in lichen planus lesions adjacent to keratinocyte, which was undergoing apoptosis [150]. Increased apoptotic index in lichen planus lesions together with higher expression of intraepithelial T-cell-restricted intracellular antigen (TIA-1), granzyme B and perforin were found by Santoro A. et al. as well [107]. These cells seem to induce apoptosis of keratinocytes in lichen planus and formation of colloid bodies [107].

There have been some interesting reports on changes in the population of cytotoxic T lymphocytes in peripheral blood as well [91,108]. CD3+P+ cells are upregulated in the phase of disease exacerbation compared to remission and healthy control, and increase is located mainly in CD8+P+ subpopulation [108]. We have also noticed a significant increase of CD8+ cells that contained perforin molecules in exacerbation phase of disease, that could suggest a possible migration of these cytotoxic cells from peripheral blood into the lesions [108].

The involvement of another cytolytic pathway mediated by Fas/FasL is also studied. Nepellberg et al. found a prominent expression of Fas/FasL in lichen planus infiltrations, more abundantly in basal cell area compared to suprabasal cell layer [109].

Both apoptotic pathways, granule egzocytosis and FasL/Fas pathway activate caspase cascade that finally leads to DNA fragmentation. Expression of caspase-3 was also studied by some authors showing significantly greater expression of this molecule in the basal layer of epidermis compared to suprabasal [110,111].

The process of apoptosis is controlled by proteins of the Bcl-2 family, i.e. proapoptotic (Bax, Bak, Bad, p53) and antiapoptoic (Bcl-2, $Bclx_L$) proteins. Bcl-2 protein prolongs cell life by inhibiting apoptosis and basal keratinocytes were negative for this protein as well as for some proapoptotic proteins like Bcl-x and Bax [112,113]. On the other hand, p53 molecule has proapoptotic or tumor suppressor activity. A prominent p53-positive staining was seen in lichen planus biopsies especially in basal keratinocyte layer [113,114,115]. Some authors also found that the number of proliferation marker Ki-67 was highly expressed in skin samples in which the proportion of p53-positive cells was high, mainly in basal and parabasal layers [114,115]. The accumulation of p53 is possibly a response to an increased proliferation rate of keratinocytes [114,115].

These findings clearly show intensive apoptotic activity in the lesional skin of lichen planus, especially in the basal and suprabasal layer, thus leading to destruction of keratinocytes and colloid body formation. Based on these observations, a novel therapy of this autoimmune disorder could be pointed at the blockage of the molecules included in the apoptotic process.

Acknowledgement

This work is financially supported by the Ministry of Science and Technology of the Republic of Croatia.

References

[1] Kerr JFR, Wyllie AH, Currie AR. Apoptosis. A basic biological phenomenon with widearanging implications in tissue kinetics. *Br J Cancer* 1972; 26:239-257.

[2] Afford S, Randhawa S. Apoptosis. *Mol Pathol* 2000; 53:55-63.

[3] Thompson CB. Apoptosis in the pathogenesis and treatment of disease. *Science* 1995; 267: 1456-1462.

[4] Raskin CA. Apoptosis and cutaneous biology. *J Am Acad Dermatol* 1997; 36:885-896.

[5] Bowen AR, Hanks AN, Allen SM, Alexander A, Diedrich MJ, Grossman D. Apoptosis regulators nad responses in human melanocytic and keratinocytic cells. *J Invest Dermatol* 2003; 120:48-55.

[6] Raj D, Brash DE, Grossman D. Keratinocyte apoptosis in epidermal development and disease. *J Invest Dermatol* 2006; 126; 243-257.

[7] Allombert-Blaise C, Tamiji S, Mortier L, Fauvel H, Tual M, Delaporte E, Piette F, DeLassale EM, Formstecher P, Marchetti P, Polakowska R. Terminal differentiation of human epidermal keratinocytes involves mitochondria- and caspase-dependent cell pathway. *Cell Death Differ* 2003; 10:850-852.

[8] Eckhart L, Declerq W, Ban J, Rendl M, Lengauer B, Mayer C, Lippens S, Vandenabeele P, Tschachler E. Terminal differentiation of human keratinocytes and stratum corneum formation is associated with caspase-14 activation. *J Invest Dermatol* 2000; 115:1148-1151.

[9] Lippens S, Kockx M, Knaapen M, Mortier L, Polakowska R, Verheyen A,Garmyn M, Zwiysen A, Formstecher P, Huylebroeck D, Vandenabeele P, Declerq W. Epidermal differentiation does not involve the pro-apoptotic executioner caspases, but is associated with caspase-14 induction and processing. *Cell Death Differ* 2000; 7:1218-1224.

[10] Gandarillas A, Goldsmith LA, Gschmeisser S, Leigh IM, Watt FM. Evidence that apoptosis and terminal differentiation of epidermal keratinocytes are distinct processes. *Exp Dermatol* 1999; 8:71-79.

[11] Curtin JF, Cotter TG. Live or let die: regulatory mechanisms in Fas-mediated apoptosis. *Cell Signal* 2003; 15983-992.

[12] Peter ME, Krammer PH. The CD95 (Apo-1/Fas) DISC and beyond. *Cell Death Differ* 2003; 10:26-35.

[13] Van Gurp M, Festjens N, van Loo G, Saelens X, Vandenabeele P. Mitochondrial intermembrane proteins in cell death. *Biochem Biophys Res Commun* 2003; 304:487-497.

[14] Riedl SJ, Shi Y. Molecular mechanisms of caspase regulation during apoptosis. *Nat Rev Mol Cell Biol* 2004; 5:897-907.

[15] Bowen AR, Hanks AN, Murphy KJ, Florell SR, Grossman D. Proliferation, apoptosis, and survivin expression in keratinocytic neoplasms and hyperplasia. *Am J Dermatopathol* 2004; 26:177-181.

[16] Harris SL, Levine AJ. The p53 pathway; positive and negative feedback loops. *Oncogene* 2005; 24:2899-2908.

[17] Yu J, Zhang L. The transcriptional targets of p 53 in apoptosis control. *Biochem Biophys Res Commun* 2005; 331:851-858.

[18] Wehrli P, Viard I, Bullani R, Tschopp J, French LE. Death receptors in cutaneous biology and disease. *J Invest Dermatol* 2000; 115:141-148.

[19] Lebwohl M. Psoriasis. *Lancet* 2003; 361:1197-1204.

[20] Prinz JC. Psoriasis vulgaris – a sterile antibacterial skin reaction mediated by cross-reactive T cells? An immunological view of the pathophysiology of psoriasis. *Clin Exp Dermatol* 2001; 26:326-332.

[21] Prinz JC. The role of T cells in psoriasis. *J Eur Acad Dermatol Venereol* 2003; 17:257-270.

[22] Krueger JG, Bowcock A. Psoriasis pathophysiology: current concepts of pathogenesis. *Ann Rheum Dis* 2005; 64 (Suppl II):ii30-ii36.

[23] Karasek MA. Progress in our understanding of the biology of psoriasis. *Cutis* 1999; 64:319-322.

[24] Krueger JG. The immunologic basis for the treatment of psoriasis with new biologic agents. *J Am Acad Dermatol* 2002; 46:1-23.

[25] Griffiths CE. The immunological basis of psoriasis. *J Eur Acad Dermatol Venereol* 2003; 17 (suppl 2):1-5.

[26] Mehlis SL, Gordon KB. The immunology of psoriasis and biologic immunotherapy. *J Am Acad Dermatol* 2003; 49:S44-S50.

[27] Leung DJ, Walsh P, Giorno R, Norris DA. A potential role for superantigens in the pathogenesis of psoriasis. *J Invest Dermatol* 1993; 100:225-228.

[28] Leung DYM, Travers JB, Norris DA. The role of superantigens in skin diseases. *J Invest Dermatol* 1995; 105:37S-42S.

[29] Valdimarsson H, Sigmundsdottir H, Jonsdottir I. Is psoriasis induced by streptococcal superantigens and maintained by M-protein-specific T cells that cross-react with keratin? *Clin Exp Immunol* 1997; 107:21-24.

[30] Boss JD. The pathomechanisms of psoriasis: the skin immune system and cyclosporin. *Br J Dermatol* 1988; 118:141-155.

[31] Gottlieb SL, Gilleaudeau P, Johnson R, Estes L, Woodworth TG, Gotlieb AB, Krueger JG. Response of psoriasis to a lymphocyte-selective toxin (DAB389IL-2) suggest a primary immune, but not keratinocyte, pathogenic basis. *Nat Med* 1995; 1:442-447.

[32] Wrone-Smith T, Nickoloff BJ. Dermal injection of immunocytes induces psoriasis. *J Clin Invest* 1996; 98:1878-1887.

[33] Prinz JC. Which T cells cause psoriasis? *Clin and Exp Dermatol* 1999; 24:291-295.

[34] Nickoloff BJ. The immunologic and genetic basis of psoriasis. *Arch Dermatol* 1999; 135:1104-1110.

[35] Schultz DR, Harrington WJ. Apoptosis: programmed cell death at a molecular level. *Semin Arthritis Rheum* 2003; 32:345-369.

[36] Yagita H, Nakata M, Kawasaki A, Shinkai J, Okumura K. Role of perforin in lymphocytes-mediated cytolysis. *Adv Immunol* 1992; 51:215-241.

[37] Teraki Y, Shiohara T. Apoptosis and the skin. *Eur J Dermatol* 1999; 9:413-426.

[38] Rubesa G, Podack ER, Sepcic J, Rukavina D. Increased perforin expression in peripheral blood lymphocytes in multiple sclerose patients in exacerbation of disease. *J Neuroimmunol* 1997; 74:198-204.

[39] Gulan G, Ravlic-Gulan J, Strbo N, Sotosek V, Nemec B, Matovinovic D. Systemic and local expression of perforin in lymphocyte subsets in acute and chronic rheumatoid arthritis. *J Rheumatol* 2003; 30:660-670.

[40] Yasukawa M, Ohminami H, Arai J, Kasahara Y, Ishida Y, Fujita S. Granule exocytosis and not the Fas/Fas ligand system, is the main pathway of cytotoxicity mediated by alloantigen-specific CD4(+) as well as CD8(+) cytotoxic T lymphocytes in humans. *Blood* 2000; 95:2352-2355.

[41] Schnyder B, Frutig K, Yawalkar N, Limat A, Pichler WJ. T cell cytotoxicity in sulfamethoxazole induced skin reaction. *Clin Exp Allergy* 1998; 28:1412-1417.

[42] Prpić L, Štrbo N, Sotošek V, Gruber F, Podack ER, Rukavina D. Assesment of perforin expression in peripheral blood lymphocytes in psoriatic patients during exacerbation of disease. *Acta Derm Venereol Stockh* 2000; suppl 211:14-16.

[43] Podack ER. Perforin: structure, function and regulation. *Curr Top Microbiol Immunol* 1992; 178:175-184.

[44] Podack ER. Functional significance of two cytolytic pathways of cytotoxic T lymphocytes. *J Leukoc Biol* 1995; 57:548-552.

[45] Valdimarsson H, Baker BS, Jonsdottir I, Powles A, Fry L. Psoriasis: a T-cell mediated autoimmune disease induced by streptococcal superantigens? *Immunol Today* 1995; 16:145-149.

[46] Kaštelan M, Prpić-Massari L, Gruber F, Zamolo G, Žauhar G, Čoklo M, Rukavina D. Perforin expression is up-regulated in the epidermis of psoriatic lesions. *Br J Dermatology* 2004; 151:831-836.

[47] Austin LM, Coven TR, Bhardway N, Steinman R, Krueger JG. Intraepidermal lymphocytes in psoriatic lesions are activated GMP-17 (TIA-1)+CD8+CD3+ CTLs as determined by phenotypic analysis. *J Cutan Pathol* 1998; 25:79-88.

[48] Yawalkar N, Schmid S, Braathen LR, Pichler WJ. Perforin and granzyme B may contribute to skin inflammation in atopic dermatitis and psoriasis. *Br J Dermatol* 2001; 144:1133-1139.

[49] Kaštelan M, Prpić-Massari L, Brajac I. Apoptosis mediated by cytolytic molecules might be responsible for maintenance of psoriatic plaques. *Medical Hypotheses* 2006; 67:336-337.

[50] Wrone-Smith T, Johnson T, Nelson B, Boise LH, Thompson CB, Nunez G, Nickoloff BJ. Discordant expression of Bcl-x and Bcl-2 by keratinocytes in vitro and psoriatic keratinocytes in vivo. *Am J Pathol* 1995; 146:1079-1088.

[51] Takahashi H, Manabe A, Ishida-Yamamoto A, Hashimoto Y, Iizuka H. Aberrant expression of apoptosis-related molecules in psoriatic epidermis. *J Dermatol Sci* 2002; 28: 187-197.

[52] Gutierrez-Steil C, Wrone-Smith T, Sun X, Krueger JG, Coven T, Nickoloff BJ. Sunlight-induced basal cell carcinoma tumor cells tumor cells and ultraviolet-B-irradiated psoriatic plaques express Fas ligand (CD95L) *J Clin Invest* 1998; 101.33-39.

[53] Lee SH, Jang JJ, Lee JY, Kim SY, Park WS, Shin MS, Dong SM, Na EY, kim KM, Kim CS, Kim SH, Yo J. Fas ligand is expressed in normal skin and some cutaneous malignancies. *Br J Dermatol* 1998; 139.186-191.

[54] Hehemtulla A, Hamilton CA, Cinnaiyan AM, Dixit VM. Ultraviolet radiation-induced apoptosis is mediated by activation of CD95 (Fas/Apo-1). *J Biol Chem* 1997; 272:25783-25786.

[55] Aragame Y, Kulms D, Metze D, Wilkers G Poppelman B, Luger TA, Schwarz T. Ultraviolet light induces apoptosis via direct activation of CD95 (Fas/Apo-1) independently of its ligand CD95L. *J Cell Biol* 1998; 140:171-182.

[56] Takahashi H, Nakamura S, Asano K, Kinouchi M, Ishida-Yamamoto A, Iizuka H. Fas modulates ultraviolet B-induced apoptosis of SVHK cells: Sequential activation of caspase 8, 3 and 1 in the apoptotic process. *Exp Cell Res* 1999; 249291.298.

[57] Bianchi L, Farrace MG, Nini G, Piacentini M. Abnormal bcl-2 and tissue transglutaminase expression in psoriatic skin. *J Invest Dermatol* 1994; 103:829-833.

[58] Berthou C, Michel L, Soulie A, Jean-Louis F, Flageul B, Dubertret L, Sigaux F, Zhang Y, Sasportes M. Acquisition of granzyme B and Fas ligand proteins by human keratinocytes contributes to epidermal cell defense. *J Immunol* 1997; 159:5293-5300.

[59] Mckay IA, Leigh IM. Altered keratinocyte growth and differentiation in psoriasis. *Clin Dermatol* 1995; 13:105-114.

[60] Wrone-Smith T, Mitra RS, Thompson CB, Jasty R, Castle VP, Nickoloff BJ. Keratinocytes derived from psoriatic plaques are resistant to apoptosis compared with normal skin. *Am J Pathol* 1997; 151:1321-1329.

[61] Adams JM, Cory S. The bcl-2 protein family: arbiters of cell survival. *Science* 1998; 281:1322-1326.

[62] Sedlak TW, Oltvai ZN, Yang E, Wang K, Boise LH, Thompson CB, Korsmeyer S. Multiple Bcl-2 family members demonstrate selective dimerization with Bax. *Proc Natl Acad Sci USA* 1995; 92: 7834-7838.

[63] Boise LH, Gonzalez-Garcia M, Postmen CE, Ding L, Lindsten T, Turka LA, Mao X, Nunez G, Thompson CB. Bcl-x, a bcl-2 related gene that functions as a dominant regulator of apoptosis cell death. *Cell* 1993; 74:597-608.

[64] Nakagawa K, Yamamura K, Maeda S, Ichihashi M. Bcl-2 expression in epidermal keratinocytic diseases. *Cancer* 1994; 74:1720-1724.

[65] Reinhartz J, Bechtel MJ, Kramer MD. Tumor necrosis factor-α-induced apoptosis in a human keratinocyte cell line (HaCaT) is counteracted by transforming growth factor-α. *Exp Cell Res* 1996; 228:334-340.

[66] Doi H, Shibata MA, Kiyokane K, Otsuki Y. Downregulation of TGF-β isoforms and their receptors contributes to keratinocyte hyperproliferation in psoriasis vulgaris. *J Dermatol Sci* 2003; 33:7-16.

[67] Tomkova H, Fujimoto W, Arata J. Expression of the bcl-2 homologue Bax in normal human skin, psoriasis vulgaris and non-melanoma skin cancers. *Eur J Dermatol* 1998; 8:256-260.

[68] Fukuya Y, Higaki M, Kawashima M. Effect of vitamin D_3 on the increased expression of Bcl-x_L in psoriasis. *Arch Dermatol Res* 2002; 293:620-625.

[69] Kawashima, Doi H, Ito Y, Shibata MA, Yoshinaka R, Otsuki Y. Evaluation of cell death and proliferation in psoriatic epidermis. *J Dermatol Sci* 2004; 35:207-214.

[70] Krüger-Krasagakis S, Galanopoulus VK, Giannikaki L, Stefanidou M, Tosca AD. Programmed cell death of keratinocytes in infliximab-treated plaque-type psoriasis. *Br J Dermatol* 2006; 154:460-466.

[71] Gaspari AA. Innate and adaptive immunity and the pathophysiology of psoriasis. *J Am Acad Dermatol* 2006; 54:S67-70.

[72] Gottlieb AB, Evans R, Li S, Dooley T, Guzzo CA, Baker D, Bala M, Marano CW, Menter A. Infliximab induction therapy for patients with severe plaque-type psoriasis: a randomized, double-blind, placebo-controlled trial. *J Am Acad Dermatol* 2004; 51:534-542.

[73] Gottlieb AB, Masud S, Ramamurthi R, Abdulghani A, romano P, Chaudhari U, Dooley LT, fasanmade AA, Wagner CL. Pharmacodynamic and pharmacokinetic response to anti-tumor necrosis factor-α monoclonal antibody (infliximab) treatment of moderate to severe psoriasis vulgaris. *J Am Acad Dermatol* 2003; 48:68-75.

[74] Goedkoop AY, Kraan MC, Teunissen MB, Picavet DI, de Rie MA, Bos JD, Tak PP. Early effects of tumor necrosis factor-α blockade on skin and synovial tissue in patients with active psoriasis and psoriasis arthritis. *Ann Rheum Dis* 2004; 63:769-773.

[75] Chen SH, Istvan A, Apisarnthanarax N, Rajaraman S, Tyring SK, Horikoshi T, Brysk H, Brysk MM. Response of keratinocytes from normal and psoriatic epidermis to interferon-α differs in the expression of zinc- α_2-glycoprotein and cathepsin D. *FASEB J* 2000; 14:565-571.

[76] Kancko F, Suzuki M, Takiguchi J, Itoh N, Minagawa T. Immunohistopathologic studies in the development of psoriatic lesions influenced by gamma-interferon and the producing cells. *J Dermatol Sci* 1990; 1:425-434.

[77] Saunders NA, Jetten MA. Control of growth regulatory and differentiation-specific genes in human epidermal keratinocytes by interferon-γ. Antagonism by retinoic acid and transforming growth factor β1. *J Biol Chem* 1994; 269:2016-2022.

[78] Baker BS, Powles AV, Valdimarsson H, Fry L. An altered response by psoriatic keratinocytes to gamma-interferon. *Scand J Immunol* 1988; 28:735-740.

[79] Barker JNWN, Goodlad JR, Ross EL, Yu CC, Groves RW, MacDonald DM. Increased epidermal cell proliferation in normal human skin in vivo following local administration of interferon-γ. *Am J Pathol* 1993; 142:1091-1097.

[80] Brysk MM, Selvanayagam P, Arany I, Brysk H, Tyring SK, Rajaraman S. Induction of apoptotic nuclei by interferon-γ and by predesquamin in cultured keratinocytes. *J Interferon Cytokine Res* 1995; 15:1029-1035.

[81] Rückert R, Asadullah K, Seifert M, Budagian VM, Arnold R, Trombotto C, Paus R, Bulfone-Paus S. Inhibition of keratinocyte apoptosis by IL-15: a new parameter in the pathogenesis of psoriasis. *J Immunol* 2000; 165:2240-2250.

[82] Bulfone-Paus S, Ungureanu D, Pohl T, Lindner G, Paus R, Rückert R, Krause H, Kunzendorf U. Interleukin-15 protects from lethal apoptosis in vivo. *Nat Med* 1997; 3:1224-1228.

[83] Nieto M, del Pozo MA, Sanchez Madrid F. Interleukin-15 induces adhesion receptor redistribution in T lymphocytes. *Eur J Immunol* 1996; 26:1302-1307.

[84] Bulfone-Paus S, Dürkop R, Paus H, Krause H, Pohl T, Onu A. Differential regulation of human T lymphoblast functions by IL-2 and IL-15. *Cytokine* 1997; 9:507-513.

[85] Kennedy MK, Glacum M, Brown SN, Butz EA, Viney JL, Embers M, Matsuki N, Charrier K, Sedger L, Willis CR et al. Reversible defects in natural killer and memory CD8 T cell lineage in interleukin 15-deficient mouse. *J Exp Med* 2000; 191:771-780.

[86] Tomkova H, Fujimoto W, Arata J. Expression pattern of the bcl-2 homologous protein bad in normal skin, psoriasis vulgaris and keratinocytic tumors. *J Dermatol Sci* 2000; 22:132-137.

[87] Franke TF, Cantley LC. A bad kinase makes good. *Nature* 1997; 390:116-117.

[88] Gavrieli Y, Sherman Y, Ben-Saason SA. Identification of programmed cell death in situ via specific labeling of nuclear DNA fragmentation. *J Cell Biol* 1992; 119:493-501.

[89] Laporte M, Galand P, Fokan D, de Graef C, Heene M. Apoptosis in established and healing psoriasis. *Dermatology* 2000; 200:314-316.

[90] Collins RJ, Harmon BV, Gove GC, Kerr JF. Internucleosomal DNA cleavage should not be sole criterion for identifying apoptosis. *Int J Radiat biol* 1992; 61:451-453.

[91] Sugerman PB, Voltz MJ, Savage NW, et al. Phenotypic and functional analysis of peripheral blood lymphocytes in oral lichen planus. *J Oral Pathol Med.* 1992; 21: 445-50.

[92] Walton LJ, Macey MG, Thornhill MH, Farthing PM. Intra-epithelial subpopulations of T lymphocytes and Langerhans cells in oral lichen planus. *J Oral Pathol Med.* 1998; 27: 116-23.

[93] Mozzanica N, Cattaneo A, Legori A, et al. Immunohistologic evaluation of the effect of cyclosporine treatment on the lichen planus immune infiltrate. *J Am Acad Dermatol.* 1991: 24: 550-554.

[94] Akasu R, From L, Kahn HJ. Lymphocyte and macrophage subsets in active and inactive lesions of lichen planus. *Am J Dermatopathol* 1993; 15:217-223.

[95] Villarroel Dorrego M, Correnti M, Delgado R, Tapia FJ. Oral lichen planus: immunohistology of mucosal lesions. *J Oral Pathol Med* 2002; 31:410-414.

[96] Khan A, Farah CS, Savage NW, Walsh LJ, Harbrow DJ, Sugerman PB. Th1 cytokines in oral lichen planus. *J Oral Pathol Med* 2003; 32:77-83.

[97] Kaštelan M, Prpić Massari L, Gruber F, Zamolo G, Žauhar G, Čoklo M, Rukavina D. The role of perforin-mediated apoptosis in lichen planus lesions. *Arch Dermatol Res* 2004; 296: 226-230.

[98] Zhou XJ, Sugerman PB, Savage NW, Walsh LJ, Seymour GJ. Intra-epithelial CD8+ T cells and basement membrane disruption in oral lichen planus. *J Oral Pathol Med* 2002; 31:23-27.

[99] Thornhill MH. Immune mechanisms in oral lichen planus. *Acta Odontol Scand* 2001; 59:174-177.

[100] Iijima W, Ohtani H, Nakayama T, Sugawara Y, Sato E, Nagura H, Yoshie O, Sasano T. Infiltrating CD8+ T cells in oral lichen planus predominantly express CCR5 and CXCR3 and carry respective chemokine ligands RANTES/CCL5 and IP-10/CXCL10 in their cytolytic granules: a potential self-recruiting mechanism. *Am J Pathol* 2003; 163:261-8.

[101] Lever WF, Lever GS. Noninfectious erythematous, papular, and squamous diseases. In: Elder D, Elenitsas R, Jaworsky C, Johnson B, eds. *Histopathology of the skin*, 7th edition. Lippincott, Philadelphia, 1990, pp 152-184.

[102] Ebner H, Gerhart W. Light and electron microscopic studies on colloid and other cytoid bodies. *Clin Exp Dermatol*. 1977; 2: 311-322.

[103] Hashimoto K. Apoptosis in lichen planus and several other dermatoses. *Acta Derm Venereol (Stockh)*, 1976; 56: 187-210.

[104] Weedon D. Apoptosis in lichen planus. *Clin Exp Dermatol*, 1980; 5: 425-430.

[105] Shimizu M, Higaki Y, Higaki M, Kawashima M. The role of granzyme B-expressing CD8-positive T cells in apoptosis of keratinocytes in lichen planus. *Arch Dermatol Res* 1997; 289: 527-532.

[106] Griffiths GM, Mueller C. Expresion of perforin and granzymes in vivo: Potential diagnostic markers for activated cytotoxic cells. *Immunol Today* 1991; 12:415-419.

[107] Santoro A, Majorana A, Bardellini E, Gentili F, Festa S, Sapelli P, Facchetti F. Cytotoxic molecule expression and epithelial cell apoptosis in oral and cutaneous lichen planus. *Am J Clin Pathol* 2004; 121:758-764.

[108] Prpić Massari L, Kaštelan M, Gruber F, Laškarin G, Sotošek Tokmadžić V, Štrbo N, Zamolo G, Žauhar G and Rukavina D. Perforin expression in peripheral blood lymphocytes and skin infiltrating cells in lichen planus patients. *Br J Dermatology* 2004; 151: 433-439.

[109] Neppelberg E, Johannessen AC, Jonsson R. Apoptosis in oral lichen planus. *Eur J Oral Sci* 2001; 109:361-364.

[110] Bascones C, Gonzales-Moles MA, Esparza G, Bravo M, Acevedo A, Gil-Montoya JA, Bascones A. Apoptosis and cell cycle arrest in oral lichen planus. Hypothesis on their possible influence on its malignant transformation. *Arch Oral Biol* 2005; 50:873-881.

[111] Tobon-Arroyave SI, Villegas-Acosta FA, Ruiz-Restrepo SM, Vieco-Duran B, Restrepo-Misas M, Londono-Lopez ML. Expression of caspase-3 and structural changes associated with apoptotic cell death of keratinocytes in oral lichen planus. *Oral Dis* 2004; 10:173-178.

[112] Dummer R, Michie SA, Kell D, Gould JW, Haeffner AC, Smoller BR, Warnke RA, Wood GS. Expression of bcl-2 protein and Ki-67 nuclear proliferation antigen in benign and malignant cutaneous T-cell infiltrates. *J Cutan Pathol* 1995; 22:11-17.

[113] Dekker NP, Lozada-Nur F, Lagenaur LA, MacPhail LA, Bloom CY, Regezi JA. Apoptosis-associated markers in oral lichen planus. *J Oral Pathol Med* 1999; 26:170-175.

[114] Soini Y, Kamel D, Paakko P, Lehto VP, Oikarinen A, Vahakangas KV. Aberrant accumulation of p53 associates with Ki67 and mitotic count in benign skin lesions. *Br J Dermatol* 1994; 131:514-520.

[115] Hirota M, Ito T, Okudela K, Kawabe R, Yazawa T, Hayashi H, Nakatani Y, Fujita K, Kitamura H. Cell proliferation activity and the expression of cell cycle regulatory proteins in oral lichen planus. *J oral Pathol Med* 2002; 31:204-212.

INDEX

A

D

E

F

G

H

I

J

K

L

M

N

O

P

Q

R

S

T

U

V